VOLUME NINETY NINE

CURRENT TOPICS IN DEVELOPMENTAL BIOLOGY

microRNAs in Development

VOLUME NINETY NINE

Current Topics in DEVELOPMENTAL BIOLOGY

microRNAs in Development

Edited by

ERAN HORNSTEIN
Department of Molecular Genetics
Weizmann Institute of Science
Rehovot, Israel

ELSEVIER

AMSTERDAM • BOSTON • HEIDELBERG • LONDON
NEW YORK • OXFORD • PARIS • SAN DIEGO
SAN FRANCISCO • SINGAPORE • SYDNEY • TOKYO
Academic Press is an imprint of Elsevier

Academic Press is an imprint of Elsevier
525 B Street, Suite 1900, San Diego, CA 92101-4495, USA
225 Wyman Street, Waltham, MA 02451, USA
32, Jamestown Road, London NW1 7BY, UK
Linacre House, Jordan Hill, Oxford OX2 8DP, UK

First edition 2012

ISBN: 978-0-12-387038-4
ISSN: 0070-2153

For information on all Academic Press publications
visit our website at elsevierdirect.com

Printed and bound by CPI Group (UK) Ltd, Croydon, CR0 4YY

Transferred to Digital Print 2012

Contents

Contributors

David Baltimore
Division of Biology, California Institute of Technology, Pasadena, California, USA

Torsten U. Banisch
Center for Molecular Biology of Inflammation, Institute of Cell Biology, University of Münster, Münster, Germany

Richard W. Carthew
Department of Molecular Biosciences, Northwestern University, Evanston, and Chicago Center for Systems Biology, Chicago, Illinois, USA

Luisa Cochella
Department of Biochemistry and Molecular Biophysics, Howard Hughes Medical Institute, Columbia University Medical Center, New York, USA

Qi Dai
Department of Developmental Biology, Sloan-Kettering Institute, New York, USA

Elaine Fuchs
Laboratory of Mammalian Cell Biology and Development, Howard Hughes Medical Institute, The Rockefeller University, New York, USA

Mehdi Goudarzi
Center for Molecular Biology of Inflammation, Institute of Cell Biology, University of Münster, Münster, Germany

Oliver Hobert
Department of Biochemistry and Molecular Biophysics, Howard Hughes Medical Institute, Columbia University Medical Center, New York, USA

Eric C. Lai
Department of Developmental Biology, Sloan-Kettering Institute, New York, USA

Jennifer H. Mansfield
Department of Biology, Barnard College, New York, USA

Edwina McGlinn
EMBL Australia, Australian Regenerative Medicine Institute, Monash University, Clayton, Australia

Vanessa Mondol
Division of Biology, University of California, San Diego, La Jolla, California, USA

Ryan M. O'Connell
Division of Microbiology and Immunology, Department of Pathology, University of Utah, Salt Lake City, Utah, USA

Amy E. Pasquinelli
Division of Biology, University of California, San Diego, La Jolla, California, USA

Nicolás Peláez
Interdepartmental Program in Biological Sciences, and Department of Molecular Biosciences, Northwestern University, Evanston; Chicago Center for Systems Biology, Chicago, Illinois, USA

Erez Raz
Center for Molecular Biology of Inflammation, Institute of Cell Biology, University of Münster, Münster, Germany

Peter Smibert
Department of Developmental Biology, Sloan-Kettering Institute, New York, USA

Nicholas S. Sokol
Department of Biology, Indiana University, Bloomington, Indiana, USA

Rui Yi
Department of Molecular, Cellular, and Developmental Biology, University of Colorado at Boulder, Boulder, Colorado, USA

Preface

Understanding how miRNA genes function in metazoan development is an important focus of study, ever since the laboratories of Victor Ambros and Gary Ruvkun discovered these tiny RNAs, almost 20 years ago. This volume of *Current Topics in Developmental Biology* provides an overview of the tremendous progress in our understanding of microRNAs (miRNAs) and their function.

Animal development is extremely dynamic and complex, yet it results in reproducible creation of an embryo. miRNAs are integrally involved in regulatory networks during embryonic development, which are required for its successful accomplishment. Often the role of miRNAs is to ensure that temporal dynamics and spatially restricted gene expression programs will be maintained during development. Thus, miRNAs impart precision and coordinated interpretation of genetic information.

Contributions from leading investigators in the field of miRNA biology revise our current knowledge and predict what we should expect to emerge of miRNA studies in the future.

In Chapter 1, Vanessa Mondol and Amy E. Pasquinelli describe let-7, the miRNA that historically ignited broader scientific interest in small RNAs, when Gary Ruvkun, Pasquinelli, and colleagues discovered its conservation throughout bilateria. Later, let-7 became a model for understanding the complex mechanisms that control miRNA biogenesis and miRNA function in development and disease.

Then, in Chapter 2, Jennifer H. Mansfield and Edwina McGlinn describe the function of Hox-embedded miRNAs and their intricate coregulation with numerous targeted Hox genes. This fundamental Hox network also provides insight into the mechanisms underlying miRNA gene evolution.

Protein expression dynamics during cell-lineage specification, morphogenesis, and organogenesis involve extensive regulation by tissue-specific miRNAs. This volume provides specific examples related to the development of central organ systems.

Thus, Chapter 3, written by Nicholas S. Sokol, focuses on miRNA regulation of myogenesis, one of the best characterized developmental systems for miRNA activity.

The pivotal functions of a wider variety of small RNAs including miRNAs, piRNA, and endo-siRNA in germ-cell development are reviewed in Chapter 4 by Torsten U. Banisch, Mehdi Goudarzi, and Erez Raz.

In Chapter 5, Luisa Cochella and Oliver Hobert review miRNAs' function in neuronal differentiation and diversity. Importantly, miRNA subcellular localization and capacity to act as posttranscriptional regulators in neuritis and perisynaptically make them particularly suited to control the dynamic molecular milieu of adult neurons as well.

Ryan M. O'Connell and David Baltimore, in Chapter 6, discuss how miRNAs fit into our current understanding of hematopoietic development in mammals and provide striking examples whereby breakdowns in miRNA-dependent pathways can trigger disease.

In the context of another adult niche, miRNAs in skin stem cells are described in Chapter 7 by Rui Yi and Elaine Fuchs, who further draw general conclusions about miRNA function in stemness and pluripotency.

The two last chapters come from fly geneticists: Chapter 8, by Qi Dai, Peter Smibert, and Eric C. Lai, reviews genetic strategies to study miRNAs in *Drosophila Melanogaster*. It provides a critical handle on the platforms used for miRNA/developmental studies. Finally, Chapter 9, by Nicolás Peláez and Richard W. Carthew, explores the landscape of miRNA interactions from a "systems" point of view, revisiting our understanding of miRNAs as means for conferring robustness to developmental genetic programs.

Together, these chapters touch upon many facets of miRNA biology. Thus, the overall perception the reader may rightfully have is that homology-based miRNA-dependent silencing creates a profound posttranscriptional regulatory layer, which affects every facet of embryogenesis and the adult life.

However, current research still awaits the resolution of critical questions in this field. For example, the extent to which miRNA loss of function has detectable phenotypic consequences may reflect a genuine fine-tuning property of miRNAs, or merely our limited ability to devise appropriate assays that will uncover the function of specific miRNAs. In addition, it seems that novel methodologies will be needed in order to address "network-level" questions, including miRNAs' preference for regulating a handful of key targets or to coordinately regulate many targets, as suggested by many bioinformatics studies. In addition, the capacity of coexpressed miRNAs to regulate simultaneously coherent sets of targets toward a particular developmental process is not yet clear. Answering these questions will provide fascinating advances in our knowledge of the roles of miRNA-based regulatory modules in development.

Eran Hornstein

CHAPTER ONE

Let's Make It Happen: The Role of let-7 MicroRNA in Development

Vanessa Mondol *and* Amy E. Pasquinelli

Contents

Abstract

Noncoding RNAs have emerged as an integral part of posttranscriptional gene regulation. Among that class of RNAs are the microRNAs (miRNAs), which posttranscriptionally regulate target mRNAs containing complementary sequences. The broad presence of miRNAs in lower eukaryotes, plants, and mammals highlights their importance throughout evolution. MiRNAs have been shown to regulate many pathways, including development, and disruption of miRNA function can lead to disease (Ivey and Srivastava, 2010; Jiang *et al.*, 2009). Although the first miRNA genes were discovered in the nematode, *Caenorhabditis elegans*, almost 20 years ago, the field of miRNA research began when they were found in multiple organisms a little over a decade ago

Division of Biology, University of California, San Diego, La Jolla, California, USA

Current Topics in Developmental Biology, Volume 99
ISSN 0070-2153, DOI: 10.1016/B978-0-12-387038-4.00001-X

(Lagos-Quintana *et al.*, 2001; Lau *et al.*, 2001; Lee and Ambros, 2001; Lee *et al.*, 1993; Pasquinelli *et al.*, 2000; Wightman *et al.*, 1993). Here, we review one of the first characterized miRNAs, let-7, and describe its role in development and the intricacies of its biogenesis and function.

1. Introduction

MicroRNAs (MiRNAs) distinguish themselves from other small noncoding RNAs by several unique features. Since miRNAs and the small interfering RNAs (siRNAs) that function in RNA interference (RNAi) associate with common Argonaute proteins and are functionally indistinct in some species, these RNAs are primarily classified based upon their biogenesis pathways. MiRNA genes are scattered throughout the genome, in intra- and intergenic positions, and are transcribed as single-stranded RNAs capable of forming stem loops that contain mismatches and bulges (Fig. 1.1; Kim *et al.*, 2009b). The stem-loop structures are processed from the nascent transcripts into ~65 nt hairpin shaped RNAs that undergo final processing to the ~21 nt mature miRNAs (Krol *et al.*, 2010). In contrast, siRNAs are usually produced from convergent transcripts forming long double-stranded RNAs that serve as substrates for RNase processing to the eventual 20–25 nt forms (Czech and Hannon, 2011). Perfect double-stranded RNAs from exogenous and endogenous sources generate exo- and endo-siRNAs, respectively. In animals, miRNAs and siRNAs are also often characterized by distinct mechanisms for regulating gene expression. The miRNAs typically bind to the 3′-untranslated regions (3′UTRs) of target messenger RNAs (mRNAs). Binding is usually imperfect, and target mRNAs are either translationally repressed or deadenylated and degraded (Bartel, 2009; Fabian *et al.*, 2010). This is in contrast to siRNAs, which form perfectly complementary bonds to any region in target mRNAs and cause degradation by cleavage at the binding site. This distinction means that one miRNA can regulate multiple mRNAs with nonidentical target sites, while siRNAs would be limited to targeting mRNAs with sites of perfect complementarity (Lim *et al.*, 2005). However, off target RNAi effects are often attributable to partial pairing between siRNAs and unintended target sites, and miRNAs can direct mRNA cleavage when presented with perfectly complementary target sequences (Alemán *et al.*, 2007; Doench *et al.*, 2003; Zeng *et al.*, 2003).

Typically, miRNA biogenesis begins with transcription by RNA polymerase II, either through an independent promoter or as part of a host gene where the miRNA is embedded within an intron of a protein-coding gene (Fig. 1.1; Kim *et al.*, 2009b). Some miRNAs are closely arranged in the genome and considered a "cluster" when they are synthesized as part of a

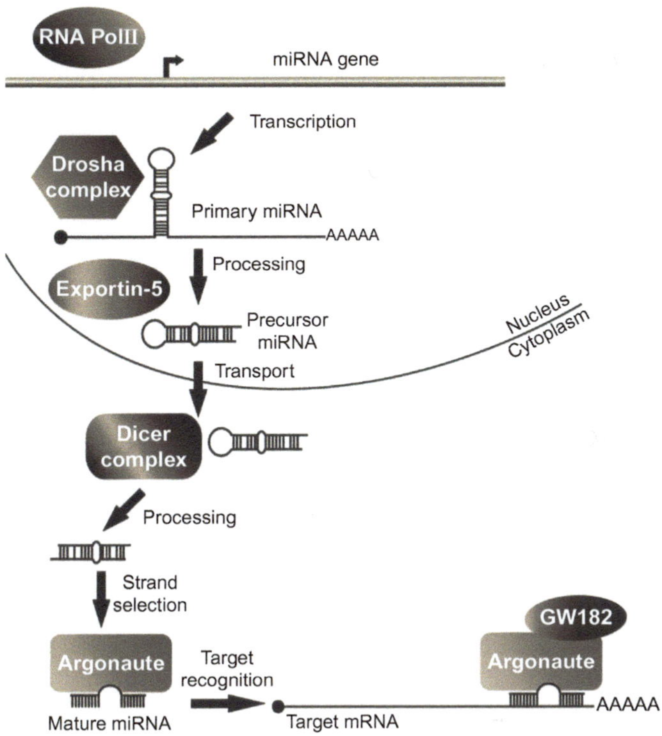

Figure 1.1 General miRNA biogenesis pathway. RNA polymerase II (RNA PolII) transcribes miRNA genes to produce a capped and polyadenylated primary miRNA (pri-miRNA). A complex containing the RNase III enzyme Drosha excises the ~65nt precursor miRNA (pre-miRNA), leaving a 2nt overhang at the 3′ end. In flies and vertebrates, nuclear transport protein Exportin-5, exports the pre-miRNA to the cytoplasm, where it is subsequently cleaved by the RNase III enzyme Dicer to produce an ~21nt double-stranded product that features 2nt overhangs at both ends. The mature miRNA strand is then loaded into Argonaute (Ago), where it works as part of a complex with GW182, and other proteins to target mRNAs for repression.

common transcript (Lau *et al.*, 2001; Lee and Ambros, 2001; Lim *et al.*, 2003). The capped and polyadenylated transcripts, known as primary miRNAs (pri-miRNAs), contain the stem–loop structure that houses the functional ~21nt miRNA sequence (Kim *et al.*, 2009b). The RNase III enzyme Drosha, working in a complex with Pasha (also known as DGCR8), excises the ~65nt hairpin, forming what is known as the precursor miRNA (pre-miRNA). A noncanonical Drosha-independent mechanism for producing pre-miRNAs was recently discovered in mammalian cells, *Drosophila melanogaster*, and *Caenorhabditis elegans* (Berezikov *et al.*, 2007; Okamura *et al.*, 2007; Ruby *et al.*, 2007a). In this pathway, pre-miRNAs derive from debranching of short introns excised from pre-mRNAs. The spliced introns have characteristics of miRNA precursors: they are ~65 nt and can fold into hairpins that contain 5′ monophosphate and 3′ hydroxyl residues. Computational and experimental analyses of short RNA transcripts in flies, nematodes, and mammals have so far revealed few mirtrons, with 14 found in *C. elegans* (Berezikov *et al.*, 2007; Chung *et al.*, 2011). In vertebrates and flies, pre-miRNAs are transported from the nucleus to the cytoplasm by a nucleocytoplasmic transport factor, Exportin-5, where the RNase III enzyme Dicer removes the loop region, leaving ~21 nt double-stranded RNA fragments with 2 nt 3′ end overhangs (Kim *et al.*, 2009b; Krol *et al.*, 2010). Curiously, *C. elegans* lacks an obvious homolog of Exportin-5, and the cellular location of miRNA processing events is yet to be determined in nematodes (Bohnsack *et al.*, 2004; Murphy *et al.*, 2008).

The mature miRNA strand, also known as the guide strand, is then loaded onto an Argonaute (Ago) protein, a key factor in the RNA-induced silencing complex (RISC; Fig. 1.1; Kim *et al.*, 2009b; Krol *et al.*, 2010). The leftover strand, called the star strand, is degraded by an unclear mechanism. Historically, the more abundant mature miRNA species has been named the guide and the less frequently detected one the star strand. However, the abundance of one strand versus the other has been found to reverse under different conditions and, thus, the term star strand does not always indicate a nonfunctional processing by-product. In fact, both halves of the Dicer cleavage product, albeit at unequal levels, are often incorporated into Ago complexes (Okamura *et al.*, 2008; Zisoulis *et al.*, 2010). The bias for loading one strand versus the other is affected by the thermodynamic stability of the duplex termini, with the strand containing the weaker paired 5′ end being favored, and the identity of the first nucleotide (Czech and Hannon, 2011).

The 27 Ago family proteins in *C. elegans* exert different roles in small RNA pathways (Yigit *et al.*, 2006). The Argonaute-Like-Genes 1 and 2 (ALG-1, ALG-2) are loaded with miRNAs and direct translational silencing or deadenylation of target mRNAs (Ding and Groszhans, 2009; Steiner *et al.*, 2007; Wu *et al.*, 2010; Zhang *et al.*, 2007; Zisoulis *et al.*, 2010). Most of the other Ago proteins function in the exo- and endo-siRNA pathways, which involve target mRNA cleavage (Yigit *et al.*, 2006). The choice of which Ago is loaded with particular small RNA species is not entirely clear but seems to depend on the origin and structure of the small RNA duplex (Jannot *et al.*, 2008; Steiner *et al.*, 2007).

MiRNAs serve as guides to lead the RISC complex to regulatory targets. The idea that miRNAs use limited base-pairing complementarity to recognize target sites was originally described for the first discovered miRNA, lin-4, and its target, *lin-14* (Lee *et al.*, 1993; Wightman *et al.*, 1993). Several partially complementary sites in well-conserved regions of the *lin-14* 3′UTR were predicted to pair with the lin-4 miRNA. The next identified miRNA, let-7, and 3′UTR sequences in its target, *lin-41*, also exhibited partial base-pairing capacity, further implicating the functionality of imperfect duplexes between miRNAs and target sites (Reinhart *et al.*, 2000). This pattern holds true for the hundreds of animal miRNAs now recognized but differs from the capacity of plant miRNAs to typically base pair perfectly with target mRNAs (Bartel, 2009). Although the inability of animal miRNAs to form antisense pairs with mRNAs complicates the assignment of miRNAs to specific targets, several parameters have emerged as useful predictors of these interactions. A common motif is the ability of the first 2–7 nt from the 5′ end of the miRNA to perfectly pair with mRNA sequences (Brennecke *et al.*, 2005; Krek *et al.*, 2005; Lewis *et al.*, 2003, 2005). This part of the miRNA is known as the "seed" sequence, and the complementary site in the mRNA is called the "seed match" (Lewis *et al.*, 2003). The requirements for conservation, 3′UTR position, and structural accessibility are also often used to restrict predicted target sites (Bartel, 2009). There are examples of targets that lack seed matches and instead utilize 3′ supplementary, where extensive complementarity with the 3′ end of the miRNA compensates, or centralized pairing conformations (Bartel, 2009; Shin *et al.*, 2010).

The first characterized miRNA target sites were all found to reside in 3′UTR sequences (Lee *et al.*, 1993; Moss *et al.*, 1997; Reinhart *et al.*, 2000; Slack *et al.*, 2000; Wightman *et al.*, 1993). Computational analyses often focus on conserved regions in 3′UTRs and take into account features like the number of target sites, their position relative to the stop codon and polyA signal, structural accessibility and whether they exist in A/U rich areas, to better predict miRNA target sites (Bartel, 2009). Surprisingly, an experimentally based genome-wide analysis of *C. elegans* Argonaute binding sites revealed that ~50% occur in coding exons and ~35% in 3′UTRs of bound mRNAs (Zisoulis *et al.*, 2010). This is consistent with similar analyses in mammalian cells, where about half of the Argonaute binding sites were found in coding exon regions (Chi *et al.*, 2009; Hafner *et al.*, 2010). While some target sites in coding exons have been shown to confer miRNA function, the general efficacy of regulation in translated regions of mRNAs may be reduced compared to that in 3′UTRs (Gu *et al.*, 2009; Kloosterman *et al.*, 2004). Recent studies trying to understand the functionality of miRNA target sites in coding regions reveal a synergistic effect, where there are stronger effects on mRNAs bound by miRNAs in the 3′UTR and coding exons, though they are not as strong as the effects when there are two target sites in the 3′UTR (Fang and Rajewsky, 2011).

Association of the miRNA complex with a target mRNA results in downregulation of the protein expression through mechanisms that are not entirely understood. There is an ongoing debate about whether the primary mode of regulation is translational repression of the mRNA, or deadenylation followed by degradation of the mRNA (Fig. 1.2; Djuranovic *et al.*, 2011; Huntzinger and Izaurralde, 2011; Krol *et al.*, 2010). Original studies of *lin-14* repression by lin-4 miRNA detected little change in target mRNA levels or polysome loading and concluded that regulation involved a

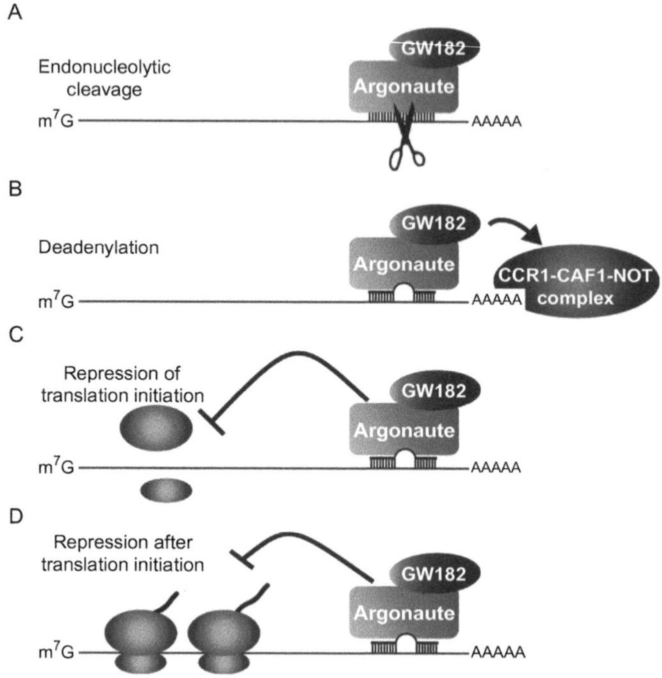

Figure 1.2 Mechanisms of miRNA target regulation. (A) When miRNAs bind to target sites with a high level of complementarity, it triggers endonucleolytic cleavage of the mRNA by Argonaute (Ago). (B) Ago interacts with GW182 proteins, which bind to the CCR1-CAF1-NOT complex to trigger deadenylation of bound target mRNAs. (C, D) Through unclear mechanisms, the Ago–GW182 complex represses translation (C) by inhibiting translation initiation and (D) by blocking protein accumulation after translation has initiated.

posttranscriptional mechanism acting at a stage after translational initiation (Olsen and Ambros, 1999; Wightman *et al.*, 1993). Subsequent work showed that mRNA levels for the *lin-14*, *lin-28*, and *lin-41* target mRNAs are downregulated in response to miRNA regulation, and there is also a shift on polysomes indicative of inhibition of translation initiation (Bagga *et al.*, 2005; Ding and Groszhans, 2009). It remains to be determined if target mRNA degradation is a cause or consequence of halted translation. Argonaute interacts with GW182 proteins (AIN-1 and AIN-2 in *C. elegans*), which recruit factors capable of interfering with translation or promoting deadenylation of the target mRNA (Tritschler *et al.*, 2010). Thus, depending on the cell type or conditions, different factors may be available to regulate miRNA targets through diverse mechanisms, ultimately resulting in diminished protein production.

2. The Discovery of miRNAs

2.1. lin-4 miRNA

Development of *C. elegans* progresses through four larval stages before reaching adulthood. The temporal fate of each cell division has been mapped and genes that regulate the timing of these events are part of the heterochronic pathway (Sulston and Horvitz, 1977). Mutations in heterochronic genes cause either precocious development, where later larval cell fates occur prematurely, or retarded development, where early larval cell fates reoccur in later stages (Ambros, 1989; Ambros and Horvitz, 1984; Chalfie *et al.*, 1981). These types of mutations cause various developmental abnormalities, some of which result in lethality. While investigating the genes important for larval development, researchers found that *lin-4* loss-of-function (lf) mutations recapitulated phenotypes observed in worms with *lin-14* gain-of-function (gf) mutations (Ambros, 1989). These mutant worms repeat L1 cell fates, which cause the absence of adult structures like the vulva and differentiated hypodermal seam cells (Chalfie *et al.*, 1981). Loss of *lin-14* activity has the opposite phenotype, where later larval fates are precociously expressed in the first larval stage (Ambros and Horvitz, 1984). Additional genetic and molecular experiments led to the model that LIN-14 protein promotes the first larval stage cell fates and then is downregulated by *lin-4* activity so that later larval fates can proceed (Arasu *et al.*, 1991; Ruvkun and Giusto, 1989).

The surprising identity of the *lin-4* gene product revealed a novel mechanism for regulation of *lin-14*. Through arduous mapping and genetic rescue experiments, researchers in the Ambros lab narrowed down the location of *lin-4* to a 693 nt region, which was later shown to contain the primary lin-4 transcripts, that lacked apparent protein-coding potential (Bracht *et al.*, 2010; Lee *et al.*, 1993). Instead, this region was found to express 22 and 61 nt RNA products that are now recognized as the mature and precursor forms of lin-4 miRNA, respectively. Since the Ruvkun lab had established that *lin-14* is regulated at the posttranscriptional level through elements in its 3′UTR, the possibility of lin-4 base-pairing to these regions was realized by both labs (Lee *et al.*, 1993; Wightman *et al.*, 1993). There are seven potential lin-4 target sites in the *lin-14* 3′UTR that are predicted to form nonidentical partial duplexes (Wightman *et al.*, 1993). Using genetic mutants, molecular experiments, and reporter assays, the Ambros and Ruvkun labs defined a novel form of gene regulation whereby the lin-4 RNA base pairs to specific sites in the 3′UTR of the *lin-14* mRNA, resulting in downregulation of LIN-14 protein expression (Lee *et al.*, 1993; Wightman *et al.*, 1993). Whether or not this mechanism of gene regulation would be an isolated example was an open question since the small RNA, lin-4, and its target gene, *lin-14*, were not obviously conserved in species beyond *Caenorhabditis*.

2.2. let-7 miRNA

Four years after the discovery of lin-4 and its target, another protein-coding gene in the heterochronic pathway was found to be under the direct control of *lin-4*. The *lin-28* 3′UTR contains a single *lin-4* binding site that mediates repression of protein expression at the posttranscriptional level (Moss *et al.*, 1997). With two different genes regulated in a similar manner by the same RNA, researchers suspected that other small RNA regulators might exist in *C. elegans* to control developmental timing. Characterization of mutations that cause delayed temporal fates in the last larval stages led to the *let-7* gene (Fig. 1.3A; Reinhart *et al.*, 2000). The *let-7* gene was originally named based on the lethal phenotype where worms rupture through the vulva and die before reaching adulthood (Fig. 1.3B; Meneely and Herman, 1979). Genetic mapping, rescue experiments, and RNA expression analyses established that *let-7* was the second example of a tiny RNA gene in the developmental timing pathway (Reinhart *et al.*, 2000). Like *lin-4*, the *let-7* gene expressed an ~21 nt RNA that could potentially derive from a processed intramolecular hairpin structure (Lee *et al.*, 1993; Reinhart *et al.*, 2000). Coincidentally, studies of the *lin-41* gene in the same lab showed that mutations in this heterochronic gene resulted in precocious adoption of differentiated cells fates, a phenotype opposite to that of *let-7* mutants (Fig. 1.3A; Reinhart *et al.*, 2000; Slack *et al.*, 2000). Using a variety of biochemical and genetic techniques, the let-7 RNA was demonstrated to regulate expression of *lin-41* through two sites of partial complementarity in its 3′UTR (Fig. 1.4; Reinhart *et al.*, 2000; Slack *et al.*, 2000; Vella *et al.*, 2004a). Thus, the accumulation of mature let-7 RNA at the end of the third larval stage causes downregulation of LIN-41 protein expression, allowing the adoption of later larval and adult fates.

2.3. let-7 miRNA in other organisms

Unlike *lin-14*, homologs of *lin-41* were found to exist in Drosophila, mouse, and other animals, raising the question of whether its small RNA regulator would also be conserved (Slack *et al.*, 2000). With whole-genome databases becoming available, the Ruvkun lab identified sequences in the Drosophila and human genomes that matched the mature let-7 sequence (Pasquinelli *et al.*, 2000). Northern blot analyses of a wide range of animal species showed that let-7 RNAs are present broadly in bilaterians, but apparently not in basal metazoans or nonanimal organisms. Remarkably, potential let-7 binding sites exist in the 3′UTRs of *lin-41* homologs in several species and, in some cases, regulation by let-7 has been demonstrated

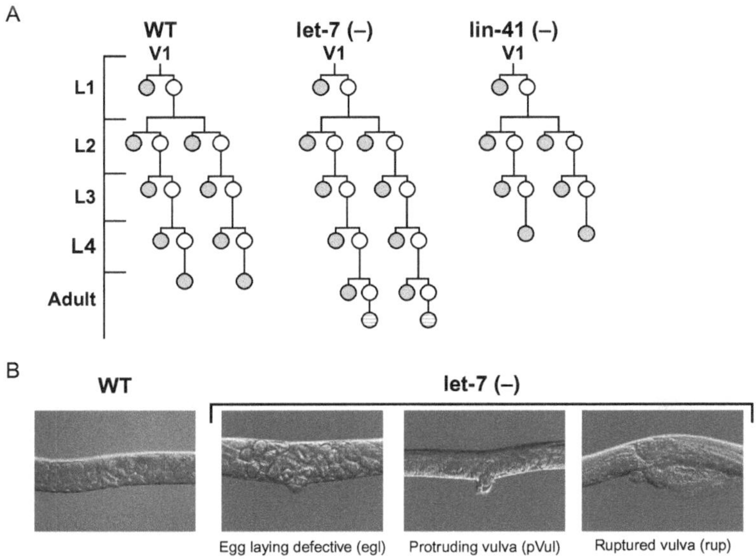

Figure 1.3 *C. elegans let-7* mutant phenotypes. (A) The lineage division patterns of wild-type (WT) V1 hypodermal seam cells are shown for the larval (L1–L4) and adult stages. In WT worms, V1 cells divide at each molt, with one cell differentiating (gray) and the other continuing to divide at the next stage (white), until the cells terminally differentiate at adulthood (gray). Worms with loss-of-function mutations in *let-7* (let-7 (−)) have retarded heterochronic phenotypes where V1 cells can undergo one or more additional divisions after L4, sometimes failing to execute the adult differentiated fate (striped). The opposite phenotype, precocious, is expressed by worms with loss of *lin-41* (lin-41(−)), where the V1 cells terminate dividing and take on the differentiated adult fate at the L3 to L4 molt. (B) Pictures of adult worms at 200× magnification; the anterior is to the left and the vulva is on the ventral side. WT adult worms form eggs, which they lay through their vulva, a narrow opening that is smoothly in line with the ventral side of the worm. Worms with mutations in *let-7* display a variety of easily identifiable vulval phenotypes. Egg laying defective (egl) is characterized by a bloated, egg-filled worm. Protruding vulva (pVul) is defined by a protrusion due to abnormal development of the vulva. Ruptured vulva (rup) occurs during the L3 to L4 molt, when the worm innards burst through the weak unstructured vulva, resulting in lethality.

(Fig. 1.4; Kanamoto *et al.*, 2006; Maller-Schulman *et al.*, 2008; Pasquinelli *et al.*, 2000). The discovery of let-7 across species fueled the search for other such RNA genes, and soon many small RNAs were identified in worms, flies, human cell lines, mice, and plants through cloning methods (Lagos-Quintana *et al.*, 2001, 2002; Lau *et al.*, 2001; Lee and Ambros, 2001; Reinhart *et al.*, 2002). A signature of these ~21 nt RNAs is that they are processed from ~65 nt stem–loop precursors, and this feature distinguishes the class of miRNAs from other tiny ncRNAs (Ambros *et al.*, 2003).

While there is one *let-7* gene in *C. elegans*, multiple let-7 genes exist in most vertebrates. The human genome includes three let-7 genes that

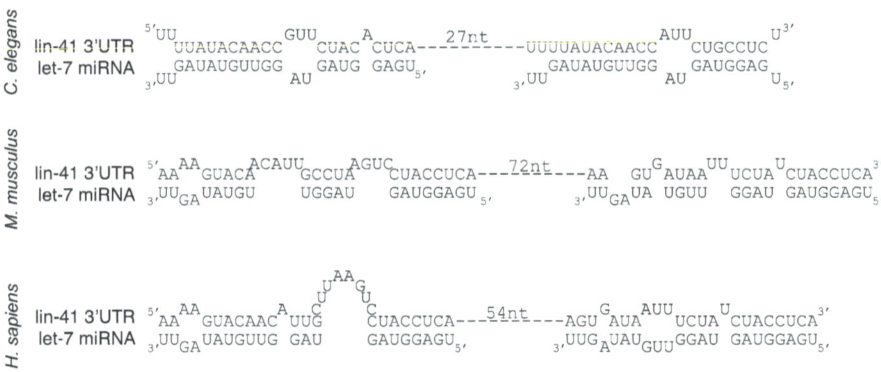

Figure 1.4 Examples of let-7 target site interactions. Illustrated here is the imperfect pairing between let-7 and 3′UTR sequences in lin-41 (also called Trim71 in mammals) genes in *C. elegans*, *M. musculus*, and *H. sapiens*.

C. elegans

let-7	UGAGGUAGUAGGUUGUAUAGUU
mir-48	UGAGGUAGGCUCAGUAGAUGCGA
mir-84	UGAGGUAGUAUGUAAUAUUGUA
mir-241	UGAGGUAGGUGCGAGAAAUGA
mir-793	UGAGGUAUCUUAGUUAGACAGA
mir-794	UGAGGUAUUCAUCGUUGUCACU
mir-795	UGAGGUAGAUUGAUCAGCGAGCUU

H. sapiens

let-7a (3)	UGAGGUAGUAGGUUGUAUAGUU
let-7b	UGAGGUAGUAGGUUGUGUGGUU
let-7c	UGAGGUAGUAGGUUGUAUGGUU
let-7d	UGAGGUAGUAGGUUGCAUAGUU
let-7e	UGAGGUAGGAGGUUGUAUAGUU
let-7f (2)	UGAGGUAGUAGAUUGUAUAGUU
let-7g	UGAGGUAGUAGUUUGUACAGUU
let-7i	UGAGGUAGUAGUUUGUGCUGUU
miR-98	UGAGGUAGUAAGUUGUAUUGUU
miR-202	AGAGGUAUAGGGCAUGGGAAAA

Figure 1.5 Homologs of let-7 in worms and humans. The let-7 family of genes is defined by the conserved seed sequence (nt 2–7). In *C. elegans*, let-7 has six nonidentical "sister" miRNAs. The human genome contains three genes that produce mature miRNAs identical to the worm let-7 (let-a-1, let-7a-2, let-a-3), two copies of let-7f and eight other miRNAs that share the let-7 seed sequence.

produce mature RNAs of identical sequence (let-7a-1, let-a-2, let-7a-3) as well as nine others that differ by one or more nucleotides (Fig. 1.5). Since the seed sequence, nucleotides 2–7, of the miRNA plays a key role in target recognition, miRNAs with identical seeds are often considered part of a family (Lim *et al.*, 2005). MiRNAs within the same family potentially regulate common targets, thus explaining why mutation of one member sometimes results in no discernible phenotypes (Miska *et al.*, 2007). In *C. elegans*, *let-7* has six "sister" genes that share 5′ end sequences (Fig. 1.5). Nonetheless, mutation of *let-7* alone is sufficient to cause developmental abnormalities and lethality in *C. elegans* (Fig. 1.3; Reinhart *et al.*, 2000). Thus, members of a miRNA family do not necessarily compensate for each other. In mammals, some of the let-7 family members exhibit specific expression patterns and have distinct targets, indicating that they may play different roles in diverse biological pathways (Boyerinas *et al.*, 2010; Chiang *et al.*, 2010).

3. The Role of let-7 in *C. elegans* Development

3.1. Biogenesis of let-7

Consistent with its role in promoting later larval cell fates, mature let-7 miRNA starts to accumulate midway through the third larval stage of development (Fig. 1.6; Reinhart *et al.*, 2000; Van Wynsberghe *et al.*, 2011). The *let-7* gene expresses three let-7 primary transcripts, two unspliced variants with differing transcriptional start sites and one SL1 *trans*-spliced isoform (Bracht *et al.*, 2004). The transcripts are capped and polyadenylated via a conserved polyA signal located ~670 nt downstream of the mature let-7 sequence. Surprisingly, expression of the let-7 primary transcripts initiates at the end of the first larval stage and oscillates during each subsequent stage (Fig. 1.6; Van Wynsberghe *et al.*, 2011). This expression pattern is regulated at the transcriptional level, as indicated by reporter genes containing the let-7 promoter fused to GFP (Esquela-Kerscher *et al.*, 2005; Johnson *et al.*, 2003; Martinez *et al.*, 2008; Van Wynsberghe *et al.*, 2011). These reporter studies also show that let-7 appears to be transcribed in most somatic tissues, consistent with the broad expression pattern of let-7 in differentiated cell types across animal species.

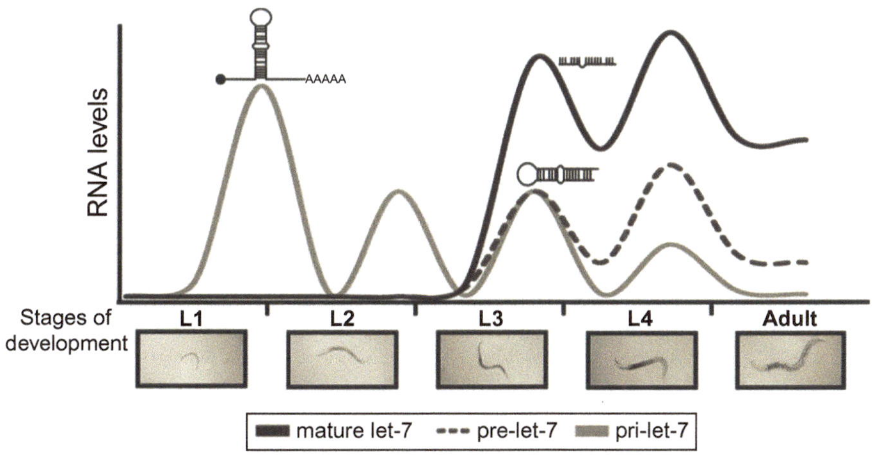

Figure 1.6 Expression of let-7 throughout *C. elegans* development. Graphical depiction of the accumulation of primary (pri-let-7), precursor (pre-let-7), and mature let-7 RNAs during the larval (L1–L4) and adult stages of worm development based on data from Van Wynsberghe *et al.* (2011). Worm pictures were captured at 100× magnification.

The search for *let-7* transcription elements identified a *cis*-acting sequence called the temporal regulatory element, TRE, located 223 and 1064nt upstream from the two transcriptional start sites (Fig. 1.7; Bracht

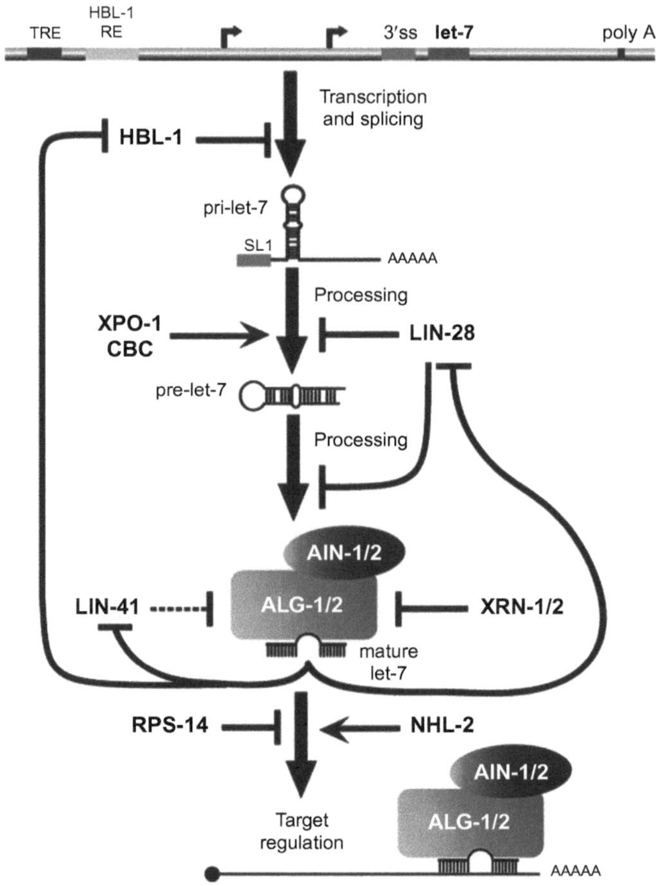

Figure 1.7 Regulation of let-7 biogenesis and function in *C. elegans*. Two *cis*-acting elements in the let-7 promoter, the temporal regulatory element (TRE) and putative HBL-1 response element, repress transcription in hypodermal seam cells. Two transcription start sites produce primary transcripts that are polyadenylated and subject to *trans*-splicing by the spliced leader 1 RNA (SL1) at the 3′ splice site (3′ss) found upstream of the mature let-7 sequence. The nuclear transport factors XPO-1 and CBC promote conversion of primary to precursor and mature, while LIN-28 blocks processing of primary and precursor RNAs. The 3′→5′ exonucleases XRN-1/2 degrade mature miRNAs upon release from the Ago complex. LIN-41, a target of let-7, indirectly regulates mature let-7 levels by targeting Argonaute for degradation in mouse cells (a dashed line represents this step since this LIN-41 activity has not yet been demonstrated in worms). As interactors with the Ago complex, NHL-2 enhances repression of let-7 targets, while RPS-14 antagonizes let-7 function.

et al., 2004; Johnson *et al.*, 2003). Internal deletions of the let-7 promoter in a GFP reporter assay identified this 116nt region, which contains a 9nt inverted repeat and is conserved in *C. briggsae*, as necessary for expression of GFP in the hypodermal seam cells (Johnson *et al.*, 2003). Compared to wild type, *let-7* transgenes lacking the TRE exhibit reduced rescue activity. The proteins that regulate transcription of let-7 through the TRE, as well as the elements responsible for the oscillating transcription pattern, and expression in other tissues are yet to be identified.

So far, two transcription factors have been found to regulate the expression of *let-7* in *C. elegans*. The *hbl-1* gene encodes a zinc-finger transcription factor that shares homology with the *Drosophila* Hunchback gene (Abrahante *et al.*, 2003; Fay *et al.*, 1999; Lin *et al.*, 2003). HBL-1 is predicted to bind to an A-rich sequence 18nt downstream of the TRE and repress transcription of *let-7* in the hypodermal seam cells (Fig. 1.7; Roush and Slack, 2009). The nuclear hormone receptor DAF-12 regulates the transcription of *let-7* in a hormone dependent manner (Bethke *et al.*, 2009; Hammell *et al.*, 2009a). In the absence of ligand, DAF-12 represses the expression of *let-7* and several of its sister miRNAs. When bound to ligand, DAF-12 activates expression of some *let-7* family members. DAF-12 response elements have been identified in the promoters of *mir-241* and *mir-84,* but direct interaction with the *let-7* promoter remains to be demonstrated (Bethke *et al.*, 2009). Interestingly, both *hbl-1* and *daf-12* are targets of regulation by *let-7* family miRNAs (Abbott *et al.*, 2005; Abrahante *et al.*, 2003; Großhans *et al.*, 2005; Lin *et al.*, 2003). Through multiple 3′UTR complementary sites, the let-7 sisters, mir-48, mir-84, and mir-241, initiate repression of *hbl-1* expression during the transition from the second to the third larval stage (Abbott *et al.*, 2005). These miRNAs are expressed by the second larval stage, one stage earlier than let-7, thus, providing a mechanism to reduce HBL-1 levels and allow for transcription of let-7 in the seam cells (Abbott *et al.*, 2005; Roush and Slack, 2009). Likewise, the let-7 sisters also target *daf-12* for downregulation at the L3 stage, which may promote transcription of let-7 in some tissues (Bethke *et al.*, 2009; Hammell *et al.*, 2009a). Expression of mature let-7 in L3 augments repression of *hbl-1* and *daf-12*, adding to the feedback loop of transcriptional and miRNA-mediated control (Abbott *et al.*, 2005; Abrahante *et al.*, 2003; Bethke *et al.*, 2009; Großhans *et al.*, 2005; Hammell *et al.*, 2009a; Lin *et al.*, 2003).

In *C. elegans*, ~70% of all mRNAs are *trans*-spliced to one of two 22nt splice leader (SL) sequences, SL1 or SL2 (Blumenthal, 2005). The *trans*-splicing reaction is carried out by a spliceosome that functions similarly to the *cis*-splicing complex, which removes introns, except that the 5′ end of the SL RNA acts as the 5′ exon and is ligated to a 3′ splice site downstream of the 5′ cap in the acceptor mRNA (Blumenthal, 2005). The role of *trans*-splicing is not fully understood, but is thought to aid in nuclear export and translation.

Thus, it was surprising to find SL1-spliced let-7 primary transcripts (Bracht *et al.*, 2004). The 3′ splice site required for *trans*-splicing is conserved in other nematodes, and disruption of this signal impairs Drosha processing of the primary transcripts (Bracht *et al.*, 2004). Secondary structure predictions by mfold suggest that splicing alters the context around the precursor hairpin, perhaps making it a better substrate for subsequent processing. Although SL-mediated *trans*-splicing is not found in most vertebrates, the primary transcripts of several human let-7 genes are part of spliced transcripts. The mechanistic role of splicing in let-7 biogenesis in worms, and perhaps other species, is yet to be resolved.

In vertebrates and Drosophila, Exportin-5 delivers miRNA precursors from the nucleus to the cytoplasm for Dicer processing (Fig. 1.1; Bohnsack *et al.*, 2004; Lund *et al.*, 2004; Yi *et al.*, 2003). The apparent absence of an Exportin-5 homolog in *C. elegans* suggests that other cellular transport factors are involved in miRNA biogenesis in this organism (Bohnsack *et al.*, 2004; Murphy *et al.*, 2008). The nuclear export receptor XPO-1, as well as the nuclear cap-binding complex (CBC), have been implicated in the *let-7* biogenesis pathway (Fig. 1.7; Bussing *et al.*, 2010). Depletion of XPO-1 or either of the two subunits of the CBC results in reduced levels of precursor and mature let-7 and accumulated levels of pri-let-7. Considering the role of XPO-1 and the CBC in mediating nuclear export of m^7G-capped U snRNAs (Hutten and Kehlenbach, 2007), one possibility is that capped primary miRNA transcripts are also substrates for transport to the cytoplasm. Since the cellular location of Drosha is not known in worms, nuclear export of pri-let-7 transcripts by XPO-1 and CBC could be important for processing (Bussing *et al.*, 2010).

The temporal expression of pri-let-7 suggests complex regulation at both the transcriptional and posttranscriptional levels (Fig. 1.7). Production of primary let-7 in the L1 and L2 stages is not coupled to accumulation of precursor and mature miRNA (Fig. 1.6; Van Wynsberghe *et al.*, 2011). The RNA binding protein LIN-28 mediates this phase of posttranscriptional regulation (Lehrbach *et al.*, 2009; Van Wynsberghe *et al.*, 2011). LIN-28 binds endogenous let-7 pri-miRNAs co-transcriptionally and blocks Drosha processing (Fig. 1.7; Van Wynsberghe *et al.*, 2011). Since *lin-28* is controlled by lin-4 miRNA, expression of this miRNA at the end of L1 results in a steady decline of LIN-28 protein and, thus, relief of let-7 processing inhibition by the third larval stage (Lehrbach *et al.*, 2009; Moss *et al.*, 1997; Van Wynsberghe *et al.*, 2011). Posttranscriptional regulation of let-7 by LIN-28 was originally discovered in mammalian cells (Heo *et al.*, 2008; Newman *et al.*, 2008; Viswanathan and Daley, 2010; Viswanathan *et al.*, 2008; Wulczyn *et al.*, 2007). In addition to preventing Drosha processing of let-7 pri-miRNAs, LIN-28 also recruits TUT4 (PUP-2 in *C. elegans*) to add a 3′ end U-tail to let-7 precursors, which blocks Dicer processing and promotes destabilization of the RNAs (Hagan *et al.*, 2009; Heo *et al.*, 2008, 2009). Comparable to the role of LIN-28 in blocking expression of mature let-7 early in worm development, high levels of Lin28 in mammalian stem cells also prevent accumulation of let-7 miRNAs.

For many miRNAs, including let-7, only one-half of the duplex that results from Dicer processing accumulates as the mature miRNA, while the other strand is presumably degraded. One factor that appears to influence miRNA stability is the association with Argonaute proteins (Kai and Pasquinelli, 2010). The ability to base pair with target sequences influences the accumulation of mature miRNAs through a mechanism dubbed target-mediated miRNA protection (TMMP; Chatterjee *et al.*, 2011). Thus, miRNA passenger strands that lack target sites are released from Argonaute and subject to degradation. The 5′→3′ exonucleases XRN-1 and XRN-2 degrade miRNAs that lose association with Argonaute (Fig. 1.7; Chatterjee and Groszhans, 2009; Chatterjee *et al.*, 2011). Although the accumulation of let-7 miRNA can be regulated by the availability of target site interactions in *C. elegans*, whether TMMP functions in other organisms is yet to be determined.

While Argonaute and the AIN-1/-2 (GW182-related) proteins are required for miRNA function, several cofactors have been found to modulate miRNA activity. The TRIM-NHL family of proteins, which contain TRIM (tripartite-containing motif; RING, B-Box, coiled-coil) and NHL domains (named after the first three proteins discovered to contain this motif; NCL-1, Ht2A, and LIN-41), includes a broadly conserved class of proteins involved in diverse biological pathways (Slack and Ruvkun, 1998). Two members of this class, NHL-2 in *C. elegans* and TRIM32 in mice, enhance the ability of let-7 miRNA to regulate target genes (Hammell *et al.*, 2009b; Schwamborn *et al.*, 2009). These proteins associate with Argonaute complexes and stimulate the repressive activity of miRNAs on certain targets through an unknown mechanism (Fig. 1.7). The small ribosome subunit protein, RPS-14, also co-precipitates with ALG-1 but, instead, seems to negatively regulate let-7 function in *C. elegans* (Fig. 1.7; Chan and Slack, 2009). Thus, the effectiveness of let-7 in target regulation is influenced not only by the level of the miRNA but also by the presence of specific miRNA complex accessory proteins.

3.2. Identification of let-7 targets

The first miRNA targets were identified as genetic suppressors of miRNA mutant phenotypes (Lee *et al.*, 1993; Moss *et al.*, 1997; Reinhart *et al.*, 2000; Slack *et al.*, 2000; Wightman *et al.*, 1993). Since miRNAs negatively regulate target gene expression, lf mutations that reduce miRNA target levels can compensate for the absence of the miRNA. For example, the reiteration of hypodermal seam cell divisions and rupturing vulva phenotypes displayed by *let-7* mutants are suppressed in worms that also contain mutations in *lin-41*, a direct target of *let-7* regulation (Reinhart *et al.*, 2000; Slack *et al.*, 2000).

In addition to the isolation of miRNA suppressors through traditional forward genetic screens, RNAi has enabled high-throughput screening to identify potential miRNA targets. Several transcription factors, including *daf-12,* emerged as *let-7* targets from an RNAi screen of candidate genes that contain predicted binding sites (Großhans *et al.*, 2005). The ability to detect single miRNA targets through genetic approaches suggests that miRNAs have a few key targets that are responsible for phenotypes associated with loss of the miRNA or that the targets are part of interconnected pathways that are sensitive to changes in expression of individual genes.

Various computational approaches have predicted dozens to hundreds of let-7 targets in *C. elegans* (Betel *et al.*, 2008; Enright *et al.*, 2003; Hammell *et al.*, 2008; John *et al.*, 2004; Kertesz *et al.*, 2007; Lall *et al.*, 2006; Miranda *et al.*, 2006; Ruby *et al.*, 2006, 2007b). While most programs incorporate seed pairing, RNA structure, free energy, and conservation of target sites, the stringency of these parameters accounts for some of the variation among the predictions (Bartel, 2009). Computational predictions are often tested by fusing the target 3′UTR to a reporter gene and determining if it is regulated in the presence of the miRNA. In many cases, disruption of seed pairing between the miRNA and target is sufficient and necessary to block regulation (Brennecke *et al.*, 2005; Doench and Sharp, 2004; Kloosterman *et al.*, 2004; Lai, 2005). However, some target sites appear more complicated. For example, systematic analysis of the *lin-41* 3′UTR revealed that multiple elements within and around the two let-7 binding sites are required for regulation (Vella *et al.*, 2004a,b).

Genome wide experimental approaches that analyze changes in gene expression in response to specific miRNAs have proven effective for identifying miRNA targets. After the recognition that established miRNA targets are regulated at the level of mRNA stability (Bagga *et al.*, 2005; Giraldez *et al.*, 2006; Wu and Belasco, 2005), microarray technology was embraced as a high-throughput method to detect potential targets (Huang *et al.*, 2007; Lim *et al.*, 2005; Schmitter *et al.*, 2006; Sood *et al.*, 2006). In the first of these studies, introduction of miR-1 or miR-124 into HeLa cells resulted in downregulation of genes that were enriched for complementary binding sites to the appropriate miRNA (Lim *et al.*, 2005). Microarray analysis of gene expression changes during early mouse development, as let-7 miRNA starts to accumulate, coupled with target prediction programs, indicates that several early embryonic genes, including HMGA2 and IMP-1, are directly repressed by this miRNA (Boyerinas *et al.*, 2008). Some of these targets are reexpressed in cancer cells when regulation by let-7 is lost (Boyerinas *et al.*, 2010).

Monitoring global changes in protein expression has also provided strong evidence for target regulation by specific miRNAs. Two studies that employed stable isotope labeling with amino acids in cell culture (SILAC) in conjunction with mass spectrometry identified hundreds of targets regulated by specific miRNAs (Baek *et al.*, 2008; Selbach *et al.*, 2008). A comparable proteomics approach performed with extracts from wild-type and *let-7* mutant worms focused on a set of predicted let-7 targets to analyze for changes in protein levels (Jovanovic *et al.*, 2010). Each of these proteomic studies also noted varying degrees of target mRNA destabilization associated with the changes in protein levels. Recently, ribosome profiling was used to compare changes in mRNA levels to polysome association, as an indicator of translational efficiency (Guo *et al.*, 2010). This study showed that changes in translation largely reflected the changes in target mRNA levels, leading to the conclusion that miRNA regulation is predominantly through target mRNA destabilization at least in some cell types (Guo *et al.*, 2010).

Since miRNAs form regulatory complexes with Argonaute and AIN-1/2 (GW182) proteins, target mRNAs can be detected in association with these factors (Beitzinger *et al.*, 2007; Easow *et al.*, 2007; Hendrickson *et al.*, 2008; Karginov *et al.*, 2007; Landthaler *et al.*, 2008; Zhang *et al.*, 2007, 2009). For example, AIN-1/2 co-precipitates with the majority of established miRNA targets in *C. elegans*, suggesting that many of the other isolated mRNAs are good candidates for regulation by the miRNA pathway (Zhang *et al.*, 2007). An advancement over the isolation of entire mRNA targets is the ability to detect the sequence of the mRNA fragment directly bound by the miRNA complex through a method called CLIP (cross-linking immunoprecipitation) This technique uses UV-irradiation to covalently bond proteins to nucleic acids, followed by immunoprecipitation of Argonaute complexes and deep sequencing of directly associated sequences (Zisoulis *et al.*, 2011). CLIP studies in mouse brain, mammalian tissue culture cells, and whole worms have revealed miRISC (miRNA-induced silencing complex) binding sites on a genome-wide scale (Chi *et al.*, 2009; Hafner *et al.*, 2010; Leung *et al.*, 2010; Zisoulis *et al.*, 2010). This method narrows the miRNA binding site to 50–100nt and provides biochemical evidence that an mRNA is bound by miRISC. In *C. elegans,* many of the well-established let-7 target sites, including those in *lin-41, daf-12,* and *hbl-1*, were detected by the CLIP method (Zisoulis *et al.*, 2010). While the exact miRNA recognition site is yet to be determined in most of the Argonaute bound sequences, several general features of miRNA targeting emerged from these studies; sites are predominantly in coding exons and 3′UTRs, seed pairing capacity is enriched within Ago bound regions, and association of Ago with 3′UTR sites is associated with target mRNA destabilization (Chi *et al.*, 2009; Hafner *et al.*, 2010; Zisoulis *et al.*, 2010).

3.3. Developmental role of let-7 and its targets

A common theme for targets of *let-7* regulation is a role in promoting cellular division and self-renewal (Büssing *et al.*, 2008). This is consistent with the phenotype of *let-7* mutants where the seam cells fail to terminally differentiate at the appropriate time and instead continue dividing (Fig. 1.3; Reinhart *et al.*,

2000). In *C. elegans* loss of *lin-41*, a direct let-7 target, results in premature cellular differentiation (Fig. 1.3; Reinhart *et al.*, 2000; Slack *et al.*, 2000). Consistent with its role in promoting the undifferentiated state in worms, the mouse homolog is concentrated in stem and early embryonic cells (Rybak *et al.*, 2009). Although the molecular function of the *C. elegans* LIN-41 protein is not yet known, recent work in mouse cells shows that mLin41 acts as an E3 ubiquitin ligase that modifies Ago2, reducing its stability (Rybak *et al.*, 2009). Thus, mLin41 can indirectly regulate the levels and function of mature miRNAs by antagonizing Argonaute (Fig. 1.7).

As mentioned above, several targets of let-7 are transcription factors. Loss of *hbl-1* or *daf-12* activity results in precious expression of adult fates in larval stage worms (Abrahante *et al.*, 2003; Großhans *et al.*, 2005; Lin *et al.*, 2003). In addition to regulation of *let-7* family members, several protein-coding genes have been identified as potential direct targets of DAF-12 and HBL-1 transcriptional control (Niwa *et al.*, 2009; Shostak *et al.*, 2004). How these transcriptional networks contribute to the maintenance of undifferentiated cell fates prior to expression of *let-7* is yet to be fully explored.

Mis-regulation of *let-60/RAS* contributes to the lethal phenotype of *let-7* mutants that rupture through the vulva (Fig. 1.3; Johnson *et al.*, 2005). The 3′UTR of *let-60/RAS* contains multiple conserved sites of complementarity to let-7 family miRNAs (Johnson *et al.*, 2005). These miRNAs act to repress expression of *let-60/RAS* in specific vulval precursor cells to restrict RAS signaling and promote adoption of appropriate cell fates. Remarkably, functional let-7 sites were also detected in the 3′UTRs of human RAS genes. Reduced expression of mature let-7 in several types of cancers, especially lung, is often linked to increased RAS protein (Boyerinas *et al.*, 2010; Johnson *et al.*, 2005). By regulating key signaling molecules and transcription factors, let-7 family miRNAs control a large network of genes that determine the correct timing of cell fates during worm development, making this regulatory pathway indispensible.

4. The Role of let-7 in Development and Disease Across Species

4.1. Stem cells and differentiation

Since several recent reviews detail the roles of mammalian let-7 miRNAs in regulating development and disease in vertebrates, only a few select highlights are presented here (Boyerinas *et al.*, 2010; Ivey and Srivastava, 2010; Jiang *et al.*, 2009; Nimmo and Slack, 2009; Osada and Takahashi, 2011). Consistent with its role in promoting differentiated cell fates in *C. elegans,* let-7 miRNAs are depleted from mammalian embryonic stem cells (ESCs; Fig. 1.8; Bar *et al.*, 2008; Hinton *et al.*, 2010; Laurent, 2008; Suh *et al.*, 2004). This expression

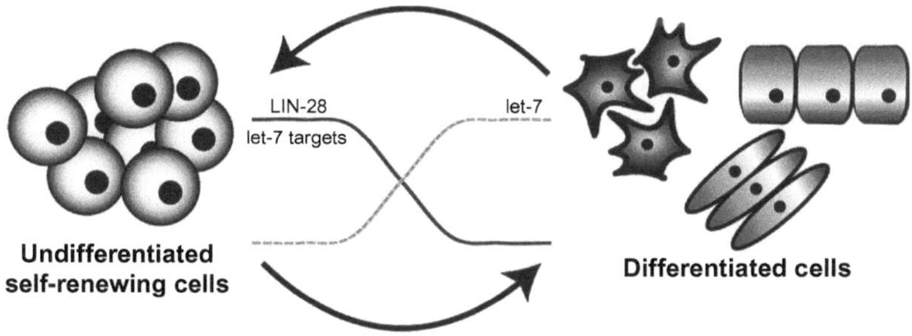

Figure 1.8 Regulation of cellular differentiation by let-7. Many types of self-renewing cells, such as embryonic stem cells and tumor cells, express high levels of LIN-28 and other targets of let-7 regulation that promote cell division. When cells are induced to differentiate, levels of these genes drop as let-7 expression increases, resulting in relatively high levels of let-7 in most differentiated cell types and repressed cell division programs.

pattern seems to be largely accomplished by LIN28, which is abundant in ESCs (Viswanathan and Daley, 2010). As described above, the LIN28 protein inhibits the accumulation of mature let-7, thus preventing cellular differentiation pathways. During stem cell differentiation, miR-125 and let-7 downregulate expression of LIN28 through complementary sites in its 3′UTR (Rybak *et al.*, 2008; Wu and Belasco, 2005). Thus, let-7 promotes its own expression by targeting its negative regulator LIN28. Given that let-7 is widely expressed in most somatic tissues, the role of LIN28 in inducing pluripotent stem (iPS) cells from fibroblasts may depend largely on its ability to repress let-7 and, thus, differentiation pathways. Additionally, aberrant activation of LIN28 occurs in some tumors, resulting in downregulation of let-7 and its ability to repress targets that promote cell division (Viswanathan and Daley, 2010).

The default pathway for ESCs depleted of miRNA processing factors is self-renewal (Kanellopoulou *et al.*, 2005; Murchison *et al.*, 2005; Wang *et al.*, 2007). However, introduction of mature let-7 miRNAs into such cells results in silencing of the self-renewal program, demonstrating the potent ability of these miRNAs to inhibit stem cell properties (Melton and Blelloch, 2010). This effect is not observed in normal ESCs because an opposing set of miRNAs is sufficient to maintain the self-renewal properties in the presence of let-7. Given the fundamental role of let-7 in promoting differentiation pathways, it is perhaps not surprising that inhibition of these miRNAs in mouse fibroblasts enhances the ability of reprogramming factors to produce iPS cells (Melton and Blelloch, 2010).

4.2. Cancer

Cancer results when cells fail to either differentiate properly or acquire the ability to self-renew, like stem cells. In many cases, this requires the activation of genes repressed by let-7 miRNAs. This is accomplished by downregulation of mature let-7 miRNA levels or loss of target regulation through deletion of 3′UTR target sites, for example. Some or all of the let-7 family miRNAs are depleted in tumor compared to normal tissues for many types of cancer (Boyerinas *et al.*, 2010). While reexpression of LIN28 explains some cases of decreased let-7 levels, other mechanisms for reducing let-7 in cancer cells are yet to be determined. The HMGA2 gene is an example of a let-7 target that escapes regulation in some types of tumors through truncation of its 3′UTR, which removes the let-7 binding sites (Lee and Dutta, 2007; Mayr *et al.*, 2007; Shell *et al.*, 2007). HMGA2 encodes a chromatin-associated nonhistone protein that is highly expressed in embryonic tissues, and undetectable is most differentiated adult cells. Chromosomal rearrangements can separate the HMGA2 3′UTR from its open reading frame, allowing for unregulated expression of the protein (Lee and Dutta, 2007; Mayr *et al.*, 2007; Shell *et al.*, 2007). The truncated HMGA2 gene has been shown to promote oncogenesis, underscoring the importance of regulation through its 3′UTR (Mayr *et al.*, 2007).

The first indication that let-7 may act as a tumor suppressor came with the discovery that it has a conserved role in regulating the expression of RAS (Johnson *et al.*, 2005). The three human RAS oncogenes each contain multiple let-7 binding sites in their 3′UTRs. The RAS proteins are frequently upregulated in lung tumors, where let-7 family miRNAs tend to be downregulated (Johnson *et al.*, 2005; Takamizawa *et al.*, 2004). A direct role for let-7 in regulating RAS and preventing lung tumor growth has been demonstrated in cell culture and mouse model systems (Esquela-Kerscher *et al.*, 2008; Johnson *et al.*, 2005, 2007; Kumar *et al.*, 2008; Takamizawa *et al.*, 2004). In addition to RAS, several oncogenes, such as IMP and MYC, have been identified as direct let-7 targets, further supporting its classification as a tumor suppressor (Boyerinas *et al.*, 2008; Kim *et al.*, 2009a). Although downregulation of let-7 family miRNAs has been associated with many types of cancers, there are a few examples of increased let-7 expression in certain tumors (Boyerinas *et al.*, 2010). Thus, the role of let-7 as a tumor suppressor will likely depend on the cellular environment and targets available for regulation.

4.3. Neuronal functions

The let-7 family miRNAs are highly expressed in neuronal cells across species, suggesting important roles in neurologic development and function (Caygill and Johnston, 2008; Johnson *et al.*, 2003; Sempere *et al.*, 2004; Smirnova *et al.*, 2005; Sokol *et al.*, 2008). Consistent with increasing expression of let-7 miRNAs during mammalian brain development, these miRNAs repress neuronal proliferation and promote differentiation programs (Nishino *et al.*, 2008; Rybak *et al.*, 2008; Schwamborn *et al.*, 2009;

Sokol *et al.*, 2008; Zhao *et al.*, 2010). In Drosophila, let-7 regulates the timing of neuromuscular remodeling events and adult behaviors (Caygill and Johnston, 2008; Sokol *et al.*, 2008). Additional work in Drosophila has revealed a potential link between disruption of let-7 function and Parkinson's disease. The Parkinson's disease related protein LRRK2 antagonizes the ability of let-7 to repress target genes (Gehrke *et al.*, 2010). In particular, mis-regulation of the let-7 target *dp* results in loss of fly dopaminergic neurons, exemplifying the importance of let-7 activity in neuronal health. It is likely that other targets of let-7 that promote cell division, like *dp*, also become mis-regulated by pathogenic LRRK2, which results in inappropriate activation of cell division programs in postmitotic neurons, leading to cell death.

5. Conclusions

Despite only about a decade's worth of research, miRNAs have emerged as key regulators of development. In particular, the let-7 miRNAs have been found to play pivotal roles in cell differentiation pathways. The strict conservation of let-7 sequences across animal species suggests an ancient function in regulating core genes that control cell division programs. The essential role of let-7 in regulating temporal development in model organisms relates to the importance of this miRNA in human disease phenotypes. In some cases, not only the general pathways but also the targets of let-7 are conserved across phylogeny. The let-7 miRNA has also been a model for understanding the complex transcriptional and post-transcriptional mechanisms that control miRNA biogenesis. The past decade introduced us to miRNAs and the basic steps for their synthesis and control of target expression. Future challenges include the elucidation of mechanisms controlling the levels of specific miRNAs, determination of features that faithfully predict miRNA target recognition in the endogenous context, and analysis of gene networks under normal and perturbed miRNA conditions. Given its conserved and essential role in animal development, it is likely that let-7 will continue to be a focus for understanding the complexities of miRNA expression and function.

Acknowledgments

We thank members of the Pasquinelli lab for critical reading of this chapter. The Pasquinelli lab is supported by the US National Institutes of Health (GM071654), Keck, Searle, V, Emerald and Peter Gruber Foundations.

REFERENCES

Abbott, A. L., Alvarez-Saavedra, E., Miska, E. A., Lau, N. C., Bartel, D. P., Horvitz, H. R., and Ambros, V. (2005). The let-7 MicroRNA family members mir-48, mir-84, and mir-241 function together to regulate developmental timing in Caenorhabditis elegans. *Dev. Cell* **9,** 403–414.

Abrahante, J. E., Daul, A. L., Li, M., Volk, M. L., Tennessen, J. M., Miller, E. A., and Rougvie, A. E. (2003). The Caenorhabditis elegans hunchback-like Gene lin-57/hbl-1 controls developmental time and is regulated by microRNAs. *Dev. Cell* **4,** 625–637.

Alemán, L. M., Doench, J., and Sharp, P. A. (2007). Comparison of siRNA-induced off-target RNA and protein effects. *RNA* **13,** 385–395.

Ambros, V. (1989). A hierarchy of regulatory genes controls a larva-to-adult developmental switch in C. elegans. *Cell* **57,** 49–57.

Ambros, V., and Horvitz, H. R. (1984). Heterochronic mutants of the nematode *Caenorhabditis elegans*. *Science* **226,** 409–416.

Ambros, V., Bartel, B., Bartel, D. P., Burge, C. B., Carrington, J. C., Chen, X., Dreyfuss, G., Eddy, S. R., Griffiths-Jones, S. A. M., Marshall, M., Matzke, M., Ruvkun, G., *et al.* (2003). A uniform system for microRNA annotation. *RNA* **9,** 277–279.

Arasu, P., Wightman, B., and Ruvkun, G. (1991). Temporal regulation of lin-14 by the antagonistic action of two other heterochronic genes, lin-4 and lin-28. *Genes Dev.* **5,** 1825–1833.

Baek, D., Villén, J., Shin, C., Camargo, F. D., Gygi, S. P., and Bartel, D. P. (2008). The impact of microRNAs on protein output. *Nature* **455,** 64–71.

Bagga, S., Bracht, J., Hunter, S., Massirer, K., Holtz, J., Eachus, R., and Pasquinelli, A. E. (2005). Regulation by let-7 and lin-4 miRNAs results in target mRNA degradation. *Cell* **122,** 553–563.

Bar, M., Wyman, S. K., Fritz, B. R., Qi, J., Garg, K. S., Parkin, R. K., Kroh, E. M., Bendoraite, A., Mitchell, P. S., Nelson, A. M., Ruzzo, W. L., Ware, C., *et al.* (2008). MicroRNA discovery and profiling in human embryonic stem cells by deep sequencing of small RNA libraries. *Stem Cells* **26,** 2496–2505.

Bartel, D. P. (2009). MicroRNAs: Target recognition and regulatory functions. *Cell* **136,** 215–233.

Beitzinger, M., Peters, L., Zhu, J. Y., Kremmer, E., and Meister, G. (2007). Identification of human microRNA targets from isolated argonaute protein complexes. *RNA Biol.* **4,** 76–84.

Berezikov, E., Chung, W. J., Willis, J., Cuppen, E., and Lai, E. C. (2007). Mammalian mirtron genes. *Mol. Cell* **28,** 328–336.

Betel, D., Wilson, M., Gabow, A., Marks, D. S., and Sander, C. (2008). The microRNA.org resource: Targets and expression. *Nucleic Acids Res.* **36,** D149–D153.

Bethke, A., Fielenbach, N., Wang, Z., Mangelsdorf, D. J., and Antebi, A. (2009). Nuclear hormone receptor regulation of microRNAs controls developmental progression. *Science* **324,** 95–98.

Blumenthal, T. (2005). Trans-splicing and operons. *In* "WormBook," (L. R. Girard, ed.)., The *C. elegans* Research Community, WormBook.

Bohnsack, M. T., Czaplinski, K., and Gorlich, D. (2004). Exportin 5 is a RanGTP-dependent dsRNA-binding protein that mediates nuclear export of pre-miRNAs. *RNA* **10,** 185–191.

Boyerinas, B., Park, S. M., Shomron, N., Hedegaard, M. M., Vinther, J., Andersen, J. S., Feig, C., Xu, J., Burge, C. B., and Peter, M. E. (2008). Identification of let-7-regulated oncofetal genes. *Cancer Res.* **68,** 2587–2591.

Boyerinas, B., Park, S. M., Hau, A., Murmann, A. E., and Peter, M. E. (2010). The role of let-7 in cell differentiation and cancer. *Endocr. Relat. Cancer* **17,** F19–F36.
Bracht, J., Hunter, S., Eachus, R., Weeks, P., and Pasquinelli, A. E. (2004). Trans-splicing and polyadenylation of let-7 microRNA primary transcripts. *RNA* **10,** 1586–1594.
Bracht, J. R., Van Wynsberghe, P. M., Mondol, V., and Pasquinelli, A. E. (2010). Regulation of lin-4 miRNA expression, organismal growth and development by a conserved RNA binding protein in C. elegans. *Dev. Biol.* **348,** 210–221.
Brennecke, J., Stark, A., Russell, R. B., and Cohen, S. M. (2005). Principles of microRNA-target recognition. *PLoS Biol.* **3,** e85.
Bussing, I., Yang, J.-S., Lai, E. C., and Grosshans, H. (2010). The nuclear export receptor XPO-1 supports primary miRNA processing in C. elegans and Drosophila. *EMBO J.* **29,** 1830–1839.
Büssing, I., Slack, F. J., and Großhans, H. (2008). let-7 microRNAs in development, stem cells and cancer. *Trends Mol. Med.* **14,** 400–409.
Caygill, E. E., and Johnston, L. A. (2008). Temporal regulation of metamorphic processes in Drosophila by the let-7 and miR-125 heterochronic microRNAs. *Curr. Biol.* **18,** 943–950.
Chalfie, M., Horvitz, H. R., and Sulston, J. E. (1981). Mutations that lead to reiterations in the cell lineages of C. elegans. *Cell* **24,** 59–69.
Chan, S. P., and Slack, F. J. (2009). Ribosomal protein RPS-14 modulates let-7 microRNA function in Caenorhabditis elegans. *Dev. Biol.* **334,** 152–160.
Chatterjee, S., and Groszhans, H. (2009). Active turnover modulates mature microRNA activity in Caenorhabditis elegans. *Nature* **461,** 546–549.
Chatterjee, S., Fasler, M., Büssing, I., and Großhans, H. (2011). Target-mediated protection of endogenous microRNAs in C. elegans. *Dev. Cell* **20,** 388–396.
Chi, S. W., Zang, J. B., Mele, A., and Darnell, R. B. (2009). Argonaute HITS-CLIP decodes microRNA-mRNA interaction maps. *Nature* **460,** 479–486.
Chiang, H. R., Schoenfeld, L. W., Ruby, J. G., Auyeung, V. C., Spies, N., Baek, D., Johnston, W. K., Russ, C., Luo, S., Babiarz, J. E., Blelloch, R., Schroth, G. P., *et al.* (2010). Mammalian microRNAs: Experimental evaluation of novel and previously annotated genes. *Genes Dev.* **24,** 992–1009.
Chung, W. J., Agius, P., Westholm, J. O., Chen, M., Okamura, K., Robine, N., Leslie, C. S., and Lai, E. C. (2011). Computational and experimental identification of mirtrons in Drosophila melanogaster and Caenorhabditis elegans. *Genome Res.* **21,** 286–300.
Czech, B., and Hannon, G. J. (2011). Small RNA sorting: Matchmaking for Argonautes. *Nat. Rev. Genet.* **12,** 19–31.
Ding, X. C., and Groszhans, H. (2009). Repression of C. elegans microRNA targets at the initiation level of translation requires GW182 proteins. *EMBO J.* **28,** 213–222.
Djuranovic, S., Nahvi, A., and Green, R. (2011). A parsimonious model for gene regulation by miRNAs. *Science* **331,** 550–553.
Doench, J. G., and Sharp, P. A. (2004). Specificity of microRNA target selection in translational repression. *Genes Dev.* **18,** 504–511.
Doench, J. G., Petersen, C. P., and Sharp, P. A. (2003). siRNAs can function as miRNAs. *Genes Dev.* **17,** 438–442.
Easow, G., Teleman, A. A., and Cohen, S. M. (2007). Isolation of microRNA targets by miRNP immunopurification. *RNA* **13,** 1198–1204.
Enright, A., John, B., Gaul, U., Tuschl, T., Sander, C., and Marks, D. (2003). MicroRNA targets in Drosophila. *Genome Biol.* **5,** R1.
Esquela-Kerscher, A., Johnson, S. M., Bai, L., Saito, K., Partridge, J., Reinert, K. L., and Slack, F. J. (2005). Post-embryonic expression of C. elegans microRNAs belonging to the lin-4 and let-7 families in the hypodermis and the reproductive system. *Dev. Dyn.* **234,** 868–877.

Esquela-Kerscher, A., Trang, P., Wiggins, J. F., Patrawala, L., Cheng, A., Ford, L., Weidhaas, J. B., Brown, D., Bader, A. G., and Slack, F. J. (2008). The *let-7* microRNA reduces tumor growth in mouse models of lung cancer. *Cell Cycle* **7,** 759–764.

Fabian, M. R., Sonenberg, N., and Filipowicz, W. (2010). Regulation of mRNA translation and stability by microRNAs. *Annu. Rev. Biochem.* **79,** 351–379.

Fang, Z., and Rajewsky, N. (2011). The impact of miRNA target sites in coding sequences and in 3′UTRs. *PLoS One* **6,** e18067.

Fay, D. S., Stanley, H. M., Han, M., and Wood, W. B. (1999). A *Caenorhabditis elegans* homologue of hunchback is required for late stages of development but not early embryonic patterning. *Dev. Biol.* **205,** 240–253.

Gehrke, S., Imai, Y., Sokol, N., and Lu, B. (2010). Pathogenic LRRK2 negatively regulates microRNA-mediated translational repression. *Nature* **466,** 637–641.

Giraldez, A. J., Mishima, Y., Rihel, J., Grocock, R. J., Van Dongen, S., Inoue, K., Enright, A. J., and Schier, A. F. (2006). Zebrafish MiR-430 promotes deadenylation and clearance of maternal mRNAs. *Science* **312,** 75–79.

Großhans, H., Johnson, T., Reinert, K. L., Gerstein, M., and Slack, F. J. (2005). The temporal patterning microRNA let-7 regulates several transcription factors at the larval to adult transition in C. elegans. *Dev. Cell* **8,** 321–330.

Gu, S., Jin, L., Zhang, F., Sarnow, P., and Kay, M. A. (2009). Biological basis for restriction of microRNA targets to the 3′ untranslated region in mammalian mRNAs. *Nat. Struct. Mol. Biol.* **16,** 144–150.

Guo, H., Ingolia, N. T., Weissman, J. S., and Bartel, D. P. (2010). Mammalian microRNAs predominantly act to decrease target mRNA levels. *Nature* **466,** 835–840.

Hafner, M., Landthaler, M., Burger, L., Khorshid, M., Hausser, J., Berninger, P., Rothballer, A., Ascano, M., Jr., Jungkamp, A.-C., Munschauer, M., Ulrich, A., Wardle, G. S., *et al.* (2010). Transcriptome-wide Identification of RNA-binding protein and microRNA target sites by PAR-CLIP. *Cell* **141,** 129–141.

Hagan, J. P., Piskounova, E., and Gregory, R. I. (2009). Lin28 recruits the TUTase Zcchc11 to inhibit let-7 maturation in mouse embryonic stem cells. *Nat. Struct. Mol. Biol.* **16,** 1021–1025.

Hammell, M., Long, D., Zhang, L., Lee, A., Carmack, C. S., Han, M., Ding, Y., and Ambros, V. (2008). mirWIP: MicroRNA target prediction based on microRNA-containing ribonucleoprotein-enriched transcripts. *Nat. Methods* **5,** 813–819.

Hammell, C. M., Karp, X., and Ambros, V. (2009a). A feedback circuit involving let-7-family miRNAs and DAF-12 integrates environmental signals and developmental timing in Caenorhabditis elegans. *Proc. Natl. Acad. Sci. USA* **106,** 18668–18673.

Hammell, C. M., Lubin, I., Boag, P. R., Blackwell, T. K., and Ambros, V. (2009b). nhl-2 modulates microRNA activity in Caenorhabditis elegans. *Cell* **136,** 926–938.

Hendrickson, D. G., Hogan, D. J., Herschlag, D., Ferrell, J. E., and Brown, P. O. (2008). Systematic identification of mRNAs recruited to argonaute 2 by specific microRNAs and corresponding changes in transcript abundance. *PLoS One* **3,** e2126.

Heo, I., Joo, C., Cho, J., Ha, M., Han, J., and Kim, V. N. (2008). Lin28 mediates the terminal uridylation of let-7 precursor microRNA. *Mol. Cell* **32,** 276–284.

Heo, I., Joo, C., Kim, Y. K., Ha, M., Yoon, M. J., Cho, J., Yeom, K. H., Han, J., and Kim, V. N. (2009). TUT4 in concert with Lin28 suppresses microRNA biogenesis through pre-microRNA uridylation. *Cell* **138,** 696–708.

Hinton, A., Afrikanova, I., Wilson, M., King, C. C., Maurer, B., Yeo, G. W., Hayek, A., and Pasquinelli, A. E. (2010). A distinct microRNA signature for definitive endoderm derived from human embryonic stem cells. *Stem Cells Dev.* **19,** 797–807. Doi: 10.1089/scd.2009.0224.

Huang, J. C., Babak, T., Corson, T. W., Chua, G., Khan, S., Gallie, B. L., Hughes, T. R., Blencowe, B. J., Frey, B. J., and Morris, Q. D. (2007). Using expression profiling data to identify human microRNA targets. *Nat. Methods* **4,** 1045–1049.

Huntzinger, E., and Izaurralde, E. (2011). Gene silencing by microRNAs: Contributions of translational repression and mRNA decay. *Nat. Rev. Genet.* **12,** 99–110.

Hutten, S., and Kehlenbach, R. H. (2007). CRM1-mediated nuclear export: To the pore and beyond. *Trends Cell Biol.* **17,** 193–201.

Ivey, K. N., and Srivastava, D. (2010). MicroRNAs as regulators of differentiation and cell fate decisions. *Cell Stem Cell* **7,** 36–41.

Jannot, G., Boisvert, M. E. L., Banville, I. H., and Simard, M. J. (2008). Two molecular features contribute to the Argonaute specificity for the microRNA and RNAi pathways in C. elegans. *RNA* **14,** 829–835.

Jiang, Q., Wang, Y., Hao, Y., Juan, L., Teng, M., Zhang, X., Li, M., Wang, G., and Liu, Y. (2009). miR2Disease: A manually curated database for microRNA deregulation in human disease. *Nucleic Acids Res.* **37,** D98–D104.

John, B., Enright, A. J., Aravin, A., Tuschl, T., Sander, C., and Marks, D. S. (2004). Human microRNA targets. *PLoS Biol.* **2,** e363.

Johnson, S. M., Lin, S. Y., and Slack, F. J. (2003). The time of appearance of the C. elegans let-7 microRNA is transcriptionally controlled utilizing a temporal regulatory element in its promoter. *Dev. Biol.* **259,** 364–379.

Johnson, S. M., Grosshans, H., Shingara, J., Byrom, M., Jarvis, R., Cheng, A., Labourier, E., Reinert, K. L., Brown, D., and Slack, F. J. (2005). RAS is regulated by the let-7 microRNA family. *Cell* **120,** 635–647.

Johnson, C. D., Esquela-Kerscher, A., Stefani, G., Byrom, M., Kelnar, K., Ovcharenko, D., Wilson, M., Wang, X., Shelton, J., Shingara, J., Chin, L., Brown, D., *et al.* (2007). The let-7 microRNA represses cell proliferation pathways in human cells. *Cancer Res.* **67,** 7713–7722.

Jovanovic, M., Reiter, L., Picotti, P., Lange, V., Bogan, E., Hurschler, B. A., Blenkiron, C., Lehrbach, N. J., Ding, X. C., Weiss, M., Schrimpf, S. P., Miska, E. A., *et al.* (2010). A quantitative targeted proteomics approach to validate predicted microRNA targets in C. elegans. *Nat. Methods* **7,** 837–842.

Kai, Z. S., and Pasquinelli, A. E. (2010). MicroRNA assassins: Factors that regulate the disappearance of miRNAs. *Nat. Struct. Mol. Biol.* **17,** 5–10.

Kanamoto, T., Terada, K., Yoshikawa, H., and Furukawa, T. (2006). Cloning and regulation of the vertebrate homologue of lin-41 that functions as a heterochronic gene in Caenorhabditis elegans. *Dev. Dyn.* **235,** 1142–1149.

Kanellopoulou, C., Muljo, S. A., Kung, A. L., Ganesan, S., Drapkin, R., Jenuwein, T., Livingston, D. M., and Rajewsky, K. (2005). Dicer-deficient mouse embryonic stem cells are defective in differentiation and centromeric silencing. *Genes Dev.* **19,** 489–501.

Karginov, F. V., Conaco, C., Xuan, Z., Schmidt, B. H., Parker, J. S., Mandel, G., and Hannon, G. J. (2007). A biochemical approach to identifying microRNA targets. *Proc. Natl. Acad. Sci. USA* **104,** 19291–19296.

Kertesz, M., Iovino, N., Unnerstall, U., Gaul, U., and Segal, E. (2007). The role of site accessibility in microRNA target recognition. *Nat. Genet.* **39,** 1278–1284.

Kim, H. H., Kuwano, Y., Srikantan, S., Lee, E. K., Martindale, J. L., and Gorospe, M. (2009a). HuR recruits let-7/RISC to repress c-Myc expression. *Genes Dev.* **23,** 1743–1748.

Kim, V. N., Han, J., and Siomi, M. C. (2009b). Biogenesis of small RNAs in animals. *Nat. Rev. Mol. Cell Biol.* **10,** 126–139.

Kloosterman, W. P., Wienholds, E., Ketting, R. F., and Plasterk, R. H. A. (2004). Substrate requirements for let-7 function in the developing zebrafish embryo. *Nucleic Acids Res.* **32,** 6284–6291.

Krek, A., Grun, D., Poy, M. N., Wolf, R., Rosenberg, L., Epstein, E. J., MacMenamin, P., da Piedade, I., Gunsalus, K. C., Stoffel, M., and Rajewsky, N. (2005). Combinatorial microRNA target predictions. *Nat. Genet.* **37,** 495–500.

Krol, J., Loedige, I., and Filipowicz, W. (2010). The widespread regulation of microRNA biogenesis, function and decay. *Nat. Rev. Genet.* **11,** 597–610.

Kumar, M. S., Erkeland, S. J., Pester, R. E., Chen, C. Y., Ebert, M. S., Sharp, P. A., and Jacks, T. (2008). Suppression of non-small cell lung tumor development by the let-7 microRNA family. *Proc. Natl. Acad. Sci. USA* **105,** 3903–3908.

Lagos-Quintana, M., Rauhut, R., Lendeckel, W., and Tuschl, T. (2001). Identification of novel genes coding for small expressed RNAs. *Science* **294,** 853–858.

Lagos-Quintana, M., Rauhut, R., Yalcin, A., Meyer, J., Lendeckel, W., and Tuschl, T. (2002). Identification of tissue-specific microRNAs from mouse. *Curr. Biol.* **12,** 735–739.

Lai, E. C. (2005). miRNAs: Whys and wherefores of miRNA-mediated regulation. *Curr. Biol.* **15,** R458–R460.

Lall, S., Grün, D., Krek, A., Chen, K., Wang, Y. L., Dewey, C. N., Sood, P., Colombo, T., Bray, N., MacMenamin, P., Kao, H.-L., Gunsalus, K. C., *et al.* (2006). A genome-wide map of conserved microRNA targets in C. elegans. *Curr. Biol.* **16,** 460–471.

Landthaler, M., Gaidatzis, D., Rothballer, A., Chen, P. Y., Soll, S. J., Dinic, L., Ojo, T., Hafner, M., Zavolan, M., and Tuschl, T. (2008). Molecular characterization of human Argonaute-containing ribonucleoprotein complexes and their bound target mRNAs. *RNA* **14,** 2580–2596.

Lau, N. C., Lim, L. P., Weinstein, E. G., and Bartel, D. P. (2001). An abundant class of tiny RNAs with probable regulatory roles in Caenorhabditis elegans. *Science* **294,** 858–862.

Laurent, L. C. (2008). MicroRNAs in embryonic stem cells and early embryonic development. *J. Cell. Mol. Med.* **12,** 2181–2188.

Lee, R. C., and Ambros, V. (2001). An extensive class of small RNAs in Caenorhabditis elegans. *Science* **294,** 862–864.

Lee, Y. S., and Dutta, A. (2007). The tumor suppressor microRNA let-7 represses the HMGA2 oncogene. *Genes Dev.* **21,** 1025–1030.

Lee, R. C., Feinbaum, R. L., and Ambros, V. (1993). The C. elegans heterochronic gene lin-4 encodes small RNAs with antisense complementarity to lin-14. *Cell* **75,** 843–854.

Lehrbach, N. J., Armisen, J., Lightfoot, H. L., Murfitt, K. J., Bugaut, A., Balasubramanian, S., and Miska, E. A. (2009). LIN-28 and the poly(U) polymerase PUP-2 regulate let-7 microRNA processing in Caenorhabditis elegans. *Nat. Struct. Mol. Biol.* **16,** 1016–1020.

Leung, A. K. L., Young, A. G., Bhutkar, A., Zheng, G. X., Bosson, A. D., Nielsen, C. B., and Sharp, P. A. (2010). Genome-wide identification of Ago2 binding sites from mouse embryonic stem cells with and without mature microRNAs. *Nat. Struct. Mol. Biol.* **18,** 237–244.

Lewis, B. P., Shih, I.h., Jones-Rhoades, M. W., Bartel, D. P., and Burge, C. B. (2003). Prediction of mammalian microRNA targets. *Cell* **115,** 787–798.

Lewis, B. P., Burge, C. B., and Bartel, D. P. (2005). Conserved seed pairing, often flanked by adenosines, indicates that thousands of human genes are microRNA targets. *Cell* **120,** 15–20.

Lim, L. P., Lau, N. C., Weinstein, E. G., Abdelhakim, A., Yekta, S., Rhoades, M. W., Burge, C. B., and Bartel, D. P. (2003). The microRNAs of Caenorhabditis elegans. *Genes Dev.* **17,** 991–1008.

Lim, L. P., Lau, N. C., Garrett-Engele, P., Grimson, A., Schelter, J. M., Castle, J., Bartel, D. P., Linsley, P. S., and Johnson, J. M. (2005). Microarray analysis shows that some microRNAs downregulate large numbers of target mRNAs. *Nature* **433,** 769–773.

Lin, S. Y., Johnson, S. M., Abraham, M., Vella, M. C., Pasquinelli, A., Gamberi, C., Gottlieb, E., and Slack, F. J. (2003). The C. elegans hunchback homolog, hbl-1, controls temporal patterning and is a probable microRNA target. *Dev. Cell* **4,** 639–650.

Lund, E., Güttinger, S., Calado, A., Dahlberg, J. E., and Kutay, U. (2004). Nuclear export of microRNA precursors. *Science* **303,** 95–98.

Maller-Schulman, B. R., Liang, X., Stahlhut, C., DelConte, C., Stefani, G., and Slack, F. J. (2008). The let-7 microRNA target gene, Mlin41/Trim71 is required for mouse embryonic survival and neural tube closure. *Cell Cycle* **7,** 3935–3942.

Martinez, N. J., Ow, M. C., Reece-Hoyes, J. S., Barrasa, M. I., Ambros, V. R., and Walhout, A. J. M. (2008). Genome-scale spatiotemporal analysis of Caenorhabditis elegans microRNA promoter activity. *Genome Res.* **18,** 2005–2015.

Mayr, C., Hemann, M. T., and Bartel, D. P. (2007). Disrupting the pairing between let-7 and Hmga2 enhances oncogenic transformation. *Science* **315,** 1576–1579.

Melton, C., and Blelloch, R. (2010). MicroRNA regulation of embryonic stem cell self-renewal and differentiation. *In* "The Cell Biology of Stem Cells," (E. Meshorer and K. Plath, eds.), pp. 105–117. Springer, USA.

Meneely, P. M., and Herman, R. K. (1979). Lethals, steriles and deficiencies in a region of the X chromosome of Caenorhabditis elegans. *Genetics* **92,** 99–115.

Miranda, K. C., Huynh, T., Tay, Y., Ang, Y. S., Tam, W. L., Thomson, A. M., Lim, B., and Rigoutsos, I. (2006). A pattern-based method for the identification of microRNA binding sites and their corresponding heteroduplexes. *Cell* **126,** 1203–1217.

Miska, E. A., Alvarez-Saavedra, E., Abbott, A. L., Lau, N. C., Hellman, A. B., McGonagle, S. M., Bartel, D. P., Ambros, V. R., and Horvitz, H. R. (2007). Most *Caenorhabditis elegans* microRNAs are individually not essential for development or viability. *PLoS Genet.* **3,** e215.

Moss, E. G., Lee, R. C., and Ambros, V. (1997). The cold shock domain protein LIN-28 controls developmental timing in C. elegans and is regulated by the lin-4 RNA. *Cell* **88,** 637–646.

Murchison, E. P., Partridge, J. F., Tam, O. H., Cheloufi, S., and Hannon, G. J. (2005). Characterization of Dicer-deficient murine embryonic stem cells. *Proc. Natl. Acad. Sci. USA* **102,** 12135–12140.

Murphy, D., Dancis, B., and Brown, J. (2008). The evolution of core proteins involved in microRNA biogenesis. *BMC Evol. Biol.* **8,** 92.

Newman, M. A., Thomson, J. M., and Hammond, S. M. (2008). Lin-28 interaction with the Let-7 precursor loop mediates regulated microRNA processing. *RNA* **14,** 1539–1549.

Nimmo, R., and Slack, F. (2009). An elegant miRror: MicroRNAs in stem cells, developmental timing and cancer. *Chromosoma* **118,** 405–418.

Nishino, J., Kim, I., Chada, K., and Morrison, S. J. (2008). Hmga2 promotes neural stem cell self-renewal in young but not old mice by reducing p16Ink4a and p19Arf expression. *Cell* **135,** 227–239.

Niwa, R., Hada, K., Moliyama, K., Ohniwa, R. L., Tan, Y. M., Carter, K. O., Chi, W., Reinke, V., and Slack, F. J. (2009). *C. elegans sym-1* is a downstream target of the Hunchback-like-1 developmental timing transcription factor. *Cell Cycle* **8,** 4147–4154.

Okamura, K., Hagen, J. W., Duan, H., Tyler, D. M., and Lai, E. C. (2007). The mirtron pathway generates microRNA-class regulatory RNAs in Drosophila. *Cell* **130,** 89–100.

Okamura, K., Phillips, M. D., Tyler, D. M., Duan, H., Chou, Y.t, and Lai, E. C. (2008). The regulatory activity of microRNA* species has substantial influence on microRNA and 3′ UTR evolution. *Nat. Struct. Mol. Biol.* **15,** 354–363.

Olsen, P. H., and Ambros, V. (1999). The lin-4 regulatory RNA controls developmental timing in Caenorhabditis elegans by blocking LIN-14 protein synthesis after the initiation of translation. *Dev. Biol.* **216,** 671–680.

Osada, H., and Takahashi, T. (2011). let-7 and miR-17-92: Small-sized major players in lung cancer development. *Cancer Sci.* **102,** 9–17.

Pasquinelli, A. E., Reinhart, B. J., Slack, F., Martindale, M. Q., Kuroda, M. I., Maller, B., Hayward, D. C., Ball, E. E., Degnan, B., Muller, P., Spring, J., Srinivasan, A., *et al.* (2000). Conservation of the sequence and temporal expression of let-7 heterochronic regulatory RNA. *Nature* **408,** 86–89.

Reinhart, B. J., Slack, F. J., Basson, M., Pasquinelli, A. E., Bettinger, J. C., Rougvie, A. E., Horvitz, H. R., and Ruvkun, G. (2000). The 21-nucleotide let-7 RNA regulates developmental timing in Caenorhabditis elegans. *Nature* **403,** 901–906.

Reinhart, B. J., Weinstein, E. G., Rhoades, M. W., Bartel, B., and Bartel, D. P. (2002). MicroRNAs in plants. *Genes Dev.* **16,** 1616–1626.

Roush, S. F., and Slack, F. J. (2009). Transcription of the C. elegans let-7 microRNA is temporally regulated by one of its targets, hbl-1. *Dev. Biol.* **334,** 523–534.

Ruby, J. G., Jan, C., Player, C., Axtell, M. J., Lee, W., Nusbaum, C., Ge, H., and Bartel, D. P. (2006). Large-scale sequencing reveals 21U-RNAs and additional microRNAs and endogenous siRNAs in C. elegans. *Cell* **127,** 1193–1207.

Ruby, J. G., Jan, C. H., and Bartel, D. P. (2007a). Intronic microRNA precursors that bypass Drosha processing. *Nature* **448,** 83–86.

Ruby, J. G., Stark, A., Johnston, W. K., Kellis, M., Bartel, D. P., and Lai, E. C. (2007b). Evolution, biogenesis, expression, and target predictions of a substantially expanded set of Drosophila microRNAs. *Genome Res.* **17,** 1850–1864.

Ruvkun, G., and Giusto, J. (1989). The Caenorhabditis elegans heterochronic gene lin-14 encodes a nuclear protein that forms a temporal developmental switch. *Nature* **338,** 313–319.

Rybak, A., Fuchs, H., Smirnova, L., Brandt, C., Pohl, E. E., Nitsch, R., and Wulczyn, F. G. (2008). A feedback loop comprising lin-28 and let-7 controls pre-let-7 maturation during neural stem-cell commitment. *Nat. Cell Biol.* **10,** 987–993.

Rybak, A., Fuchs, H., Hadian, K., Smirnova, L., Wulczyn, E. A., Michel, G., Nitsch, R., Krappmann, D., and Wulczyn, F. G. (2009). The let-7 target gene mouse lin-41 is a stem cell specific E3 ubiquitin ligase for the miRNA pathway protein Ago2. *Nat. Cell Biol.* **11,** 1411–1420.

Schmitter, D., Filkowski, J., Sewer, A., Pillai, R. S., Oakeley, E. J., Zavolan, M., Svoboda, P., and Filipowicz, W. (2006). Effects of Dicer and Argonaute down-regulation on mRNA levels in human HEK293 cells. *Nucleic Acids Res.* **34,** 4801–4815.

Schwamborn, J. C., Berezikov, E., and Knoblich, J. A. (2009). The TRIM-NHL protein TRIM32 activates microRNAs and prevents self-renewal in mouse neural progenitors. *Cell* **136,** 913–925.

Selbach, M., Schwanhausser, B., Thierfelder, N., Fang, Z., Khanin, R., and Rajewsky, N. (2008). Widespread changes in protein synthesis induced by microRNAs. *Nature* **455,** 58–63.

Sempere, L., Freemantle, S., Pitha-Rowe, I., Moss, E., Dmitrovsky, E., and Ambros, V. (2004). Expression profiling of mammalian microRNAs uncovers a subset of brain-expressed microRNAs with possible roles in murine and human neuronal differentiation. *Genome Biol.* **5,** R13.

Shell, S., Park, S. M., Radjabi, A. R., Schickel, R., Kistner, E. O., Jewell, D. A., Feig, C., Lengyel, E., and Peter, M. E. (2007). Let-7 expression defines two differentiation stages of cancer. *Proc. Natl. Acad. Sci. USA* **104,** 11400–11405.

Shin, C., Nam, J.-W., Farh, K. K.-H., Chiang, H. R., Shkumatava, A., and Bartel, D. P. (2010). Expanding the microRNA targeting code: Functional sites with centered pairing. *Mol. Cell* **38,** 789–802.

Shostak, Y., Van Gilst, M. R., Antebi, A., and Yamamoto, K. R. (2004). Identification of C. elegans DAF-12-binding sites, response elements, and target genes. *Genes Dev.* **18,** 2529–2544.

Slack, F. J., and Ruvkun, G. (1998). A novel repeat domain that is often associated with RING finger and B-box motifs. *Trends Biochem. Sci.* **23,** 474–475.

Slack, F. J., Basson, M., Liu, Z., Ambros, V., Horvitz, H. R., and Ruvkun, G. (2000). The lin-41 RBCC gene acts in the C. elegans heterochronic pathway between the let-7 regulatory RNA and the LIN-29 transcription factor. *Mol. Cell* **5,** 659–669.

Smirnova, L., Gräfe, A., Seiler, A., Schumacher, S., Nitsch, R., and Wulczyn, F. G. (2005). Regulation of miRNA expression during neural cell specification. *Eur. J. Neurosci.* **21,** 1469–1477.

Sokol, N. S., Xu, P., Jan, Y.-N., and Ambros, V. (2008). Drosophila let-7 microRNA is required for remodeling of the neuromusculature during metamorphosis. *Genes Dev.* **22,** 1591–1596.

Sood, P., Krek, A., Zavolan, M., Macino, G., and Rajewsky, N. (2006). Cell-type-specific signatures of microRNAs on target mRNA expression. *Proc. Natl. Acad. Sci. USA* **103,** 2746–2751.

Steiner, F. A., Hoogstrate, S. W., Okihara, K. L., Thijssen, K. L., Ketting, R. F., Plasterk, R. H. A., and Sijen, T. (2007). Structural features of small RNA precursors determine Argonaute loading in Caenorhabditis elegans. *Nat. Struct. Mol. Biol.* **14,** 927–933.

Suh, M.-R., Lee, Y., Kim, J. Y., Kim, S.-K., Moon, S.-H., Lee, J. Y., Cha, K.-Y., Chung, H. M., Yoon, H. S., Moon, S. Y., Kim, V. N., and Kim, K.-S. (2004). Human embryonic stem cells express a unique set of microRNAs. *Dev. Biol.* **270,** 488–498.

Sulston, J. E., and Horvitz, H. R. (1977). Post-embryonic cell lineages of the nematode, Caenorhabditis elegans. *Dev. Biol.* **56,** 110–156.

Takamizawa, J., Konishi, H., Yanagisawa, K., Tomida, S., Osada, H., Endoh, H., Harano, T., Yatabe, Y., Nagino, M., Nimura, Y., Mitsudomi, T., and Takahashi, T. (2004). Reduced expression of the let-7 microRNAs in human lung cancers in association with shortened postoperative survival. *Cancer Res.* **64,** 3753–3756.

Tritschler, F., Huntzinger, E., and Izaurralde, E. (2010). Role of GW182 proteins and PABPC1 in the miRNA pathway: A sense of deja vu. *Nat. Rev. Mol. Cell Biol.* **11,** 379–384.

Van Wynsberghe, P. M., Kai, Z. S., Massirer, K. B., Burton, V. H., Yeo, G. W., and Pasquinelli, A. E. (2011). LIN-28 co-transcriptionally binds primary let-7 to regulate miRNA maturation in Caenorhabditis elegans. *Nat. Struct. Mol. Biol.* **18,** 302–308.

Vella, M. C., Choi, E.-Y., Lin, S.-Y., Reinert, K., and Slack, F. J. (2004a). The C. elegans microRNA let-7 binds to imperfect let-7 complementary sites from the lin-41 3′UTR. *Genes Dev.* **18,** 132–137.

Vella, M. C., Reinert, K., and Slack, F. J. (2004b). Architecture of a validated microRNA:: target interaction. *Chem. Biol.* **11,** 1619–1623.

Viswanathan, S. R., and Daley, G. Q. (2010). Lin28: A microRNA regulator with a macro role. *Cell* **140,** 445–449.

Viswanathan, S. R., Daley, G. Q., and Gregory, R. I. (2008). Selective blockade of microRNA processing by Lin28. *Science* **320,** 97–100.

Wang, Y., Medvid, R., Melton, C., Jaenisch, R., and Blelloch, R. (2007). DGCR8 is essential for microRNA biogenesis and silencing of embryonic stem cell self-renewal. *Nat. Genet.* **39,** 380–385.

Wightman, B., Ha, I., and Ruvkun, G. (1993). Posttranscriptional regulation of the heterochronic gene lin-14 by lin-4 mediates temporal pattern formation in C. elegans. *Cell* **75,** 855–862.

Wu, L., and Belasco, J. G. (2005). Micro-RNA regulation of the mammalian lin-28 gene during neuronal differentiation of embryonal carcinoma cells. *Mol. Cell. Biol.* **25,** 9198–9208.

Wu, E., Thivierge, C., Flamand, M., Mathonnet, G., Vashisht, A. A., Wohlschlegel, J., Fabian, M. R., Sonenberg, N., and Duchaine, T. F. (2010). Pervasive and cooperative deadenylation of 3′UTRs by embryonic microRNA families. *Mol. Cell* **40,** 558–570.

Wulczyn, F. G., Smirnova, L., Rybak, A., Brandt, C., Kwidzinski, E., Ninnemann, O., Strehle, M., Seiler, A., Schumacher, S., and Nitsch, R. (2007). Post-transcriptional

regulation of the let-7 microRNA during neural cell specification. *FASEB J.* **21,** 415–426.
Yi, R., Qin, Y., Macara, I. G., and Cullen, B. R. (2003). Exportin-5 mediates the nuclear export of pre-microRNAs and short hairpin RNAs. *Genes Dev.* **17,** 3011–3016.
Yigit, E., Batista, P. J., Bei, Y., Pang, K. M., Chen, C.-C. G., Tolia, N. H., Joshua-Tor, L., Mitani, S., Simard, M. J., and Mello, C. C. (2006). Analysis of the C. elegans Argonaute family reveals that distinct Argonautes act sequentially during RNAi. *Cell* **127,** 747–757.
Zeng, Y., Yi, R., and Cullen, B. R. (2003). MicroRNAs and small interfering RNAs can inhibit mRNA expression by similar mechanisms. *Proc. Natl. Acad. Sci. USA* **100,** 9779–9784.
Zhang, L., Ding, L., Cheung, T. H., Dong, M.-Q., Chen, J., Sewell, A. K., Liu, X., Yates Iii, J. R., and Han, M. (2007). Systematic identification of C. elegans miRISC proteins, miRNAs, and mRNA targets by their interactions with GW182 proteins AIN-1 and AIN-2. *Mol. Cell* **28,** 598–613.
Zhang, L., Hammell, M., Kudlow, B. A., Ambros, V., and Han, M. (2009). Systematic analysis of dynamic miRNA-target interactions during C. elegans development. *Development* **136,** 3043–3055.
Zhao, C., Sun, G., Li, S., Lang, M.-F., Yang, S., Li, W., and Shi, Y. (2010). MicroRNA let-7b regulates neural stem cell proliferation and differentiation by targeting nuclear receptor TLX signaling. *Proc. Natl. Acad. Sci. USA* **107,** 1876–1881.
Zisoulis, D. G., Lovci, M. T., Wilbert, M. L., Hutt, K. R., Liang, T. Y., Pasquinelli, A. E., and Yeo, G. W. (2010). Comprehensive discovery of endogenous Argonaute binding sites in Caenorhabditis elegans. *Nat. Struct. Mol. Biol.* **17,** 173–179.
Zisoulis, D. G., Yeo, G. W., and Pasquinelli, A. E. (2011). Comprehensive identification of miRNA target sites in live animals. *Methods Mol. Biol.* **732,** 169–185.

CHAPTER TWO

Evolution, Expression, and Developmental Function of Hox-Embedded miRNAs

Jennifer H. Mansfield* *and* Edwina McGlinn†

Contents

* Department of Biology, Barnard College, New York, USA
† EMBL Australia, Australian Regenerative Medicine Institute, Monash University, Clayton, Australia

Current Topics in Developmental Biology, Volume 99
ISSN 0070-2153, DOI: 10.1016/B978-0-12-387038-4.00002-1

Abstract

Exquisite regulation of Hox protein activity is fundamental to the regionalization of the early embryo across diverse taxa. Highlighting the critical importance of these transcription factors, an astonishing number of different mechanisms have evolved to tightly coordinate their activity both in time and in space. The recent identification of numerous microRNAs that are not only embedded within Hox clusters but also target numerous Hox genes suggests an important role for these regulatory molecules in shaping Hox protein output. Here, we discuss the positioning of these miRNAs within clusters over evolutionary time, the unexpected complexity in miRNA processing and target interactions, and the current understanding of Hox-embedded miRNA function during development.

1. Introduction

Hox genes have fascinated developmental biologists for decades given their central role in regionalization of the early embryo, as demonstrated by the striking developmental defects observed following their functional alteration. Since their discovery in *Drosophila* (Lewis 1963, 1978), Hox genes have been identified in all bilaterian lineages and unique Hox signatures in divergent species likely represent a fundamental mechanism underlying morphological diversity. Hox genes provide cells their context within the embryonic anterior–posterior (A–P) axis, allowing them to coordinate within their local environment to perform specialized functions and give rise to region-specific morphologies, for example, the choice between vertebrae that are ribless (cervical) and ribbed (thoracic) in vertebrates, or between a wing and haltere in *Drosophila*. A unique feature of Hox biology is that multiple genes are expressed and work as a coordinated unit, a feature facilitated by their genomic positioning and by the multilayered and precise control this positioning affords.

In most bilaterian genomes, Hox genes are clustered in an ordered chromosomal arrangement that is translated into ordered expression along the embryonic A–P axis, a property termed colinearity. Colinearity is usually both spatial and temporal. Genes at the 3′ ends of clusters are expressed earlier and more anteriorly in the body, and those at the 5′ ends are expressed later and more posteriorly, producing a staggered and overlapping pattern of expression. It is thought that Hox proteins act combinatorially, with a given "code" of protein expression specifying each unique segmental identity. However, in vertebrates, members of paralogous groups are often redundant and may sometimes provide quantitative rather than qualitative patterning information. Further, activity is not purely combinatorial, since the most 5′ Hox protein(s) expressed in a given segment generally make the greatest contribution to phenotype, a phenomenon known as posterior prevalence (Kessel and Gruss, 1991; Kmita and Duboule, 2003).

In *Drosophila* and in mouse, each Hox gene has been knocked out and a detailed understanding of their functions is emerging (Reviewed in McGinnis and Krumlauf, 1992; Wellik, 2009). In flies, loss-of-function mutations cause homeotic transformations, in which segmental identities are transformed to those of other (usually more anterior) segments. In vertebrates, homeotic transformations are also observed although phenotypes are not always straightforward to interpret in these terms. Individual Hox mutations often do not result in widespread defects; however, paralogous gene knockouts are uncovering redundant but dramatic patterning functions of related genes (Horan *et al.*, 1995; McIntyre *et al.*, 2007; Wellik and Capecchi, 2003; reviewed comprehensively in Wellik, 2007). With this accumulating knowledge of Hox function, a key outstanding question in the field is how Hox gene expression is so exquisitely regulated.

Precise spatial and temporal Hox expression is governed by multiple mechanisms and is best studied in *Drosophila* and mouse. Individual (sometimes multiple) promoters and enhancers exist for most Hox genes, but in vertebrates, there are also cases of enhancer and even promoter sharing to produce polycistronic transcripts. Individual enhancers often exhibit both auto- and cross-regulation by Hox proteins. Promoter competition also occurs in both vertebrates and invertebrates. (e.g., see Coulombe *et al.*, 2010; Gould *et al.*, 1997; Herault *et al.*, 1997; Ohtsuki *et al.*, 1998; Sharpe *et al.*, 1998). At least in vertebrates, in which clusters are more consolidated than in other lineages (Duboule, 2007), global enhancer elements have been found that regulate collinear expression along the body axis or nested 5′ *Hoxd* expression in appendages (reviewed in Deschamps, 2007). Long noncoding RNAs within or near clusters can regulate Hox transcription in *cis*, for example, via transcriptional interference in the *Drosophila* bithorax complex, or in *trans*, for example, through effects on chromatin packaging, in vertebrates (Rinn *et al.*, 2007; Wang *et al.*, 2011; reviewed in Brock *et al.*, 2009). Collinear expression is found even in species where Hox clusters are dispersed and thus in those cases must depend on individual enhancer elements. Nevertheless, both global regulatory elements and extensive cross-regulation of individual Hox enhancers and promoters are often cited as reasons for the maintenance of Hox clustering in most bilaterian lineages.

Extensive posttranscriptional regulation also ensures proper Hox expression, evidenced by several cases where Hox protein and mRNA expression patterns are strikingly different (Brend *et al.*, 2003; Nelson *et al.*, 1996). Novel mechanisms of posttranscriptional regulation are continually being uncovered (Kondrashov *et al.*, 2011), and in recent years, it has become clear that one important mechanism is via microRNA (miRNA) control, and in particular, via miRNA genes embedded within the Hox clusters.

miRNAs are short, approximately 22-nucleotide (nt) RNA species that act posttranscriptionally to negatively regulate gene expression (reviewed in Bartel, 2009). Nascent primary transcripts undergo a series of RNA processing and translocation steps to yield mature miRNAs within the cytoplasm (reviewed in Carthew and Sontheimer, 2009). There, miRNAs bind to target mRNAs generally, although not exclusively, in the 3′UTR, which directs the RNA-induced silencing complex (RISC) to the mRNA. Almost all animal miRNAs bind their target sequence with imperfect complementarity, and the importance of particular nucleotides in mediating this interaction has become evident (Brennecke *et al.*, 2005; Grimson *et al.*, 2007; Lim *et al.*, 2005). The "seed" sequence is critical for specificity and consists of Watson–Crick pairing between nt 2–7, with an additional complementary base at nt 8 or an A at nt 1. This interaction reduces target protein levels by multiple mechanisms leading to translational inhibition, and potentially to a large extent through mRNA destabilization (Baek *et al.*, 2008; Carthew and Sontheimer, 2009; Guo *et al.*, 2010; reviewed in Carthew and Sontheimer, 2009). One important exception is the interaction between *miR-196* and its target gene *Hoxb-8*. Here, the miRNA binds with near-perfect complementarity, initiating endonucleolytic cleavage and degradation of the mRNA (Mansfield *et al.*, 2004; Yekta *et al.*, 2004).

miRNAs have emerged recently as important regulators of embryonic development. Some of the best-studied animal miRNAs are thought to help regulate cell lineage progression and/or maintain transcription profiles of differentiated cell types, by targeting suites of mRNAs associated with progenitor cell states and by fine-tuning levels of coexpressed transcripts (reviewed in Takacs and Giraldez, 2011). More generally, there has been recent suggestion that, through subtle but extensive regulation of mRNA targets both outside and within of targets' expression domains, miRNAs may function to canalize developmental programs, buffering against stochastic variations in transcription and raising the threshold of gene expression required to alter morphology (reviewed in Hornstein and Shomron, 2006). The extent to which miRNA regulation contributes to coordinated Hox output is under intense investigation following the observation that several highly conserved miRNA families are embedded within Hox clusters (Aravin *et al.*, 2003; Lagos-Quintana *et al.*, 2001, 2003; Lim *et al.*, 2003; Ronshaugen *et al.*, 2005; Yekta *et al.*, 2004). For example, *miR-10* resides in almost all taxa between *Hox4* and *5* paralogs and arose in early bilaterians. *miR-196* is located between *Hox9* and *10* paralogs and is specific to vertebrates and urochordates. In arthropods, *miR-iab4/8* is located at the analogous position to *miR-196* and interestingly, may have a similar function.

Hox-embedded miRNA families have a dynamic evolutionary history that often parallels that of Hox clusters themselves, with duplication, gene loss, and gene transposition occurring in various lineages (Fig. 2.1). Further, lineage-specific and developmentally regulated posttranscriptional variation including arm switching, seed shifting, and RNA editing increases the

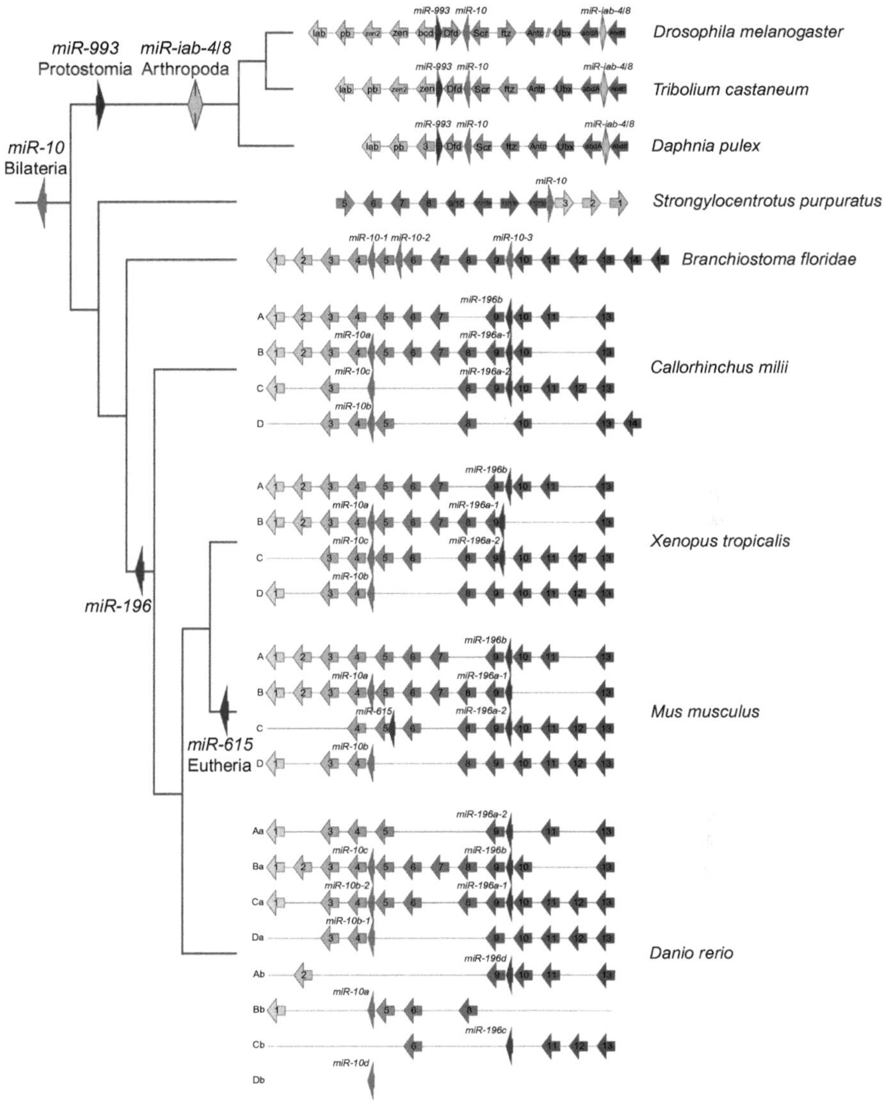

Figure 2.1 Phylogenetic distribution of Hox-embedded miRNA families. The five known families of miRNAs that have become positioned within bilaterian Hox clusters are diagrammed for several organisms. Each cluster is shown as a line, with blue arrows representing protein encoding genes and smaller arrows representing the miRNAs indicated (*miR-10*: light blue; *miR-993*: dark blue; *miR-iab-4/8*: yellow; *miR-196*: green; *miR-615*: red). The direction of transcription for each gene is indicated. Lineages within which each miRNA family became incorporated into Hox clusters are indicated on the cladogram. Homeobox genes that remain clustered but have lost Hox function (in insects) and are shown in gray. Please note that neither branch lengths nor gene or intergenic sequence lengths are drawn to scale, and this diagram does not reflect substantial variation in cluster sizes, intergenic DNA length and composition, sizes, structures, or splice isoforms of individual Hox transcripts, which are beyond the scope of this review. (See Color Insert.)

complexity of predicted target specificities. The location of these miRNAs within Hox clusters provides an extensive framework for predicting and understanding their developmental patterns of expression which have now been detailed in many model organisms (Asli and Kessel, 2010; Bender, 2008; Kloosterman *et al.*, 2006; Mansfield *et al.*, 2004; Qiu *et al.*, 2009; Ronshaugen *et al.*, 2005; Stark *et al.*, 2008; Woltering and Durston, 2006, 2008; Zhao *et al.*, 2010). This spatial information, coupled with the wealth of data being generated by deep sequencing projects, has provided important insight into the tissue restriction and relative expression levels of Hox-embedded miRNAs.

Hox-embedded miRNAs are predicted to extensively, though by no means exclusively, target protein-coding Hox mRNAs. Functional studies that have been performed to date have borne this out, confirming *in vivo* Hox mRNA regulation by *miR-196* and *miR-10* in vertebrates and by *miR-iab-4/8* in *Drosophila* (Asli and Kessel, 2010; Bender, 2008; Hornstein *et al.*, 2005; McGlinn *et al.*, 2009; Ronshaugen *et al.*, 2005; Stark *et al.*, 2008; Tyler *et al.*, 2008; Woltering and Durston, 2008).

In the following sections, we review current progress toward understanding the evolution of Hox-embedded miRNA families, on regulation of their embryonic expression patterns, and on their patterning functions. Hox genes are predicted targets of many other miRNAs, and conversely, Hox-embedded miRNAs target many non-Hox transcripts; however, here, we focus primarily on understanding the role of Hox-embedded miRNAs in Hox mRNA regulation.

2. Genomic Organization and Evolution

2.1. Hox clusters

The Hox gene clusters arose via duplication and divergence of an ancestral homeobox gene prior to the cnidarian–bilaterian split. The last common bilaterian ancestor likely possessed at least eight clustered Hox genes (Butts *et al.*, 2008; Lemons and McGinnis, 2006). Most extant bilaterians retained a single cluster of 8–9 genes, although there have been lineage-specific gene duplications, losses, and rearrangements. In chordates, *Abd-B*-class genes expanded to *Hox9–13* in vertebrates (*Hox9–15* in cephalochordates). Several whole-cluster duplications also occurred; the first was likely in stem vertebrates, and a second in basal gnathostomes, generating four clusters, *Hoxa–d* (Holland and Garcia-Fernandez, 1996; Prohaska and Stadler, 2004; Stadler *et al.*, 2004). An additional round of duplication and subsequent loss generated 7–8 clusters in teleosts (Amores *et al.*, 1998). Finally, 1–2 duplications occurred in the lineage leading to extant cyclostomes (lamprey and hagfish) (Force *et al.*, 2002; Fried *et al.*, 2003; Stadler *et al.*, 2004).

Species vary greatly in the degree of clustering, with vertebrates possessing the most compact and organized Hox clusters. In a few lineages, in contrast, clusters have become significantly rearranged or dispersed, although interestingly, axially restricted and even collinear expression of at least some Hox genes is preserved in most cases (reviewed in Duboule, 2007).

2.2. Hox-embedded miRNAs

2.2.1. *miR-10* and *miR-933*

Several miRNA gene families are embedded in Hox clusters (Fig. 2.1). *miR-10* is the most ancient and arose early in bilaterian history. Genome searches and deep sequencing of small RNA libraries have led to identification of *miR-10* sequences in species widely distributed across the bilaterian tree, but not in cnidarians or poriferans (Huang *et al.*, 2010b; Lagos-Quintana *et al.*, 2001; Lim *et al.*, 2003; Sperling *et al.*, 2009; Wheeler *et al.*, 2009). *miR-10* genes show uncommonly high sequence conservation, with universal conservation of the 5p seed sequence and near-perfect conservation outside the seed region (although considerable variation is generated via posttranscriptional processing; see below) (Ruby *et al.*, 2007; Wheeler *et al.*, 2009). The genomic position of *miR-10* between *Hox4* and *5* genes is also highly conserved (Fig. 2.1).

Whole-cluster duplications in vertebrates increased *miR-10* copy number, though multiple gene losses also occurred (Fig. 2.1). Ancestral gnathostomes likely had three copies on the *Hoxb–d* clusters. This condition is generally preserved except in mammals and birds, in which the *Hoxc* copy was lost, probably independently given its presence in the squamate *Anolis carolinensis* (Amemiya *et al.*, 2010; Kent, 2002; Ravi *et al.*, 2009; Richardson *et al.*, 2007; Tanzer *et al.*, 2005). Additional copies of *miR-10* arose via another round of Hox cluster duplication in teleosts. Fascinatingly, in zebrafish, the entire *HoxDb* cluster has been reduced to encode only *miR-10d* (Woltering and Durston, 2006). Although the functional significance of *miR-10d* preservation is unknown, it could suggest selection for a particular *miR-10* gene dosage in this lineage.

The genomic position of *miR-10* between *Hox4* and *Hox5* genes is highly conserved, but there are a few additional exceptions. *miR-10* was lost or dispersed in lineages where Hox clusters disintegrated, including *Ciona*, *Caenorhabditis elegans,* and *Caenorhabditis briggsae* (Aboobaker and Blaxter, 2003; Ikuta *et al.*, 2004). *Ceanorhabditis* genomes do, however, encode a closely related miRNA (*miR-57*), with an identical seed sequence to *miR-10* (Lau *et al.*, 2001). Further, some unusual duplications have occurred. *Amphioxus* was recently shown to have three *mir-10* copies embedded in its single cluster, one of which is in the conserved position between *Hox4* and *Hox5* (Fig. 2.1; Campo-Paysaa *et al.*, 2011). *Amphioxus miR-10* genes are also subject to particularly complex posttranscriptional regulation, generating at least five mature transcripts (discussed below;

Wheeler *et al.*, 2009). A similar genomic organization occurs in the annelid *Capitella*; based on comparison of the genomic locations of *miR-10* to those for protein-coding Hox genes (Frobius *et al.*, 2008; Wheeler *et al.*, 2009), copies are found between *Dfd* and *Scr*, between *lox5/ftz* and *Antp*, and 5′ (distal) to *lox4* (*Ubx*). It is striking that these distantly related organisms show such similar patterns of *miR-10* duplication.

Finally, *miR-993* is another *miR-10/miR-100* family member that probably arose as a duplication of *miR-10* early in the protostome lineage (Griffiths-Jones *et al.*, 2011). *miR-933* has been identified in multiple insect orders and in the annelid *Capitella*, and it is located 3′ (proximal) to *Hox4/Dfd* genes in all species examined (Marco *et al.*, 2010; Ruby *et al.*, 2007; Sperling *et al.*, 2009). No functional studies of *miR-993* have yet been reported.

2.2.2. *miR-196*

miR-196 likely arose in a common ancestor of chordates and urochordates. It has been found in *Ciona,* in an agnathan (lamprey), and in all major groups of gnathostomes (Heimberg *et al.*, 2010; Hendrix *et al.*, 2010; Tanzer *et al.*, 2005; Yekta *et al.*, 2004). All known vertebrate *miR-196* genes are located between *Hox9* and *Hox10* (Fig. 2.1) (Heimberg *et al.*, 2010; Tanzer *et al.*, 2005; Yekta *et al.*, 2004). It is notable that the *Amphioxus* genome, although lacking *miR-196,* has a *miR-10* paralog between *Hox9* and *Hox10,* and some invertebrate genomes also have a miRNA in a similar position. Together, this suggests that selection for a miRNA in this genomic location may have arisen prior to the evolution of *miR-196* itself. *Ciona miR-196* is not positioned near Hox genes; however, the *Ciona* Hox cluster has been significantly rearranged and dispersed (Hendrix *et al.*, 2010; Ikuta *et al.*, 2004).

miR-196 paralogs are located on the *Hoxa–c* clusters of most extant gnathostomes, although losses have occurred particularly in teleosts (Ravi *et al.*, 2009; Tanzer *et al.*, 2005; Yekta *et al.*, 2004). The *miR-196-5p* seed sequence is perfectly conserved across vertebrates. To date, there is no evidence for alternative posttranscriptional processing of *miR-196* as there is for *miR-10.*

One variation of the genomic position of *miR-196* has been described: in the amphibian *Xenopus tropicalis,* two of the three copies are located within introns of *Hox9* genes. As described below, *miR-196* and *Hox9* are cotranscribed in many vertebrates; however, *X. tropicalis* is only one known in which the major *Hox9* transcripts, rather than rarer variants, appear to be polycistronic (Tang and Maxwell, 2008).

2.2.3. *miR-iab-4/miR-iab-8*

miR-196 is specific to chordates; however, a miRNA unrelated in sequence occupies a similar position within the posterior Hox clusters of some arthropods. This miRNA is transcribed from both DNA strands to generate antisense *miR-iab-4* and *miR-iab-8* (also known as *miR-iab-4AS*) miRNAs. *miR-iab-4* has been identified in multiple insect orders and in the crustacean

Daphnia, and is located between *abd-A* and *Abd-B* in all cases (Fig. 2.1) (Aravin *et al.*, 2003; Miura *et al.*, 2011; Ronshaugen *et al.*, 2005). Both miRNAs show perfect conservation in the 5p and 3p seed sequences across all arthropods, and near-perfect conservation for all four mature transcripts (Miura *et al.*, 2011; Ronshaugen *et al.*, 2005).

2.2.4. *miR-615*

miR-615 is found only in eutherian mammals, in a single copy located within the second intron of *Hoxc-5*. *miR-615* may be processed from the major *Hoxc-5* transcript, as EST libraries do not suggest any additional transcript structures or alternative promoters (Kent, 2002). Mature *miR-615* transcripts processed from both arms of the pre-miRNA have been recovered, and transcripts have been isolated from multiple tissues, but to date no function for this miRNA has been reported (Chiang *et al.*, 2010; Landgraf *et al.*, 2007; Mineno *et al.*, 2006).

3. Predicted Targets of Hox-Embedded miRNAs: A Role in Posterior Prevalence, and More?

While miRNAs located within Hox clusters are predicted to target hundreds of mRNA transcripts, a striking bias toward targeting Hox genes themselves has been identified (Woltering and Durston, 2008; Yekta *et al.*, 2004, 2008). Moreover, the potential scale of this regulation is quite extensive with individual Hox 3′ UTRs having up to four (*Hoxa-7*, *Hoxc-8* in vertebrates) or even seven (*Ubx* in *Drosophila*) target binding sites. This statistical enrichment of Hox targets becomes even more intriguing when the asymmetric distribution of these targets is considered (Woltering and Durston, 2008; Yekta *et al.*, 2008). Predicted targets of Hox-embedded miRNAs are, in general, located genomically adjacent to but more 3′ (proximal) in the cluster than the miRNA (see Table 2.1). This is clearly exemplified by *miR-196* in *Mus musculus* where target sites for these miRNAs have evolved in 10 nearby Hox genes, all positioned more 3′ and expressed more anteriorly. This bias largely holds true for *miR-iab4/8* in *Drosophila* and for most of *miR-10* targets in vertebrates as well, although some targets do not follow this trend. In *Drosophila*, all *miR-10* targets are instead located genomically more 5′ (distal) though whether genomic positioning in this context strictly translates to more posterior developmental expression will be discussed in Section 4.2.1. This analysis is restricted to the 3′ UTR, where evolutionary conservation within a rapidly diverging region is important in identifying *bona fide* target sites. However, coding sequence may also represent a valid target for miRNA regulation (Woltering and Durston, 2008).

Table 2.1 Predicted targets of Hox-embedded miRNAs in *Drosophila* and mouse

Drosophila	miR-10-5p	*Scr* (light gray)
Drosophila	miR-10-3p	*Ubx, Abd-B* (light gray)
Drosophila	mir-iab-4/miR-iab-8	*Scr,* Antp, *Ubx, abd-A* (dark gray)
Drosophila	mir-iab-4/miR-iab-8	*AbdB* (light gray)
Mus musculus	miR-10	*Hoxa1, Hoxa3, Hoxb3, Hoxd1* (dark gray)
Mus musculus	miR-10	*Hoxd10* (light gray)
Mus musculus	mir-196	*Hoxa4, Hoxa5, Hoxa7, Hoxa9, Hoxb1, Hoxb6, Hoxb7, Hoxb8, Hoxc8, Hoxd8* (dark gray)

Gray shade indicates whether target genes are genomically located more 3′ (light gray) or 5′ (dark gray) in the cluster than the miRNA. (See Color Insert.)

The relative genomic positioning of Hox miRNAs and their targets has important implications for predicting their patterns of coexpression and functional interactions. Colinear Hox transcription results in staggered yet overlapping expression of adjacent genes, with each vertebral segment in mammals, for example, normally requiring input from at least two paralog groups (McIntyre *et al.*, 2007; reviewed in Wellik, 2007). The evolution of miRNA target sites in adjacent or near-adjacent genes allows at least the possibility that a miRNA is coexpressed with its target genes, where it may act as a tuning mechanism to optimize functional output (Fig. 2.2). Dependent on their relative expression levels at the posterior boundary of target Hox expression, this could essentially abolish functional output of the target mRNA, thereby sharpening boundary formation. In addition, positioning of the miRNA 5′ to its targets suggests that its expression will also extend more posteriorly than its targets, where it could act as a failsafe mechanism to repress aberrant Hox transcripts posterior to their domains (Fig. 2.2). This latter role would be expected to reinforce the phenomenon of posterior prevalence, whereby genes located more posteriorly exert a phenotypic dominance over more anterior programs. Indeed, although still poorly understood, posttranscriptional control is known to contribute to posterior prevalence (Gehring *et al.*, 2009). Experimental evidence exists for both tuning and failsafe mechanisms of miRNA–Hox interaction (Asli and Kessel, 2010; Hornstein *et al.*, 2005; McGlinn *et al.*, 2009), and understanding the relative contribution of each to a given developmental system will require detailed characterization of the level of coexpression between the miRNA and its target genes.

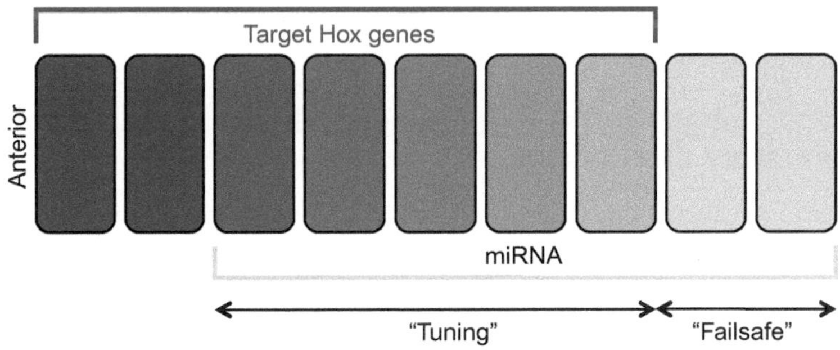

Figure 2.2 Potential functional interactions between Hox-embedded miRNAs (yellow) and their Hox targets (blue) along the anterior–posterior embryonic axis, based on anticipated extent of coexpression. Where coexpressed, the miRNA could modulate and optimize target output. The positioning of these miRNAs more 5′ in the cluster suggests their expression will extend more posteriorly and act as a failsafe mechanism to prevent output of aberrant transcripts. (See Color Insert.)

miRNA–target interactions are clearly predicted around targets' posterior boundaries, but they could be important in any region sensitive to target dosage. Vertebrate homeotic transformations often cluster at morphological transitions (e.g., cervical–thoracic, or thoracic–lumbar) (Wellik, 2009), suggesting heightened sensitivity to Hox dosage at these locations (although this may also to some degree reflect the relative ease with which transformations are identified). It is therefore, interesting to note a correlation between miRNA positions, their targets, and major morphological transition points. For example, *miR-196* forms a posterior bracket for genes involved in thoracic/trunk morphology. mir-196 is predicted or known to target 75% of the *Hox6–8* genes in mammals, which collectively specify thoracic identities and whose dosages are apparently most critical at each end of the thoracic region (McIntyre *et al.*, 2007; van den Akker *et al.*, 2001). *miR-196* negatively regulates *Hoxb-8* both anterior and posterior to the thoracic region (Asli and Kessel, 2010; Hornstein *et al.*, 2005; McGlinn *et al.*, 2009) and knockdown perturbs vertebral identities at the cervical–thoracic transition (McGlinn *et al.*, 2009; see Section 6.3). Thus, at least for this target, miRNA regulation occurs in both anterior and posterior regions of its expression.

Finally, it is interesting to note that although *miR-196* targets are genomically located more 3′, they rarely overlap with vertebrate *miR-10* targets suggesting clear division of function. However, this division is less clear in *Drosophila*, where *miR-10* and *miR-iab-4/8* share three predicted target genes.

In the following sections, we examine the developmental expression of Hox-embedded miRNAs in the context of their predicted Hox targets and review studies testing these interactions and their functional importance *in vivo*.

4. *MIR-10*

4.1. Transcript structure, processing, and regulatory mechanisms

Little is known about the structure of *miR-10* regulatory elements. A *miR-10a* proximal enhancer active during smooth muscle differentiation has been defined in mouse, but it is not clear whether it functions in other contexts (Huang *et al.*, 2010a). Interestingly, *miR-10* is frequently cotranscribed with *Hox3* or *Hox4* genes in vertebrates. In zebrafish, a polycistronic transcript originating near the *HoxB5a* transcription start site contains the *HoxB3a* coding sequence and *miR-10c* within an intron (Hadrys *et al.*, 2004; Woltering and Durston, 2008). Consistent with colinearity, this transcript has an anterior boundary of expression similar to *HoxB5a* though it is not translated since its expression does not overlap with HoxB3 protein. Interestingly, the transcript is upregulated following *miR-10* knockdown suggesting autoregulation by *miR-10* (Woltering and Durston, 2008). EST libraries further reveal cotranscription of human *HoxB-3* with *miR-10a* and *HoxD-3* with *miR-10b,* and mouse *miR-10b* with *Hoxd-3* or *Hoxd-4* (Benson *et al.*, 2004; Mainguy *et al.*, 2007). It has been proposed that via polycistronic transcription, *miR-10* (and perhaps other Hox-embedded miRNAs) may be important to block expression of aberrant Hox transcripts that arise due to the proximity of regulatory elements within Hox clusters (Woltering and Durston, 2008). Defining the number and usage of *miR-10* promoters and enhancers, and testing the functional significance of polycistronic transcripts, will be an important area of future study.

In *C. elegans,* the *miR-10*-related gene *miR-57* is not clustered with Hox genes. A *miR-57* enhancer that recapitulates its expression pattern was defined using reporter assays, and it is regulated by the Abd homolog Nob-1 (Zhao *et al.*, 2010; discussed further in Section 4.2.2). *miR-10* transcript structure and regulation in other invertebrates is not well characterized. In *Drosophila, miR-10* is alternatively spliced and, although the significance of this is unknown, both isoforms include the mature miRNA (Qian *et al.*, 2011).

In addition to variable transcript structures, *miR-10* is subject to extensive posttranscriptional processing. Despite the unusually high conservation of this gene (Ruby *et al.*, 2007; Wheeler *et al.*, 2009), a diversity of mature transcripts are produced, suggesting functional specialization within and across species. First, transcripts processed from both arms of the pre-miRNA (which have different target specificities) have been cloned from many organisms (Landgraf *et al.*, 2007; Lim *et al.*, 2003; Ruby *et al.*, 2007; Stark *et al.*, 2007; Wheeler *et al.*, 2009). Further, the expression ratio of the 5p and 3p products varies across species, a phenomenon termed arm switching (Griffiths-Jones *et al.*, 2011). miR-*933* similarly exhibits arm switching (Griffiths-Jones *et al.*, 2011; Marco *et al.*, 2010). Second, *miR-10* transcripts are subject to both RNA editing and seed shifting, documented in cephalochordates (which also show an unusual pattern of *miR-10* gene duplication; Fig. 2.1), echinoderms, insects, and a priapulid (lophotrochozoan) (Marco *et al.*, 2010; Wheeler *et al.*, 2009). Seed shifting allows multiple miRNAs to be processed from one primary transcript by varying the position of the 5′end of the mature miRNA. All of these modifications alter the predicted targeting specificity of processed transcripts. *miR-10* has not been functionally characterized outside of vertebrates, so the significance of alternative processing is unknown.

4.2. Developmental expression, target predictions, and function

4.2.1. *Drosophila miR-10*

In *Drosophila*, *miR-10* expression initiates at the cellular blastoderm stage, in a striped pattern within central segments. It is downregulated during gastrulation but reinitiates throughout the ventral nerve cord and in the posterior midgut and hindgut (Aboobaker *et al.*, 2005). *Drosophila miR-10* is predicted to regulate several Hox mRNAs, and the *miR-10*-5p and -3p transcripts have different predicted targets (Table 2.1). A strong, compensatory *miR-10-5p* site is predicted in the 3′UTR of *Scr* (a *Hox5* ortholog) that is conserved across *Drosophilids* and present in other insect orders (Brennecke *et al.*, 2005; Enright *et al.*, 2003; Miura *et al.*, 2011). *miR-10-3p* (the predominant product in *Drosophila* but not in other insect orders; Griffiths-Jones *et al.*, 2011) is predicted to bind the *Ubx* and *Abd-B 3′UTRs* via sites conserved across *Drosophilids* (Stark *et al.*, 2007). None of these potential interactions has been tested but predicted targeting of more distal genes within the cluster stands in contrast to vertebrates, where *miR-10* primarily targets *Hox1–3* genes (reviewed in Yekta *et al.*, 2008). However, at least for *Scr* (but not other targets), *Drosophila miR-10* expression actually extends further posteriorly. The *miR-10* primary transcript is expressed throughout central segments, but *Scr* is restricted to parasegments 2–3 (precursors of maxillary, labial, and T1 segments) (Aboobaker *et al.*, 2005; Kuroiwa *et al.*, 1985; Riley *et al.*, 1987). Thus, despite their genomic positions, this interaction appears to follow the trend of targeting more anteriorly expressed genes.

4.2.2. *C. elegans miR-57*

Although *C. elegans* lacks clustered Hox genes, *miR-57* is expressed in a Hox-like, axially restricted pattern, initiating in the posterior embryo in cells of diverse lineages. Expression persists in larvae and adults (Zhao *et al.*, 2010).

While the *miR-57* enhancer is positively regulated by the Abd-class Nob-1, *miR-57*, in turn, directly targets *nob-1*. High-resolution expression analysis and lineage mapping revealed that the two genes are coexpressed in all *nob-1*-expressing cells; however, *nob-1* expression precedes *miR-57* by approximately one cell division. This interaction suggests that *miR-57* may act as part of a switch mechanism required to inactivate *nob-1* during lineage progression. Genetic data strongly support a functional interaction *in vivo*. *miR-57* loss-of-function partially rescues defects associated with a *nob-1* hypomorphic allele. Further, *miR-57* misexpression phenocopies posterior patterning defects found in *nob-1* null embryos. *miR-57* null animals show temperature-sensitive sterility and partially penetrant embryonic or larval arrest with defects in tail morphology. Whether or which of these phenotypes are due to *nob-1* misregulation is unknown but confirms a clear, if partially redundant, patterning role for *miR-57* (Zhao *et al.*, 2010).

4.2.3. Vertebrate *mir-10*

In many vertebrate species, *miR-10* paralogs are expressed early in development with an anterior limit roughly at the level of *Hox4* paralogs (Darnell *et al.*, 2006; He *et al.*, 2011a; Kloosterman *et al.*, 2006; Mansfield *et al.*, 2004; Wienholds *et al.*, 2005; Woltering and Durston, 2006, 2008). In zebrafish, expression is observed in the posterior trunk region at early stages, including both paraxial mesoderm and the central nervous system (CNS) tissue, with expression becoming largely restricted to the CNS over time (Kloosterman *et al.*, 2006; Wienholds *et al.*, 2005; Woltering and Durston, 2008). *miR-10* paralogs exhibit differential anterior boundaries of expression within the CNS; *miR-10a* and *miR-10c* are caudal to rhombomere 6/7, while *miR-10b* and *miR-10d* are caudal to the hindbrain. It is quite striking that *miR-10d* expression is so similar to paralogs, given that all Hox genes surrounding *miR-10d* and global regulatory elements of this cluster were lost (Woltering and Durston, 2006). In chick, additional expression was observed in the developing branchial arches and forelimb bud, though absent from the hindlimb (Darnell *et al.*, 2006). High-throughput direct sequencing of miRNAs expressed in whole mouse embryos at E7.5, E9.5, and E12.5 has allowed the first glimpse of relative expression levels of Hox-embedded paralogous miRNAs during development (Chiang *et al.*, 2010). Absolute read counts for *miR-10a* and *miR-10b* are broadly comparable; however, the relative temporal expression of each miRNA is different. *miR-10b* is highest at E7.5, while *miR-10a* exhibited highest relative expression at E12.5. Deep sequencing approaches also indicate expression of *miR-10* paralogs in developing somites (Rathjen *et al.*, 2009) as well as the embryonic and adult kidney (Landgraf *et al.*, 2007; http://microrna.org)

In vivo validation of *miR-10* targeting has been most extensively characterized in zebrafish. Two predicted targets, *HoxB1a* and *HoxB3a*, are expressed in discrete rhombomeric stripes, with additional diffuse expression in the very posterior hindbrain and into the spinal cord. While anterior hindbrain expression is more anterior to/mutually exclusive with *miR-10c*, forced *miR-10* expression can reduce both target transcripts at this location and direct repression is supported by *in vivo* sensor assays (Woltering and Durston, 2008). In the spinal cord where *miR-10* paralogs are coexpressed with target genes, *miR-10* loss-of-function studies result in upregulation and posterior expansion of *HoxB1a* and *HoxB3a*, supporting the *in vivo* relevance of these target gene interactions. *HoxB1b* and *HoxA1a* are also predicted *miR-10* targets; however, *HoxA3a* did not respond to *miR-10* manipulation at least in this developmental context. The repression of anterior Hox genes by *miR-10*, which was shown to synergize with more classic Hox-mediated repressive mechanisms, lends strong support to a multifaceted approach to posterior prevalence.

With this compelling molecular evidence, and the observation that *miR-10* overexpression phenocopies loss of anterior *HoxB1a* as judged by mis-patterning of rhombomere 4 and branchiomotor neuron migration, it was perhaps surprising that no overt phenotype was observed in *miR-10* loss-of-function fish (Woltering and Durston, 2008). This suggests that *miR-10* regulation is not a principal mechanism restricting target Hox gene levels in the zebrafish embryo, or at least, that additional compensatory mechanisms limit phenotypic outcomes. To date, loss-of-function studies have not been performed in any other species and will be an important resource in dissecting *miR-10* function.

5. *MIR-IAB-4/MIR-IAB-8*

5.1. Transcript structure, processing, and upstream regulatory mechanisms

The *Drosophila iab-4* and *iab-8* loci were originally defined through fine-scale mutational analysis of the *BX-C* and mapped between *abd-A* and *Abd-B* (reviewed in Lewis, 1978). Partially overlapping transcripts are generated from opposite DNA strands within this region (Bae *et al.*, 2002; Cumberledge *et al.*, 1990), generating antisense miRNAs, *miR-iab-4* and *miR-iab-8* (Aravin *et al.*, 2003; Ruby *et al.*, 2007). The primary *miR-iab-4* transcript has alternate polyadenylation sites, one of which excludes the miRNA precursor, although differential expression of these isoforms has not been reported (Bender, 2008; Cumberledge *et al.*, 1990). *miR-iab-8* is processed from a very long (at least 120 kb) transcription unit that initiates within the *iab-8* locus and extends proximally through the *abd-A* coding sequence, thus encompassing nearly the entire *abd-A–Abd-B* intergenic region (Bae *et al.*, 2002; Bender, 2008; Zhou *et al.*, 1999). Mature transcripts processed from both arms of each miRNA have been detected; however, the 5p product predominates for *miR-iab-4* and is almost exclusively expressed for *miR-iab-8* (Aravin *et al.*, 2003; Bender, 2008; Ruby *et al.*, 2007; Stark *et al.*, 2008; Tyler *et al.*, 2008).

5.2. Integrating developmental expression with target predictions

The rostral boundaries of *Drosophila miR-iab-4* and *miR-iab-8* expression are consistent with colinearity. Their expression patterns are nonoverlapping with one another and are largely complementary to predicted and known Hox targets, which include more proximal (3′) Hox mRNAs (Table 2.1) (Bender, 2008; Ronshaugen *et al.*, 2005; Stark *et al.*, 2008; Tyler *et al.*, 2008). Further, it has been noted that the mutually exclusive expression of *miR-iab-4* and *miR-iab-8* could be generated or reinforced through mutual repression, perhaps via transcriptional interference or posttranscriptionally through annealing of transcripts to produce of double-stranded RNA (Stark *et al.*, 2008).

miR-iab-4 is expressed in precursors of segments A2–A7 (parasegments 8–12) at the cellular blastoderm stage and becomes refined to a subset of ectodermal cells during germband extension. Ectodermal expression overlaps with Abd-A, but levels are inverse to the validated target Ubx in segments where they are coexpressed, with areas of highest *miR-iab-4* corresponding to areas of lowest Ubx protein. Although *miR-iab-4* expression is highest in embryos, processed transcripts are present in larvae and adult (Bender, 2008; Ronshaugen *et al.*, 2005; Stark *et al.*, 2008).

miR-iab-8 is expressed in the precursors of segments A8–A9 (parasegments 13–14), with similar timing to *miR-iab-4*. Expression in late embryos becomes refined to the CNS within these segments. Although primary transcripts can be detected by RT-PCR at all stages, processed *miR-iab-8* appears limited to late embryos and early larvae. In contrast to *miR-iab-4*, *miR-iab-8* is expressed only in the posterior segments, overlapping with Abd-B, but where its other predicted and known targets (*Antp, Ubx,* and *abd-A*) are not translated at all (Bender, 2008; Lewis, 1998; Stark *et al.*, 2008; Tyler *et al.*, 2008).

In *D. melanogaster, the Ubx* 3′ UTR bears seven predicted binding sites for *miR-iab-4–5p*, five or six of which are also predicted to bind more strongly to *miR-iab-8-5p*. Most sites are conserved across *Drosophilids*, and some across arthropods (Bender, 2008; Miura *et al.*, 2011; Ronshaugen *et al.*, 2005; Stark *et al.*, 2003, 2008; Tyler *et al.*, 2008). Although individual target sites evolve rapidly, site variation (at least within *Drosophila*) appears compensatory, such that loss or weakening of one binding site often accompanies gain or strengthening of another. Such changes might be predicted to maintain a similar level of regulation of *Ubx* across species (Ronshaugen *et al.*, 2005; Stark *et al.*, 2008). *In vivo* reporter assays have confirmed that both *miR-iab-8* and, to a lesser extent, *miR-iab-4* can repress the *D. melanogaster Ubx* 3′UTR (Ronshaugen *et al.*, 2005; Stark *et al.*, 2008; Tyler *et al.*, 2008).

miR-iab-4 and *miR-iab-8* are also predicted to target *Antp*, *abd-A*, and *Abd-B* (Table 2.1). Between 1 and 3 binding sites for each have been predicted in the *Antp* 3′ UTRs of most *Drosophilids*. In addition, six perfectly conserved *miR-iab-8* binding sites are present in the *abd-A* 3′ UTR (Stark *et al.*, 2008; Tyler *et al.*, 2008). Hox regulation by these miRNAs appears to be a common feature of arthropods, although the number and degree of Hox mRNAs targeted is labile; varying numbers of sites in *Antp*, *Ubx*, *abd-A*, *Abd-B*, and *Scr* are predicted in non-*Drosophilid* insects and in *Daphnia* (Miura *et al.*, 2011). *In vivo* reporter assays show that *miR-iab-8*, but not *miR-iab-4*, can repress the *Drosophila abd-A* 3′UTR (Stark *et al.*, 2008; Tyler *et al.*, 2008). Similar assays have not yet been reported for *Antp*. Finally, although *miR-iab-4* and *miR-iab-8* sites have also been predicted in the *Drosophila Abd-B 3′UTR* (the only predicted *miR-iab-8* target that is coexpressed with it), regulation by *miR-iab-8* was not detected in a reporter assay in cultured cells (Miura *et al.*, 2011; Stark *et al.*, 2008; Thomsen *et al.*, 2010).

5.3. Function

miR-iab-4 or *miR-iab-8* misexpression in haltere discs induces a classic *Ubx* loss-of-function phenotype: partial transformation of halteres to wings. The *miR-iab-8*-induced homeotic transformation is more complete than that induced by *miR-iab-4*, in keeping with its stronger regulation of the *Ubx* 3′ UTR in reporter assays (Ronshaugen *et al.*, 2005; Stark *et al.*, 2008; Tyler *et al.*, 2008).

Given these interactions, the *miR-iab-4/8* loss-of-function phenotype is surprisingly subtle (Bender, 2008). Flies homozygous for a targeted deletion of both pre-miRNAs (which leaves other *iab* elements intact) are viable and lack patterning phenotypes in adult cuticles. Homozygotes are male and female sterile, but both sexes produce mature gametes; sterility seems to be due to a behavioral or locomotor defect. Interestingly, subtle upregulation of Ubx protein is observed in the neural ectoderm of late gastrula abdominal segments and is most dramatic within the normal *miR-iab-8* expression domain posterior to A8. Whether this upregulation causes the observed phenotypic defects remains to be determined. However, complementation tests show that sterility is associated specifically with loss of *miR-iab-8*; alleles that block transcription of only *miR-iab-4* fully complement the miRNA deletion (Bender, 2008). No changes in Abd-A or Abd-B protein were observed by immunohistochemical analysis. Together, these results show that miRNA interaction cannot be the primary mode for Hox mRNA repression in posterior segments. Dissecting the redundancy with other transcriptional and posttranscriptional regulatory mechanisms that operate in this system will be an important area for future study.

Another important question raised by these experiments is why Ubx levels are affected only at late gastrula stages, despite the earlier and compellingly complementary expression with both miRNAs. In fact, *Ubx* has alternate polyadenylation sites that produce transcripts with varied degrees of miRNA regulation. The shorter transcript predominates at early stages and loss of *miR-iab4/8* has no effect on its expression levels despite bearing three predicted binding sites. The inclusion of distal 3′ UTR sequence, which incorporates an additional four potential *miR-iab-4/8* binding sites and is expressed at later stages, confers miRNA regulation (Bender, 2008; Thomsen *et al.*, 2010). This principle was further extended to other Hox genes *abd-A*, *Abd-B*, and *Antp*, all of which exhibit the same striking use of alternate polyadenylation resulting in either masking or availability of these transcripts to miRNA control. It is possible that early *miR-iab-4/8* expression is not functional and could even be due simply to coregulation with Abd-A and Abd-B, respectively. Alternatively, it may be that these miRNAs can repress the shorter transcript variants, though to a lesser extent, while serving in a failsafe mode; for example, they could prevent Ubx expression that is also repressed by Abd-A and Abd-B. Such a role could be tested by generating compound mutant animals. Further, how these Hox transcripts collectively switch to the long, miRNA-regulated form at a particular stage of development is an important question for future investigation. Whether similar mechanisms exist in mammalian Hox gene regulation is unknown; however, both mouse *Hoxa-4* and *Hoxa-7* transcripts have alternative polyadenylation sites which have the potential to alter the amplitude of Hox-embedded miRNA regulation (Yekta *et al.*, 2008). The temporal and spatial division of alternate transcript expression and their *in vivo* relevance is yet to be determined.

6. *MIR-196*

6.1. Transcript structure, processing, and upstream regulatory mechanisms

miR-196 regulatory elements have not been characterized, but there is evidence that many *miR-196* paralogs are cotranscribed with protein-coding Hox genes. Polycistronic transcripts containing *miR-196* pre-miRNAs within introns have been cloned for *Hoxa-5* (*X. tropicalis*), *Hoxa-9* (human, mouse, guinea pig, and zebrafish), *HoxB-7* (human), *Hoxb-9* (*X. tropicalis*), *HoxC-5* (human), *HoxC-6* (human), *Hoxc-9* (*X. tropicalis*), and *HoxC10a* (zebrafish) (Benson *et al.*, 2004; Griffiths-Jones, 2004; Mainguy *et al.*, 2007; Tang and Maxwell, 2008). In addition, an antisense transcript containing *miR-196b* and part of *HoxA-10* is expressed in humans (Mainguy *et al.*, 2007). In chick, EST libraries reveal a transcript with alternate polyadenylation sites, in which an antisense *miR-196* is present in one isoform (Benson *et al.*, 2004; Kent, 2002). Humans and mice both transcribe antisense *miR-196a-1* as well (Mainguy *et al.*, 2007). It is currently unknown whether or what roles these transcripts play.

6.2. Integrating developmental expression with target predictions

miR-196 expression analysis has revealed a strong consensus across many species: expression is axially restricted and more caudal than *miR-10*, as is expected given its chromosomal position. In zebrafish, robust *miR-196a* expression is observed by 24hpf throughout the posterior trunk including somitic tissue and the CNS (Wienholds *et al.*, 2005), similar to adjacent *Hox9* paralog expression (http://zfin.org). With time, expression becomes more restricted to the CNS, potentially spreading more anteriorly than some adjacent 3′ genes (http://zfin.org). Pectoral fin expression is suggested at 72 hpf (Wienholds *et al.*, 2005), with *miR-196a-1* specific pronephric expression identified following primary transcript detection (He *et al.*, 2011a). This dynamic pattern of expression is mirrored quite closely in *Xenopus* (Qiu *et al.*, 2009). In chick, Asli and Kessel (2010) provide evidence that *miR-196* expression is not initiated prior to Hamburger and Hamilton Stage 15 (HH15), a time when robust expression of the validated *miR-196* target *Hoxb-8* is observed in the hindlimb field and posterior unsegmented paraxial mesoderm. As development proceeds, the expression of *Hoxb-8* diminishes in posterior tissue and *miR-196* expression was shown to increase, resolving into a mutually exclusive pattern at the level of somite 23–24 (Asli and Kessel, 2010). This expression pattern is in keeping with either a classic "switch" or "failsafe" mechanism whereby the miRNA acts at a more posterior location to either actively repress or reinforce repression of anterior Hox gene expression and anterior morphological programs. However, these expression data are not consistent with morphological and molecular observations following loss-of-function studies in chick (McGlinn *et al.*, 2009). Here, loss of cumulative *miR-196* by antagomiR-based knockdown resulted in an expansion of *Hoxb-8* at HH12 and importantly, homeotic transformations of vertebral morphology at the cervical–thoracic boundary (somite 11–12). Classic grafting experiments (Kieny *et al.*, 1972), complemented by more recent genetic strategies in mouse (Carapuco *et al.*, 2005), support the view that vertebral identity is established in the PSM prior to somite segmentation, suggesting that very low levels of *miR-196* may function at this time and place.

In mouse, combined analysis of *miR-196a-1* and *miR-196a-2* expression using LNA probes (Asli and Kessel, 2010; Kloosterman *et al.*, 2006) supports early transgenic sensor approaches (Mansfield *et al.*, 2004), highlighting *miR-196* in posterior derivatives of all three germ layers. However, detailed section *in situ* analysis remains to be performed. While the paraxial expression limits of Hox genes are quite dynamic over time, the anterior limit of *miR-196a* defined by these approaches at embryonic day 9.5 (E9.5) (Asli and Kessel, 2010; Kloosterman *et al.*, 2006) is quite caudal when compared with genomically adjacent Hox genes (e.g., *Hoxb-9* expression up to prevertebra 3; Chen and Capecchi, 1997). This could reflect differential upstream regulation or, alternatively, differential stability of the miRNA relative to Hox transcripts, a point to consider in understanding the complexity of target regulation. Microarray analysis of the developing mouse limb buds identified a more than 20-fold enrichment of *miR-196a* in the hindlimb when compared to forelimb (Hornstein *et al.*, 2005). High-throughput direct sequencing at three developmental stages (Chiang *et al.*, 2010) showed quite striking differences in the relative expression levels of individual *miR-196* family members. *miR-196a* was more abundant at E12.5 than earlier stages, while *miR-196b* exhibits very high relative levels of expression at E7.5 (Chiang *et al.*, 2010). These data suggest that *miR-196* can potentially act at presomite stages, at least in mouse.

6.3. Function

Contrary to loss-of-function observations for *miR-10* and *miR-iab-4/8*, accumulating evidence supports a dramatic, nonredundant developmental role for the *miR-196* family of miRNAs. Much of this work has centered on the regulation of one key gene, *Hoxb-8*, which in mouse has a single target binding site exhibiting extensive complementarity to *miR-196* within its 3′ UTR (Yekta *et al.*, 2004). Akin to plant miRNA regulation, this suggests *miR-196* binding will initiate endonucleolytic cleavage between nt 10 and 11 and subsequent mRNA degradation in a RNAi-like manner. Importantly, such *Hoxb-8* cleavage products have been identified in the early mouse embryo (Mansfield *et al.*, 2004; Yekta *et al.*, 2004) confirming this mode of regulation *in vivo*. Genetic manipulation of *Hoxb-8* by both gain- and loss-of-function highlights the capability of this gene to affect patterning of the skeleton, nervous system, and A–P symmetry in the forelimb (Charite *et al.*, 1994; Fanarraga *et al.*, 1997; Greer and Capecchi, 2002; Holstege *et al.*, 2008; van den Akker *et al.*, 1999), and recent evidence suggests a functional interaction between *miR-196* and *Hoxb-8* in each of these developmental contexts.

Recent studies indicate that *miR-196* acts to clear or repress unwanted *Hoxb-8* activity from both neural and mesodermal tissue caudal to the domain of *Hoxb-8* expression (Asli and Kessel, 2010; Hornstein *et al.*, 2005). In the lumbar region of the chick CNS, enforced *Hoxb-8* expression caudal to its normal domain compromises motor neuron differentiation, highlighting the need to tightly regulate posterior *Hoxb-8* boundary formation. *In ovo miR-196* antisense oligo electroporation phenocopies these defects; however, no upregulation of *Hoxb-8* could be identified in this context, and whether additional target derepression contributes to the phenotype remains unclear (Asli and Kessel, 2010). Given this potential role in spinal cord generation, it is interesting to note that *miR-196* has also been shown to regulate key stages of the spinal cord regenerative process in the axolotl, *Ambystoma mexicanum*, however, potentially not through Hox gene regulation in this case (Sehm *et al.*, 2009). In the developing forelimb, *Hoxb-8* acts downstream of retinoic acid (RA) and can act upstream of the key A–P patterning molecule Sonic hedgehog (Charite *et al.*, 1994; Stratford *et al.*, 1999). Interestingly, this genetic hierarchy is not conserved in the hindlimb (Stratford *et al.*, 1999) given the absence of *Hoxb-8* expression at this site and the inability of ectopic RA to induce hindlimb *Hoxb-8* expression. The tapering of *Hoxb-8* expression caudally is achieved primarily by transcriptional mechanisms; however, the presence of *mir-196*-directed *Hoxb-8* cleavage products in the hindlimb indicates the incomplete nature of this regulation and confirms an endogenous role for this miRNA interaction in reinforcing repression at this site (Hornstein *et al.*, 2005). In this context, *miR-196* apparently acts in a failsafe mode: it is redundant with transcriptional repression because *Hoxb-8* does not accumulate in *dicer* mutant hindlimbs unless exogenous RA is added.

These failsafe mechanisms are intuitive with the genomic positioning of *miR-196* and parallel the molecular repression of *Ubx* by *miR-iab-4/8*; however, a more direct role for *miR-196* in defining functional levels of *Hoxb-8* within its normal domain of expression has been suggested (McGlinn *et al.*, 2009). *miR-196* knockdown studies in chick, utilizing antagomiR injection into precursors of the axial skeleton at stage 10–12 of development, resulted in an anteriorly expanded domain of *Hoxb-8* and homeotic transformation of the last cervical vertebrae toward a thoracic identity. This suggests that *miR-196* acts to raise the threshold of *Hoxb-8* expression required to induce morphological change, thereby preventing small fluctuations in gene expression and increasing the robustness of this genetic program. No alterations were observed at more posterior transition points. This is likely to be due to a technical reason, since dilution of the antagomiR is possibly below the threshold of functional activity by the time more posterior vertebrae are being specified. Alternatively, in this context, ectopic expression of a more anterior Hox gene may be silenced by additional mechanisms reinforcing posterior prevalence.

7. Summary and Conclusions

The efforts over recent years to integrate a whole new layer of regulatory control within the previously established Hox genetic hierarchies have yielded fascinating insight into the complexity of these processes. The availability of sequence information from evolutionarily distant species indicates a conserved benefit to particular locations for Hox-embedded miRNAs since there have been several independent events positioning a miRNA upstream of *Hox8* or *9* paralogs. Further, the position of *miR-10* between *Hox4* and *Hox5* is almost universally conserved. Why are these miRNAs positioned here? It is perhaps notable that, as discussed above, they bracket groups of Hox genes that pattern distinct morphological regions along the body axis. At least in vertebrates, morphological transition points appear especially sensitive to quantitative changes in Hox protein and might require miRNA regulation. In insects, both miRNA families bracket and predominantly regulate central Hox genes which control thoracic patterning, considered an evolutionary "ground state" developmental program (reviewed in Gehring *et al.*, 2009). This could be beneficial because more recently evolved programs (anterior and posterior) may require robust repression of central Hox genes to support neofunctionalization. In vertebrates, *miR-10* and *miR-196* target predictions are more in line with roles in reinforcing posterior prevalence (rather than only regulating central genes) along the length of the body axis. However, blocking *miR-196* in chick does lead to posterior homeotic transformations of cervical toward thoracic (also ground state) segments and anterior expansion of *Hoxb-8* expression.

When overexpressed, these miRNAs can have quite dramatic consequences for developmental programs (e.g., *Drosophila miR-iab-4/8*, and to a lesser extent zebrafish *miR-10*) indicating the *in vivo* context for regulation exists, in terms of 3′UTR availability. It is therefore surprising that loss-of-function studies to date have revealed subtle, if any, defects. These data lend support to accumulating evidence suggesting many miRNAs do not drive certain developmental processes but may buffer genetic systems providing robustness of output (Li *et al.*, 2009; Staton *et al.*, 2011). Given the many levels of Hox network regulation, it is plausible that the molecular changes occurring following loss of miRNAs which would otherwise lead to phenotypic changes are being compensated for by other mechanisms. For a genetic network as fundamentally critical to embryogenesis as the Hox network, it is clear to see the importance of redundant and/or fine-scale regulatory mechanisms to limit potentially detrimental effects of gene expression fluctuations or misregulation.

Why have Hox-embedded miRNAs been evolutionarily maintained? These miRNAs clearly contribute to shaping Hox expression domains particularly at their posterior boundaries and likely synergize with other regulatory mechanisms to ensure timely clearance of unwanted transcripts. In addition, there has been suggestion that these miRNAs may act to restrict "transcriptional noise" that is generated as a byproduct of Hox clustering (Woltering and Durston, 2008). While not essential, clustering of Hox genes facilitates coordinated regulation. However, in some cases, this clustering, and therefore the proximity of unshared as well as shared regulatory elements, might also produce aberrant transcription of closely linked genes.

In this context, it is interesting to note extensive cotranscription of miRNAs with protein-coding Hox transcripts that at least in some cases are found in areas where translation is repressed. Finally, the extent to which Hox-embedded miRNAs "fine-tune" target gene expression in areas of coexpression remains an open question. Many miRNA knockouts have limited or no phenotypic alterations observed until the system is stressed with environmental or genetic alterations (Li *et al.*, 2009; van Rooij *et al.*, 2007; Zhao *et al.*, 2010). Given the redundant nature of Hox network regulation, compound mutant analysis (with non-miRNA-mediated regulatory mechanisms) may be required to reveal miRNA contribution.

Functional data for Hox-embedded miRNAs are still lacking in many cases, notably in mammals. Not until all mouse knockouts are generated, a complicated task due to presumed redundancy, will we know the full extent to which these miRNAs contribute to Hox output in mammalian systems. Finally, most Hox miRNA regulatory elements have yet to be defined, and although there is accumulating evidence of complex posttranscriptional processing, we know very little mechanistically regarding their transcriptional and posttranscriptional regulation. While great strides have been made, there are certainly exciting times ahead integrating miRNA function within the classic Hox genetic hierarchies.

7.1. Note added in proof

Recent functional studies in zebrafish support a critical, nonredundant role for *miR-196* in multiple developmental processes, often affecting quite anterior structures (He *et al.*, 2011b). Overexpression of *miR-196* induces homeotic transformations of anterior vertebral structures along with a striking inhibition of pectoral fin initiation, which in the latter case, appears to be achieved via attenuation of RA signaling. Complementary gain- and loss-of-function experiments indicate *miR-196* acts to restrict both pharyngeal arch number as well as vertebral number specifically of the rib-bearing precaudal vertebrae. While these results indicate a conserved function of *miR-196* across species in suppressing rib formation, it is particularly interesting to note that this is achieved by different mechanisms (homeotic transformation in chick (McGlinn *et al.*, 2009) and altered somite number in zebrafish (He *et al.*, 2011b)). Hox target gene expression was observed to be dysregulated in mutant fish; however, the exact extent to which these alterations contribute to altered skeletal patterning is currently unclear.

ACKNOWLEDGMENTS

We thank Cliff Tabin and Jessica Lehoczky for critical reading of the review. The Australian Regenerative Medicine Institute at the Monash University is supported by grants from the Australian Government and the State Government of Victoria, Australia. J. H. M. is supported in part by NSF IOS-0818223.

REFERENCES

Aboobaker, A. A., and Blaxter, M. L. (2003). *Curr. Biol.* **13,** 37–40.

Aboobaker, A. A., Tomancak, P., Patel, N., Rubin, G. M., and Lai, E. C. (2005). *Proc. Natl. Acad. Sci. USA* **102,** 18017–18022.

Amemiya, C. T., Powers, T. P., Prohaska, S. J., Grimwood, J., Schmutz, J., Dickson, M., Miyake, T., Schoenborn, M. A., Myers, R. M., Ruddle, F. H., and Stadler, P. F. (2010). *Proc. Natl. Acad. Sci. USA* **107,** 3622–3627.

Amores, A., Force, A., Yan, Y. L., Joly, L., Amemiya, C., Fritz, A., Ho, R. K., Langeland, J., Prince, V., Wang, Y. L., Westerfield, M., Ekker, M., *et al.* (1998). *Science* **282,** 1711–1714.

Aravin, A. A., Lagos-Quintana, M., Yalcin, A., Zavolan, M., Marks, D., Snyder, B., Gaasterland, T., Meyer, J., and Tuschl, T. (2003). *Dev. Cell* **5,** 337–350.

Asli, N. S., and Kessel, M. (2010). *Dev. Biol.* **344,** 857–868.

Bae, E., Calhoun, V. C., Levine, M., Lewis, E. B., and Drewell, R. A. (2002). *Proc. Natl. Acad. Sci. USA* **99,** 16847–16852.

Baek, D., Villen, J., Shin, C., Camargo, F. D., Gygi, S. P., and Bartel, D. P. (2008). *Nature* **455,** 64–71.

Bartel, D. P. (2009). *Cell* **136,** 215–233.

Bender, W. (2008). *Genes Dev.* **22,** 14–19.

Benson, D. A., Karsch-Mizrachi, I., Lipman, D. J., Ostell, J., and Wheeler, D. L. (2004). *Nucleic Acids Res.* **32,** D23–D26.

Brend, T., Gilthorpe, J., Summerbell, D., and Rigby, P. W. (2003). *Development* **130,** 2717–2728.

Brennecke, J., Stark, A., Russell, R. B., and Cohen, S. M. (2005). *PLoS Biol.* **3,** e85.

Brock, H. W., Hodgson, J. W., Petruk, S., and Mazo, A. (2009). *Biochem. Cell Biol.* **87,** 27–34.

Butts, T., Holland, P. W., and Ferrier, D. E. (2008). *Trends Genet.* **24,** 259–262.

Campo-Paysaa, F., Semon, M., Cameron, R. A., Peterson, K. J., and Schubert, M. (2011). *Evol. Dev.* **13,** 15–27.

Carapuco, M., Novoa, A., Bobola, N., and Mallo, M. (2005). *Genes Dev.* **19,** 2116–2121.

Carthew, R. W., and Sontheimer, E. J. (2009). *Cell* **136,** 642–655.

Charite, J., de Graaff, W., Shen, S., and Deschamps, J. (1994). *Cell* **78,** 589–601.

Chen, F., and Capecchi, M. R. (1997). *Dev. Biol.* **181,** 186–196.

Chiang, H. R., Schoenfeld, L. W., Ruby, J. G., Auyeung, V. C., Spies, N., Baek, D., Johnston, W. K., Russ, C., Luo, S., Babiarz, J. E., Blelloch, R., Schroth, G. P., *et al.* (2010). *Genes Dev.* **24,** 992–1009.

Coulombe, Y., Lemieux, M., Moreau, J., Aubin, J., Joksimovic, M., Berube-Simard, F. A., Tabaries, S., Boucherat, O., Guillou, F., Larochelle, C., Tuggle, C. K., and Jeannotte, L. (2010). *PLoS One* **5,** e10600.

Cumberledge, S., Zaratzian, A., and Sakonju, S. (1990). *Proc. Natl. Acad. Sci. USA* **87,** 3259–3263.

Darnell, D. K., Kaur, S., Stanislaw, S., Konieczka, J. H., Yatskievych, T. A., and Antin, P. B. (2006). *Dev. Dyn.* **235,** 3156–3165.

Deschamps, J. (2007). *Curr. Opin. Genet. Dev.* **17,** 422–427.

Duboule, D. (2007). *Development* **134,** 2549–2560.

Enright, A. J., John, B., Gaul, U., Tuschl, T., Sander, C., and Marks, D. S. (2003). *Genome Biol.* **5,** R1.

Fanarraga, M. L., Charite, J., Hage, W. J., De Graaff, W., and Deschamps, J. (1997). *J. Neurosci. Methods* **71,** 11–18.

Force, A., Amores, A., and Postlethwait, J. H. (2002). *J. Exp. Zool.* **294,** 30–46.

Fried, C., Prohaska, S. J., and Stadler, P. F. (2003). *J. Exp. Zool. B Mol. Dev. Evol.* **299,** 18–25.
Frobius, A. C., Matus, D. Q., and Seaver, E. C. (2008). *PLoS One* **3,** e4004.
Gehring, W. J., Kloter, U., and Suga, H. (2009). *Curr. Top. Dev. Biol.* **88,** 35–61.
Gould, A., Morrison, A., Sproat, G., White, R. A., and Krumlauf, R. (1997). *Genes Dev.* **11,** 900–913.
Greer, J. M., and Capecchi, M. R. (2002). *Neuron* **33,** 23–34.
Griffiths-Jones, S. (2004). *Nucleic Acids Res.* **32,** D109–D111.
Griffiths-Jones, S., Hui, J. H., Marco, A., and Ronshaugen, M. (2011). *EMBO Rep.* **12,** 172–177.
Grimson, A., Farh, K. K., Johnston, W. K., Garrett-Engele, P., Lim, L. P., and Bartel, D. P. (2007). *Mol. Cell* **27,** 91–105.
Guo, H., Ingolia, N. T., Weissman, J. S., and Bartel, D. P. (2010). *Nature* **466,** 835–840.
Hadrys, T., Prince, V., Hunter, M., Baker, R., and Rinkwitz, S. (2004). *J. Exp. Zool. B Mol. Dev. Evol.* **302,** 147–164.
He, X., Yan, Y. L., DeLaurier, A., and Postlethwait, J. H. (2011a). *Zebrafish* **8,** 1–8.
He, X., Yan, Y. L., Eberhart, J. K., Herpin, A., Wagner, T. U., Schartl, M., and Postlethwait, J. H. (2011b). *Dev. Biol.* **357,** 463–477.
Heimberg, A. M., Cowper-Sal-lari, R., Semon, M., Donoghue, P. C., and Peterson, K. J. (2010). *Proc. Natl. Acad. Sci. USA* **107,** 19379–19383.
Hendrix, D., Levine, M., and Shi, W. (2010). *Genome Biol.* **11,** R39.
Herault, Y., Fraudeau, N., Zakany, J., and Duboule, D. (1997). *Development* **124,** 3493–3500.
Holland, P. W., and Garcia-Fernandez, J. (1996). *Dev. Biol.* **173,** 382–395.
Holstege, J. C., de Graaff, W., Hossaini, M., Cano, S. C., Jaarsma, D., van den Akker, E., and Deschamps, J. (2008). *Proc. Natl. Acad. Sci. USA* **105,** 6338–6343.
Horan, G. S., Ramirez-Solis, R., Featherstone, M. S., Wolgemuth, D. J., Bradley, A., and Behringer, R. R. (1995). *Genes Dev.* **9,** 1667–1677.
Hornstein, E., and Shomron, N. (2006). *Nat. Genet.* **38**(Suppl.), S20–S24.
Hornstein, E., Mansfield, J. H., Yekta, S., Hu, J. K., Harfe, B. D., McManus, M. T., Baskerville, S., Bartel, D. P., and Tabin, C. J. (2005). *Nature* **438,** 671–674.
Huang, H., Xie, C., Sun, X., Ritchie, R. P., Zhang, J., and Chen, Y. E. (2010a). *J. Biol. Chem.* **285,** 9383–9389.
Huang, Q. X., Cheng, X. Y., Mao, Z. C., Wang, Y. S., Zhao, L. L., Yan, X., Ferris, V. R., Xu, R. M., and Xie, B. Y. (2010b). *PLoS One* **5,** e13271.
Ikuta, T., Yoshida, N., Satoh, N., and Saiga, H. (2004). *Proc. Natl. Acad. Sci. USA* **101,** 15118–15123.
Kent, W. J. (2002). *Genome Res.* **12,** 656–664.
Kessel, M., and Gruss, P. (1991). *Cell* **67,** 89–104.
Kieny, M., Mauger, A., and Sengel, P. (1972). *Dev. Biol.* **28,** 142–161.
Kloosterman, W. P., Wienholds, E., de Bruijn, E., Kauppinen, S., and Plasterk, R. H. (2006). *Nat. Methods* **3,** 27–29.
Kmita, M., and Duboule, D. (2003). *Science* **301,** 331–333.
Kondrashov, N., Pusic, A., Stumpf, C. R., Shimizu, K., Hsieh, A. C., Xue, S., Ishijima, J., Shiroishi, T., and Barna, M. (2011). *Cell* **145,** 383–397.
Kuroiwa, A., Kloter, U., Baumgartner, P., and Gehring, W. J. (1985). *EMBO J.* **4,** 3757–3764.
Lagos-Quintana, M., Rauhut, R., Lendeckel, W., and Tuschl, T. (2001). *Science* **294,** 853–858.
Lagos-Quintana, M., Rauhut, R., Meyer, J., Borkhardt, A., and Tuschl, T. (2003). *RNA* **9,** 175–179.

Landgraf, P., Rusu, M., Sheridan, R., Sewer, A., Iovino, N., Aravin, A., Pfeffer, S., Rice, A., Kamphorst, A. O., Landthaler, M., Lin, C., Socci, N. D., *et al.* (2007). *Cell* **129,** 1401–1414.
Lau, N. C., Lim, L. P., Weinstein, E. G., and Bartel, D. P. (2001). *Science* **294,** 858–862.
Lemons, D., and McGinnis, W. (2006). *Science* **313,** 1918–1922.
Lewis, E. B. (1963). *Am. Zool.* **3,** 33–56.
Lewis, E. B. (1978). *Nature* **276,** 565–570.
Lewis, E. B. (1998). *Int. J. Dev. Biol.* **42,** 403–415.
Li, X., Cassidy, J. J., Reinke, C. A., Fischboeck, S., and Carthew, R. W. (2009). *Cell* **137,** 273–282.
Lim, L. P., Glasner, M. E., Yekta, S., Burge, C. B., and Bartel, D. P. (2003). *Science* **299,** 1540.
Lim, L. P., Lau, N. C., Garrett-Engele, P., Grimson, A., Schelter, J. M., Castle, J., Bartel, D. P., Linsley, P. S., and Johnson, J. M. (2005). *Nature* **433,** 769–773.
Mainguy, G., Koster, J., Woltering, J., Jansen, H., and Durston, A. (2007). *PLoS One* **2,** e356.
Mansfield, J. H., Harfe, B. D., Nissen, R., Obenauer, J., Srineel, J., Chaudhuri, A., Farzan-Kashani, R., Zuker, M., Pasquinelli, A. E., Ruvkun, G., Sharp, P. A., Tabin, C. J., *et al.* (2004). *Nat. Genet.* **36,** 1079–1083.
Marco, A., Hui, J. H., Ronshaugen, M., and Griffiths-Jones, S. (2010). *Genome Biol. Evol.* **2,** 686–696.
McGinnis, W., and Krumlauf, R. (1992). *Cell* **68,** 283–302.
McGlinn, E., Yekta, S., Mansfield, J. H., Soutschek, J., Bartel, D. P., and Tabin, C. J. (2009). *Proc. Natl. Acad. Sci. USA* **106,** 18610–18615.
McIntyre, D. C., Rakshit, S., Yallowitz, A. R., Loken, L., Jeannotte, L., Capecchi, M. R., and Wellik, D. M. (2007). *Development* **134,** 2981–2989.
Mineno, J., Okamoto, S., Ando, T., Sato, M., Chono, H., Izu, H., Takayama, M., Asada, K., Mirochnitchenko, O., Inouye, M., and Kato, I. (2006). *Nucleic Acids Res.* **34,** 1765–1771.
Miura, S., Nozawa, M., and Nei, M. (2011). *Genome Biol. Evol.* **3,** 129–139.
Nelson, C. E., Morgan, B. A., Burke, A. C., Laufer, E., DiMambro, E., Murtaugh, L. C., Gonzales, E., Tessarollo, L., Parada, L. F., and Tabin, C. (1996). *Development* **122,** 1449–1466.
Ohtsuki, S., Levine, M., and Cai, H. N. (1998). *Genes Dev.* **12,** 547–556.
Prohaska, S. J., and Stadler, P. F. (2004). *Theory Biosci.* **123,** 89–110.
Qian, J., Zhang, Z., Liang, J., Ge, Q., Duan, X., Ma, F., and Li, F. (2011). *Genomics* **97,** 294–303.
Qiu, R., Liu, Y., Wu, J. Y., Liu, K., Mo, W., and He, R. (2009). *Brain Res. Bull.* **79,** 26–31.
Rathjen, T., Pais, H., Sweetman, D., Moulton, V., Munsterberg, A., and Dalmay, T. (2009). *FEBS Lett.* **583,** 1422–1426.
Ravi, V., Lam, K., Tay, B. H., Tay, A., Brenner, S., and Venkatesh, B. (2009). *Proc. Natl. Acad. Sci. USA* **106,** 16327–16332.
Richardson, M. K., Crooijmans, R. P., and Groenen, M. A. (2007). *Cytogenet. Genome Res.* **117,** 110–119.
Riley, P. D., Carroll, S. B., and Scott, M. P. (1987). *Genes Dev.* **1,** 716–730.
Rinn, J. L., Kertesz, M., Wang, J. K., Squazzo, S. L., Xu, X., Brugmann, S. A., Goodnough, L. H., Helms, J. A., Farnham, P. J., Segal, E., and Chang, H. Y. (2007). *Cell* **129,** 1311–1323.
Ronshaugen, M., Biemar, F., Piel, J., Levine, M., and Lai, E. C. (2005). *Genes Dev.* **19,** 2947–2952.
Ruby, J. G., Stark, A., Johnston, W. K., Kellis, M., Bartel, D. P., and Lai, E. C. (2007). *Genome Res.* **17,** 1850–1864.

Sehm, T., Sachse, C., Frenzel, C., and Echeverri, K. (2009). *Dev. Biol.* **334,** 468–480.

Sharpe, J., Nonchev, S., Gould, A., Whiting, J., and Krumlauf, R. (1998). *EMBO J.* **17,** 1788–1798.

Sperling, E. A., Vinther, J., Moy, V. N., Wheeler, B. M., Semon, M., Briggs, D. E., and Peterson, K. J. (2009). *Proc. R. Soc. B* **276,** 4315–4322.

Stadler, P. F., Fried, C., Prohaska, S. J., Bailey, W. J., Misof, B. Y., Ruddle, F. H., and Wagner, G. P. (2004). *Mol. Phylogenet. Evol.* **32,** 686–694.

Stark, A., Brennecke, J., Russell, R. B., and Cohen, S. M. (2003). *PLoS Biol.* **1,** E60.

Stark, A., Kheradpour, P., Parts, L., Brennecke, J., Hodges, E., Hannon, G. J., and Kellis, M. (2007). *Genome Res.* **17,** 1865–1879.

Stark, A., Bushati, N., Jan, C. H., Kheradpour, P., Hodges, E., Brennecke, J., Bartel, D. P., Cohen, S. M., and Kellis, M. (2008). *Genes Dev.* **22,** 8–13.

Staton, A. A., Knaut, H., and Giraldez, A. J. (2011). *Nat. Genet.* **43,** 204–211.

Stratford, T., Logan, C., Zile, M., and Maden, M. (1999). *Mech. Dev.* **81,** 115–125.

Takacs, C. M., and Giraldez, A. J. (2011). *Semin. Cell Dev. Biol.* **21,** 760–767.

Tang, G. Q., and Maxwell, E. S. (2008). *Genome Res.* **18,** 104–112.

Tanzer, A., Amemiya, C. T., Kim, C. B., and Stadler, P. F. (2005). *J. Exp. Zool. B Mol. Dev. Evol.* **304,** 75–85.

Thomsen, S., Azzam, G., Kaschula, R., Williams, L. S., and Alonso, C. R. (2010). *Development* **137,** 2951–2960.

Tyler, D. M., Okamura, K., Chung, W. J., Hagen, J. W., Berezikov, E., Hannon, G. J., and Lai, E. C. (2008). *Genes Dev.* **22,** 26–36.

van den Akker, E., Reijnen, M., Korving, J., Brouwer, A., Meijlink, F., and Deschamps, J. (1999). *Mech. Dev.* **89,** 103–114.

van den Akker, E., Fromental-Ramain, C., de Graaff, W., Le Mouellic, H., Brulet, P., Chambon, P., and Deschamps, J. (2001). *Development* **128,** 1911–1921.

van Rooij, E., Sutherland, L. B., Qi, X., Richardson, J. A., Hill, J., and Olson, E. N. (2007). *Science* **316,** 575–579.

Wang, K. C., Yang, Y. W., Liu, B., Sanyal, A., Corces-Zimmerman, R., Chen, Y., Lajoie, B. R., Protacio, A., Flynn, R. A., Gupta, R. A., Wysocka, J., Lei, M., *et al.* (2011). *Nature* **472,** 120–124.

Wellik, D. M. (2007). *Dev. Dyn.* **236,** 2454–2463.

Wellik, D. M. (2009). *Curr. Top. Dev. Biol.* **88,** 257–278.

Wellik, D. M., and Capecchi, M. R. (2003). *Science* **301,** 363–367.

Wheeler, B. M., Heimberg, A. M., Moy, V. N., Sperling, E. A., Holstein, T. W., Heber, S., and Peterson, K. J. (2009). *Evol. Dev.* **11,** 50–68.

Wienholds, E., Kloosterman, W. P., Miska, E., Alvarez-Saavedra, E., Berezikov, E., de Bruijn, E., Horvitz, H. R., Kauppinen, S., and Plasterk, R. H. (2005). *Science* **309,** 310–311.

Woltering, J. M., and Durston, A. J. (2006). *Nat. Genet.* **38,** 601–602.

Woltering, J. M., and Durston, A. J. (2008). *PLoS One* **3,** e1396.

Yekta, S., Shih, I. H., and Bartel, D. P. (2004). *Science* **304,** 594–596.

Yekta, S., Tabin, C. J., and Bartel, D. P. (2008). *Nat. Rev. Genet.* **9,** 789–796.

Zhao, Z., Boyle, T. J., Liu, Z., Murray, J. I., Wood, W. B., and Waterston, R. H. (2010). *PLoS Genet.* **6,** e1001089.

Zhou, J., Ashe, H., Burks, C., and Levine, M. (1999). *Development* **126,** 3057–3065.

CHAPTER THREE

The Role of MicroRNAs in Muscle Development

Nicholas S. Sokol

Contents

Abstract

MicroRNAs play essential roles during animal development, including in developing muscle. Many microRNAs are expressed during muscle development and some, like *miR-1* and *miR-133*, are muscle specific. Muscle microRNAs are integrated into myogenic regulatory networks: their expression is under the transcriptional and posttranscriptional control of myogenic factors, and they in turn have widespread control of muscle gene expression. This review summarizes recent work characterizing the function of microRNAs in muscle biology and specifically focuses on the genetic analysis of muscle microRNAs in a variety of model organisms including worms, flies, zebrafish, and mice.

Department of Biology, Indiana University, Bloomington, Indiana, USA

Current Topics in Developmental Biology, Volume 99

ISSN 0070-2153, DOI: 10.1016/B978-0-12-387038-4.00003-3

1. Introduction

MicroRNAs (miRNAs) are a class of small RNAs that are 20–24 nucleotides in length, and that function to regulate the expression of target mRNAs (Bartel, 2009). This class of noncoding RNAs is well represented in animal genomes and contains members that are perfectly conserved between diverse species. The metazoan origin and evolutionary conservation of miRNAs indicate that they have essential roles in the biology of multicellular animals (Grimson *et al.*, 2008), and many display conserved tissue-specific expression patterns (Wienholds *et al.*, 2005). Consistent with these expression profiles, miRNAs are known to play critical roles in diverse cells and tissues during animal development (Ambros, 2004). This review summarizes the current understanding of the roles that miRNAs play in developing muscle. I concentrate on insights from the genetic analyses of miRNA function in key model organisms including *Caenorhabditis elegans*, *Drosophila melanogaster*, *Danio rerio*, and *Mus musculus*, with apologies to those whose work could not be included due to space constraints.

2. Muscle miRNA Regulatory Networks

The existence of muscle-specific miRNAs was suggested by one of the first papers identifying miRNAs (Lee and Ambros, 2001). This paper showed that the *miR-1* miRNA was highly expressed in human heart tissue but not in human brain, liver, kidney, or lung tissue. Since then, the expression of miRNAs in muscle has been studied intensively. This section describes the identification of muscle miRNAs, as well as their placement within genetic regulatory networks that control muscle formation and function.

2.1. Identification of muscle miRNAs

Analysis of miRNA expression patterns in whole mount specimens was one of the first approaches used to identify muscle miRNAs. Such analyses found that *miR-1* is expressed specifically in muscles during the development of worms (Simon *et al.*, 2008), flies (Kwon *et al.*, 2005; Sokol and Ambros, 2005), zebrafish (Wienholds *et al.*, 2005), and mice (Zhao *et al.*, 2005) and thus displays a striking evolutionary conservation of its tissue-specific expression. Whole mount analysis identified numerous other miRNAs expressed in muscles, though most were expressed in other tissues as well. These include *miR-45*, *-51*, *-90*, *-227*, and *-794* in *C. elegans* (Martinez *et al.*, 2008), *miR-8*, *-13b-2*, *-184*, and *-316* in *D. melanogaster*

(Aboobaker *et al.*, 2005), and *miR-133*, *-206*, *-216*, and *-499* in zebrafish (Kloosterman *et al.*, 2006; Wienholds *et al.*, 2005).

Microarray analysis and deep sequencing experiments have significantly extended the number of miRNAs implicated in vertebrate muscle development. These expression-profiling approaches have focused on identifying the complete repertoire of muscle miRNAs in embryonic and adult muscle tissue from diverse species (Huang *et al.*, 2008; Li *et al.*, 2011). Microarray analysis of mouse tissue, for example, identified a total of 114 miRNAs that are expressed in the mouse heart, 25 of which are apparently heart specific (Tang *et al.*, 2007). Subsequent deep sequencing of young adult mouse heart tissue found an enrichment of a small number of miRNAs, with *miR-1* accounting for 40% of all cardiac miRNA expression (Rao *et al.*, 2009). The level of *miR-1* expression is striking: *miR-1* is ~20-fold more abundant than *miR-208*, a miRNA encoded within the intron of a cardiac myosin and therefore a reasonable proxy for a highly expressed cardiac mRNA. Other miRNAs enriched in heart tissues include members of the *let-7*, *miR-26*, *miR-29*, and *miR-30* families.

In a parallel approach to identify miRNAs involved in muscle biology, miRNA expression dynamics have been profiled in cultured mammalian cells undergoing skeletal or cardiac muscle differentiation (Chen *et al.*, 2006; Dey *et al.*, 2011; Ivey *et al.*, 2008; Naguibneva *et al.*, 2006). For example, the expression of a collection of miRNAs including *miR-1*, *-22*, *-26*, *-133*, *-181*, *-206*, and *-486* is upregulated as cultured murine C2C12 myoblasts differentiate into skeletal muscle (Chen *et al.*, 2006; Dey *et al.*, 2011; Naguibneva *et al.*, 2006). Activation of these miRNAs may be required to promote differentiation programs and is coincident with the downregulation of a second set of C2C12-expressed miRNAs including *miR-125b* (Ge *et al.*, 2011), which likely repress C2C12 differentiation. Indeed, an analogous role for *miR-125b* has recently been found during skin cell differentiation, since *miR-125b* is highly expressed in skin progenitor cells and is sharply downregulated during their differentiation (Zhang *et al.*, 2011).

Similar profiling approaches have also been used to identify miRNAs involved in cardiac muscle differentiation. For example, 17 murine miRNAs are upregulated as embryonic stem cells differentiate into cardiomyocytes, including *miR-1*, *-133*, *-143*, *-182*, *-200*, and *-295* (Ivey *et al.*, 2008). Taken together, these profiling analyses indicate that muscles have a rich palette of miRNAs that likely play myriad diverse roles during muscle development.

2.2. Genomic organization of muscle miRNAs

In many animal genomes, muscle miRNAs are found clustered together (see Fig. 3.1). In both the mouse and human genomes, for example, each of the three members of the *miR-1* family of miRNAs is located next to a

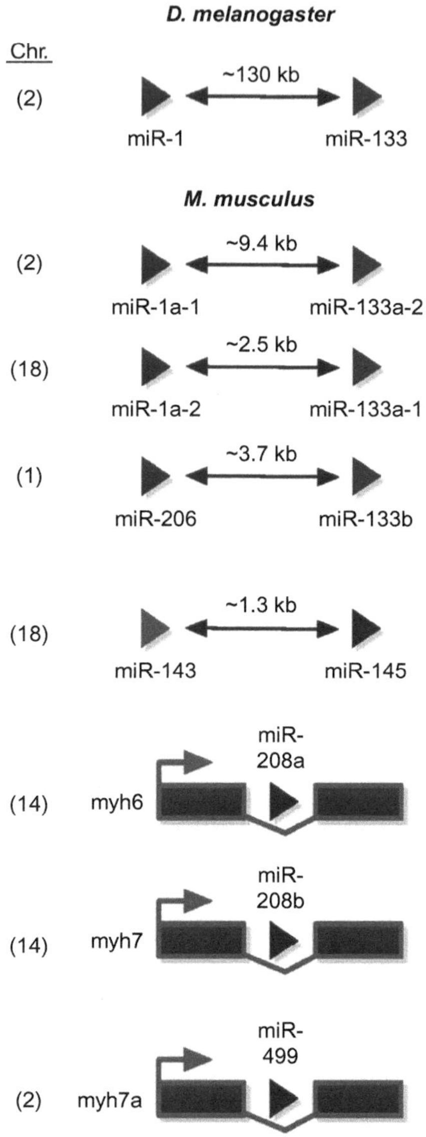

Figure 3.1 Genomic organization of muscle miRNAs in flies and mice. In both flies and mice, members of the *miR-1* and *miR-133* are clustered together. Additional mouse miRNAs are also clustered, including *miR-143* and *miR-145*. Other muscle miRNAs, like *miR-208a*, *miR-208b*, and *miR-499*, are located within the introns of myosin genes. Chromosomal locations are indicated in parentheses. (See Color Insert.)

member of the *miR-133* family (Chen *et al.*, 2006; Liu *et al.*, 2007; Rao *et al.*, 2006). Such bicistronic organization places distinct miRNAs under common transcriptional control. Indeed, one member of the *miR-1* family, *mir-206*, and its neighbor, *miR-133b*, are specifically expressed in mouse skeletal muscle but not in the heart (McCarthy, 2008). The other two *miR-1/miR-133* clusters, located on mouse chromosomes 2 and 18, are transcriptionally activated in skeletal as well as cardiac muscle (Liu *et al.*, 2007; Zhao *et al.*, 2005). Since *miR-1* miRNAs are ~10-fold more abundant in heart tissue than *miR-133* miRNAs, these cotranscribed miRNAs are likely the targets of differential posttranscriptional control (Rao *et al.*, 2009). The functional significance of the differential expression of *miR-1* and *miR-133* is currently unknown, however. The bicistronic genomic organization of vertebrate *miR-1/miR-206* and *miR-133* is reflected in the *D. melanogaster* genome as well, where the single copy of *miR-1* is located ~130 kb away from *miR-133*, relatively close in genomic terms. Although RNA profiling indicates that fly *miR-1* and *miR-133* have distinct temporal expression profiles and are not cotranscribed (Graveley *et al.*, 2011), the spatial expression pattern of *miR-133* has not been reported, so it may be expressed in muscles and under the control of enhancers it shares with *miR-1*. Other vertebrate muscle miRNAs also display bicistronic organization, including the unrelated smooth muscle miRNAs *miR-143* and *miR-145* (Cordes *et al.*, 2009).

Muscle miRNAs can also be found within the introns of myogenic loci (see Fig. 3.1). For example, *miR-208a*, *miR-208b*, and *miR-499* are located in the introns of three corresponding myosin genes: *myh6*, *myh7*, and *myh7b*, respectively (Callis *et al.*, 2009; van Rooij *et al.*, 2007, 2009). This genomic organization is intriguing, particularly because the *myh6* and *myh7* loci encode proteins that display nonoverlapping expression profiles that are shared by their intronic miRNAs. *myh6* and *myh7* encode α-cardiac muscle myosin heavy chain (α-MHC) and β-cardiac muscle myosin heavy chain (β-MHC), respectively. The ratio of α-MHC and β-MHC is under tight transcriptional control since it controls cardiac contractility: β-MHC is expressed during embryonic development and is downregulated shortly after birth when α-MHC is upregulated. The intronic miRNAs display these same patterns of temporal expression, with *miR-208a* abundantly expressed in fetal but not in adult hearts and *miR-208b* expressed in the opposite pattern (Callis *et al.*, 2009). This is due to the transcription of the intronic miRNAs from their host genes, although *miR-208b* can also be independently controlled by an intronic promoter (Monteys *et al.*, 2010). Knockout analysis of the miRNAs, described below, indicates that these miRNAs are not required to ensure the mutually exclusive expression profiles of α-MHC and β-MHC. Nevertheless, the structural relationship between *miR-208* and myosin family members is conserved from fish to humans, indicating that the coexpression of the intronic miRNAs with their myosin host genes is important for muscle function.

2.3. Transcriptional control of muscle miRNAs

Muscle miRNAs are well integrated within myogenic transcriptional regulatory networks. Intronic miRNAs like *miR-208a*, *miR-208b*, and *miR-499*, for example, are under the same transcriptional control as their myosin hosts. Intergenic miRNAs are also targeted by myogenic transcription factors. For example, transcription of *miR-1* and *miR-133* in skeletal muscle is controlled by the myogenic transcription factor MyoD, while their transcription in cardiac muscle is controlled by cardiogenic transcription factors SRF and MEF2 (Liu *et al.*, 2007; Zhao *et al.*, 2005). The transcriptional regulation of *miR-1* by MEF2 is highly conserved, since MEF2 coregulates *miR-1* expression in flies along with the mesodermal transcription factor Twist (Biemar *et al.*, 2005; Kwon *et al.*, 2005; Sokol and Ambros, 2005). In addition to the *miR-1/miR-133* cluster, vertebrate SRF activates the smooth muscle miRNA *miR-143/miR-145* cluster together with its coactivator myocardin (Cordes *et al.*, 2009; Xin *et al.*, 2009). Predicted SRF binding sites are located near another 40 miRNA loci (Niu *et al.*, 2007), so it will be interesting to see how many of these are directly controlled by SRF.

2.4. Posttranscriptional regulation of muscle miRNAs

miRNA expression is also regulated at the posttranscriptional level. miRNAs are transcribed within long primary transcripts, which are cleaved twice to release functional 21-nt RNAs. The Dicer and Drosha enzymes carry out these sequential cleavages, and their activities can be modulated by an emerging repertoire of auxiliary RNA-binding proteins including KSRP, p68, and Lin-28 (Davis *et al.*, 2008; Suzuki *et al.*, 2009; Trabucchi *et al.*, 2009; Viswanathan *et al.*, 2008). KSRP is a KH-type splicing regulatory protein that physically associates with both Dicer and Drosha to promote the biogenesis of a subset of miRNAs, including *miR-1* and *miR-206* during skeletal muscle differentiation of murine C2C12 cells (Trabucchi *et al.*, 2009). Similarly, the DEAD-box RNA helicase p68 interacts with the Drosha processing complex to enhance the posttranscriptional processing of the smooth muscle miRNAs *miR-21*, *miR-143*, and *miR-145* (Davis *et al.*, 2008; Suzuki *et al.*, 2009). p68 is a key regulatory component of the Drosha complex, since it is targeted by the TGF-B pathway to promote *miR-21* processing (Davis *et al.*, 2008) and by the p53 tumor suppressor protein to promote *miR-143* and *miR-145* processing (Suzuki *et al.*, 2009). Unlike KSRP and p68, the pluripotency factor Lin-28 is a negative regulator of miRNA biogenesis and targets *let-7* (Viswanathan and Daley, 2010; Viswanathan *et al.*, 2008). Recently, *pre-miR-1* was also identified as a target of Lin-28 (Rau *et al.*, 2011), which is strongly induced during the differentiation of adult primary myoblasts (Polesskaya *et al.*, 2007; Viswanathan *et al.*, 2008). Collectively, the studies summarized here identify a handful of muscle miRNAs whose expression is posttranscriptionally regulated. Such posttranscriptional control provides a mechanism to rapidly increase or decrease the amount of functional miRNAs in a cell, ensuring rapid transitions in miRNA expression during myogenesis as well as in postmitotic muscle cells. The functional significance of this regulation during muscle development has yet to be reported, however.

2.5. mRNA targets of muscle miRNAs

Muscle miRNAs target a diverse set of mRNAs, including transmembrane receptors, ion channels, cytoskeleton components, transcription factors, and epigenetic enzymes. Of particular interest are the targeting issues raised by the bicistronic organization of sequence divergent muscle miRNAs. Some cotranscribed muscle miRNAs, like mouse *miR-143* and *miR-145*, cotarget an overlapping set of mRNAs: reporter assays indicate that 3′UTRs of at least five genes contain functional binding sites for both *miR-143* and *miR-145*, including *klf4*, *klf5*, *adducin3*, *ssh2*, and *mrtf-B* (Xin *et al.*, 2009). In contrast, other cotranscribed miRNAs, like mouse *miR-1* and *miR-133*, regulate largely nonoverlapping sets of mRNA targets. This conclusion is supported by the fact that there is little overlap in the sets of mRNAs that are misregulated in either *miR-1* or *miR-133* mutant tissue as detected by microarray analysis (Liu *et al.*, 2008; Mishima *et al.*, 2009; Shkumatava *et al.*, 2009; Zhao *et al.*, 2007). Further, with the exception of one recently analyzed 3′UTR (Vo *et al.*, 2010), most characterized 3′UTRs contain binding sites for either *miR-1* or *miR-133* but not both (Boutz *et al.*, 2007; Chen *et al.*, 2006; Ikeda *et al.*, 2009; Li *et al.*, 2010; Liu *et al.*, 2008; Zhao *et al.*, 2005, 2007). Indeed, *miR-1* and *miR-133* have opposing functions in C2C12 cells, with *miR-1* favoring differentiation by inhibiting myoblast proliferation and *miR-133* suppressing differentiation by promoting myoblast proliferation (Chen *et al.*, 2006). Collectively, the analysis of the *miR-143*/*miR-145* and *miR-1*/*miR-133* clusters suggests that cotranscribed miRNA clusters use differing targeting strategies and can target either distinct or overlapping sets of mRNAs. The specific targets of a number of other muscle miRNAs have been identified and will be discussed below in reference to the miRNA mutant phenotype.

3. Functional Analysis of miRNAs During Muscle Development

miRNA function can be determined by analyzing the consequences of eliminating individual miRNAs on a case-by-case basis. Alternatively, the consequences of depleting many miRNAs at once can be assayed by targeting the general RNA-binding proteins required for miRNA production including Dicer and Drosha. In this section, I focus first on muscle phenotypes displayed by *dicer* mutants and then discuss the phenotypic analysis of individual miRNA mutants.

3.1. Muscle phenotypes of zebrafish and mouse *dicer* mutants

Although *dicer* mutants provide a powerful approach to analyzing the global role of miRNAs, some caveats exist. First, Dicer is required for the processing of not only miRNAs but other biologically active small RNA species like endogenous siRNAs as well (for a recent review, see Suh and Blelloch, 2011). Second, Dicer depletion does not result in the complete elimination of all miRNAs, since some stable miRNAs can still be detected well after *dicer* depletion (Giraldez *et al.*, 2005) and other miRNAs are processed by a Dicer-independent mechanism (Cheloufi *et al.*, 2010; Cifuentes *et al.*, 2010; Giraldez *et al.*, 2005). Nevertheless, the expression of many miRNAs is affected by Dicer depletion, which therefore provides a good starting point for analyzing the general roles of miRNAs in developmental processes. Below, I summarize the phenotypes displayed by zebrafish and mouse *dicer* mutants. Depleting Dicer to identify the role of miRNAs during nematode and fruit fly muscle development has not been possible, since *dicer* mutant worms arrest at an early developmental stage that is difficult to study (Grishok *et al.*, 2001), and maternal fly *dicer* is required for germline development, precluding its analysis in the early embryo (Hatfield *et al.*, 2005).

The role of miRNAs during embryogenesis was explored in zebrafish, through the analysis of embryos lacking both maternal and zygotic Dicer (Giraldez *et al.*, 2005). These MZ*dicer* mutants undergo axis formation and differentiate multiple cell types including muscles, indicating that miRNAs are not required for the initial patterning and specification of the germ layers including the mesoderm and at least some of its derivatives. MZ*dicer* mutants display defects in subsequent embryonic events, however, including morphogenesis of the somites and heart. Although somites form normally in MZ*dicer* mutants, they do not acquire their normal chevron shape. Similarly, a two-chamber heart does not form, even though contractile cardiomyocytes are specified correctly in MZ*dicer* mutants. Instead, MZ*dicer* mutants display a tubular heart and pericardial edema. These results indicate that miRNAs are not absolutely required for myogenic cell fate specification, but may play roles in modulating cell type-specific differentiation as well as influencing the behavior of myogenic cells once differentiated.

To extend this phenotypic analysis to the molecular level, global gene expression was profiled in muscle cells isolated from 24-h-old MZ*dicer* mutants (Mishima *et al.*, 2009). Hundreds of inappropriately expressed mRNAs were identified, which is not surprising given the large repertoire of muscle miRNAs. However, most of these misexpressed mRNAs contain predicted binding sites for either *miR-1* or *miR-133*, suggesting that these two muscle-specific miRNAs play a predominant role in the miRNA-dependent control of muscle gene expression. Further, these muscle miRNA targets are enriched for actin-related and acting-binding proteins, providing a mechanistic explanation for the morphological defects displayed by MZ*dicer* mutants.

Phenotypic analysis of *dicer* mutants has also been used to analyze the role of miRNAs during muscle development in mice. Unlike MZ*dicer*-deficient zebrafish embryos, however, mouse embryos that are depleted of zygotic *dicer* arrest early in development at embryonic day 7.5 (E7.5) and are shrunken and malformed (Bernstein *et al.*, 2003). These *dicer* mutant embryos express the early mesodermal marker *brachyury,* indicating that mesodermal cells form, but its expression domain is reduced, suggesting that the progression of patterning is abnormal in these embryos (Spruce *et al.*, 2010). Since cultured Dicer-deficient mouse embryonic stem cells have proliferation and general differentiation defects due to the inability to silence pluripotency factors (Kanellopoulou *et al.*, 2005; Murchison *et al.*, 2005; Sinkkonen *et al.*, 2008), they have not been useful for investigating the requirement for miRNAs in cell fate specification of the mouse mesoderm. *miR-1* and *miR-133* have been directly implicated in these early cell fate decisions, however, since forced expression of either in mouse embryonic stem cells results in enhanced mesoderm gene expression and a corresponding reduction in ectodermal and endodermal gene expression (Ivey *et al.*, 2008).

Subsequent roles for Dicer during the development of cardiac muscle, skeletal muscle, or smooth muscle have been investigated by a series of studies using a conditional knockout approach. These studies have made use of different floxed *dicer* alleles in combination with tissue-specific Cre lines that together disrupt *dicer* function in specific spatiotemporal and tissue-specific patterns during mouse development. The results of these studies are summarized below, but in general Dicer is implicated in the morphogenetic events of muscle systems rather than the initial cell fate specification of muscle subtypes. This result may be a consequence of the conditional depletion strategy, however, since early cell fate specification may have already occurred at the time when Dicer depletion has functional consequences.

3.1.1. Mouse skeletal muscle

The role of Dicer in skeletal muscle development has been investigated using a MyoD-Cre line, which is expressed in developing skeletal muscle in the mandibular arch, somitic myotomes, developing limb buds, and trunk musculature between E9.75 and E12.5 (O'Rourke *et al.*, 2007). These mice display a significant reduction in muscle mass and abnormal myofiber morphology, ultimately leading to perinatal lethality. The reduction of muscle mass is due to decreased skeletal muscle cell proliferation, rather than an inability to pattern skeletal muscle or specific myofiber type.

3.1.2. Mouse cardiac muscle

The role of Dicer in heart development has been investigated by two studies using Nkx2.5 Cre lines (Saxena and Tabin, 2010; Zhao *et al.*, 2007), which express Cre in the earliest identifiable cardiac progenitors present in the cardiac crescent. Despite this early depletion of Dicer, the general patterning and formation of the heart are not severely affected. However, later morphogenetic events are flawed, and resulting mice display a series of heart phenotypes including underdevelopment of the ventricular myocardium, pericardial edema, and defects in outflow tract alignment and chamber septation. These findings are consistent with the results from a third study that reported that later depletion of cardiac Dicer using MHC-Cre leads to progressive dilated cardiomyopathy, heart failure, and postnatal lethality (Chen *et al.*, 2008).

3.1.3. Smooth muscle

Two groups have investigated the consequence of Dicer depletion in smooth muscle during developing (Albinsson *et al.*, 2010; Pan *et al.*, 2011) by using the SM22-cre line, which is expressed in arterial, venous, and visceral smooth muscle cells. This loss of miRNA processing resulted in dilated and thin-walled blood vessels caused by a reduction in cellular proliferation and disarray of vascular architecture, leading to embryonic lethality associated with extensive internal hemorrhage.

3.2. Muscle phenotypes of miRNA mutants

Since a sizable number of genetic mutations in muscle miRNAs have been reported, I focus here on a description of their phenotypes and mention the results of miRNA knockdown experiments where appropriate (see Table 3.1 for a summary of muscle miRNA phenotypes). Consistent with the *dicer* mutant phenotypes described above, elimination of individual miRNAs leads to defects late in muscle development involving muscle morphology and function rather than early events during muscle development including mesoderm formation or muscle progenitor fate specification. However, miRNAs may be playing redundant roles either with one another or with protein-coding genes, as has been shown recently in *C. elegans* (Alvarez-Saavedra and Horvitz, 2010; Brenner *et al.*, 2010). Thus, conclusive evidence regarding the general role of miRNAs in muscle development awaits the construction of compound miRNA mutants.

Table 3.1 Muscle development phenotypes of miRNA mutant animals

Gene	Animal	Phenotype	References
miR-1	*C. elegans*	No obvious developmental phenotype	Simon *et al.* (2008)
miR-1	*D. melanogaster*	Muscle growth	Sokol and Ambros (2005)
		Muscle patterning	Kwon *et al.* (2005)
let-7-C	*D. melanogaster*	NMJ and muscle growth	Caygill and Johnston (2008); Sokol *et al.* (2008)
miR-1-2	*M. musculus*	~50% lethality, cardiac defects	Zhao *et al.* (2007)
miR-17–92	*M. musculus*	100% postnatal death, cardiac defects	Ventura *et al.* (2008); Wang *et al.* (2010)
miR-133a-1	*M. musculus*	No obvious developmental phenotype	Liu *et al.* (2008)
miR-133a-2	*M. musculus*	No obvious developmental phenotype	Liu *et al.* (2008)
miR-133a-1/-2	*M. musculus*	~50% lethality, cardiac defects	Liu *et al.* (2008)
miR-143	*M. musculus*	No obvious developmental phenotype	Xin *et al.* (2009)
miR-145	*M. musculus*	Thinner smooth muscle layer in arteries	Xin *et al.* (2009)
miR-143/145	*M. musculus*	Similar to miR-145 phenotype	Elia *et al.* (2009); Xin *et al.* (2009)
miR-206	*M. musculus*	No obvious developmental phenotype	Williams *et al.* (2009)
miR-208a	*M. musculus*	No obvious developmental phenotype	van Rooij *et al.* (2007); Callis *et al.* (2009)
miR-208b	*M. musculus*	No obvious developmental phenotype	van Rooij *et al.* (2009)
miR-499	*M. musculus*	No obvious developmental phenotype	van Rooij *et al.* (2009)
miR-208b/-499	*M. musculus*	Loss of slow myofibers in skeletal muscle	van Rooij *et al.* (2009)

3.2.1. miR-1/206 and miR-133 families in muscle development

As described above, there are three bicistronic miRNA clusters in the mouse genome that each encodes one member of the *miR-1/206* family and one member of the *miR-133* family (Chen *et al.*, 2006; Liu *et al.*, 2007; Rao *et al.*, 2006). Deletions of four of these six miRNAs have been generated and analyzed to date, including *miR-1-2* and *miR-206* as well as *miR-133a-1* and *miR-133a-2* (Liu *et al.*, 2008; Williams *et al.*, 2009; Zhao *et al.*, 2007). In each case, care was taken so that the expression of the cotranscribed neighbor was not affected. While the elimination of *miR-206* seems to have no effect on cardiac and skeletal muscle development (Williams *et al.*, 2009), *miR-1-2* is required for heart development (Zhao *et al.*, 2007): roughly half of the *miR-1-2* mutants die prior to adulthood and display defects in the formation of the ventricular septum, which is the wall that divides the left and right ventricles of the heart. This malformation is indicative of abnormal heart morphogenesis and arises late during embryogenesis; histological analysis of early embryos indicates that initial heart formation is normal. Surviving *miR-1-2* mutants have grossly normal heart morphology but display a number of adult phenotypes including an increase in cardiomyocyte number. Thus, removal of one of the two copies of *miR-1* leads to clear defects in muscle development and function (Zhao *et al.*, 2007) and indicates that the extremely high expression of *miR-1* in the heart is functionally significant (Rao *et al.*, 2009).

Although *miR-1-2* mutants display no apparent defects in mesoderm determination, embryonic myoblast fate specification, or skeletal muscle myogenesis, involvement in these processes might be compensated for by *miR-1-1* as well as the closely related *miR-206*. However, genetic depletion of the single copy of *miR-1* in both flies and worms as well as of MZ*dicer* in zebrafish indicates that *miR-1* is not required for muscle formation (Giraldez *et al.*, 2005; Simon *et al.*, 2008; Sokol and Ambros, 2005). In *Drosophila* mutants in which a 57-bp region containing the *miR-1* 21-mer has specifically been deleted, for example, the muscle system forms and is morphologically and functionally normal (Sokol and Ambros, 2005). Indeed, ~80% of *miR-1* mutant embryos hatch into motile larvae with normal body wall contraction and heart rates. Feeding triggers their death, however, indicating that *miR-1* is required for postmitotic growth of larval muscle. Interestingly, mutants homozygous for a 31-kbp deletion that removed *miR-1* and surrounding sequences displayed a much stronger muscle phenotype: ~66% of these mutants die during embryogenesis and display defects in cardiac and somitic muscle patterning (Kwon *et al.*, 2005). The difference between the phenotypes of these two *miR-1* alleles suggests that the *miR-1* phenotype might be very sensitive to genetic background, or alternatively that some of the noncoding transcripts neighboring *miR-1* and detected by profiling microarrays may coordinately control muscle formation along with *miR-1* (Graveley *et al.*, 2011). Such scenarios are consistent with the analysis of *miR-1* in *C. elegans*: *miR-1* mutants ordinarily proceed through development normally, but display developmental defects when containing mutations that partially compromise either the general miRNA production machinery or a range of other regulatory pathways (Brenner *et al.*, 2010). Taken together, the analysis of worms, flies, and zebrafish indicates that *miR-1* is not essential for myogenesis in these animals. A definitive determination of the requirement for *miR-1* family miRNAs in mouse muscle development awaits the construction of *miR-1-1*/*miR-1-2* double mutants as well as *miR-1-1*/*miR-1-2*/*miR-206* triple mutants.

Deletions of the cotranscribed partner of *miR-1-2*, *miR-133a-1*, as well as its sibling *miR-133a-2* have also been reported (Liu *et al.*, 2008). Deletion of either *miR-133a-1* or *miR-133a-2* has no apparent effect on development. However, *miR-133a-1*/*miR-133a-2* double mutants display a constellation of phenotypes that are very similar to the *miR-1-2* single mutants (Liu *et al.*, 2008). Roughly half the *miR-133a-1*/*miR-133a-2* double mutant mice die as late embryos or neonates with ventricular septum defects. Those that survive display cardiomyocyte hyperplasia but normal skeletal muscles formation and function. Gene expression analysis of *miR-133a-1*/*miR-133a-2* double mutant hearts found a disproportionate increase in smooth muscle genes regulated by SRF (Liu *et al.*, 2008). This finding is consistent with the *miR-1*/*miR-133* targets identified in zebrafish (Mishima *et al.*, 2009) and provides a mechanistic basis for the defects in heart morphology displayed by the *miR-133a-1*/*miR-133a-2* double mutants. The similarity between the *miR-133a-1*/*miR-133a-2* and *miR-1-2* mice is intriguing, given the characterization of their opposing roles in cell culture (Chen *et al.*, 2006). Understanding how bicistronic miRNAs can work together in some contexts but antagonistically in others will be interesting.

3.2.2. miR-1/206 and let-7 in neuromuscular junctions

Genetic analyses have identified postdevelopmental roles for *miR-1/206* family members in neuromuscular junction (NMJ) function in both worms (Simon *et al.*, 2008) and mice (Williams *et al.*, 2009), and a developmental role for the *let-7* miRNA in NMJ formation in flies (Caygill and Johnston, 2008; Sokol *et al.*, 2008). Muscle development in *miR-1* mutant worms (Simon *et al.*, 2008) and *miR-206* mutant mice (Williams *et al.*, 2009) proceed normally, but each of these mutant strains displays defective muscle behavior in response to external perturbation. Pharmacological treatments of *miR-1* mutant worms, for example, identified a role for *miR-1* both in controlling the postsynaptic activity of nicotinic acetylcholine receptors as well as in the presynaptic release of acetylcholine (Simon *et al.*, 2008). Double mutant and transgenic experiments indicate that *miR-1* performs its postsynaptic role by directly regulating the levels of nicotinic acetylcholine receptor subunits in muscle cells. Similarly, *miR-1* mediates its presynaptic role by directly regulating the MEF2 transcription factor, which presumably activates a retrograde signal that culminates in changes in acetylcholine release by innervating neurons.

Like worm *miR-1*, mouse *miR-206* is also involved in retrograde signaling, in its case to promote the formation of new NMJs after denervation (Williams *et al.*, 2009). Expression of *miR-206* is sharply upregulated in response to denervation, and the initial rate of reinnervation by motor axons is delayed in *miR-206* mutants relative to wild type. Roughly 90% of both wild-type and *miR-206*-deficient muscles are reinnervated after 8 weeks, though, suggesting a compensatory mechanism that may involve *miR-206*'s neighbor *miR-133b*, which is also robustly activated in response to denervation. Since *miR-206* apparently promotes retrograde signaling by modulating FGF signaling through its direct target *hdac4*, it will be interesting to see whether *miR-133b* also promotes reinnervation after injury or is implicated in FGF signaling.

While the muscle-specific *miR-1/206* family is involved in postdevelopmental NMJs, the *let-7* miRNA is required for the formation of NMJs during development (Caygill and Johnston, 2008; Sokol *et al.*, 2008). Elimination of the *let-7* miRNA results in a reduction in NMJ length and a corresponding decrease in muscle size. The *let-7* mutant phenotype mimics the effects of denervation of wild-type muscle and suggests that *let-7* may function in neurons and/or muscles. Either way, it presumably mediates its affect via direct control of the *abrupt* transcription factor (Caygill and Johnston, 2008).

3.2.3. miR-143 and miR-145 in smooth muscle

Like the *miR-1* and *miR-133* miRNAs, *miR-143* and *miR-145* are also encoded in a bicistronic locus (Cordes *et al.*, 2009). This one is present in single copy in the mouse genome and is expressed in the heart during early embryogenesis (E7.5–E16.5) (Cordes *et al.*, 2009; Xin *et al.*, 2009). In adult mice, though, expression of the *miR-143/miR-145* locus is confined to vascular and visceral smooth muscle cells (Cordes *et al.*, 2009; Xin *et al.*, 2009). Genetic deletion of *miR-143* and *miR-145*, either alone or together, has no effect on viability, cardiac structure, or cardiac gene expression (Boettger *et al.*, 2009; Elia *et al.*, 2009; Xin *et al.*, 2009). However, the smooth muscle layer of the aorta and other arteries of the *miR-145* single mutants as well as the *miR-143/miR-145* double mutants is thinner, due to a reduction in the width of smooth muscle cells (Boettger *et al.*, 2009; Elia *et al.*, 2009; Xin *et al.*, 2009). Thus, removal of *miR-143/miR-145* affects the morphology rather than the development or differentiation of smooth muscle cells, presumably due to the direct regulation of cytoskeletal components by these miRNAs. While the developmental role of these miRNAs is relatively mild, they play critical roles in mediating vascular smooth muscle cell programs in response to injury and may target additional mRNAs in this context (Boettger *et al.*, 2009; Elia *et al.*, 2009; Xin *et al.*, 2009).

3.2.4. miR-208/-499 mutants in cardiac muscle

As mentioned above, *miR-208a*, *miR-208b*, and *miR-499* are encoded in the introns of three myosin genes: *myh6*, *myh7*, and *myh7b*, respectively. Evidence in adults that these miRNAs function to control the expression levels of each other's host gene (Callis *et al.*, 2009; van Rooij *et al.*, 2007, 2009) suggested that they may also orchestrate the developmental transition from fetal cardiac β-MHC expression to adult cardiac α-MHC expression. *miR-208a* mutants as well as *miR-208b*/*miR-499* double mutants proceed through development normally, however, indicating that the function of these miRNAs to modulate the α-MHC/β-MHC ratio is limited to the adult (Callis *et al.*, 2009; van Rooij *et al.*, 2007, 2009). These *miR-208b*/*miR-499* double mutants display a substantial loss of type I myofibers in the soleus, indicating a redundant function for these two miRNAs in skeletal muscle (van Rooij *et al.*, 2009).

3.2.5. miR-17–92 in the heart

Recent analysis of a mouse mutant for the *miR-17–92* polycistronic cluster, which encodes members of the *miR-17*, *-18*, *-19*, *-20*, and *-92* families, indicates that these miRNAs play a role in heart development (Ventura *et al.*, 2008; Wang *et al.*, 2010). Despite the low expression of these miRNAs in the heart, *miR-17–92* mutant embryos are perinatal lethal and display strong ventricular septal and outflow tract defects in which the aorta rises from the right ventricle instead of the left ventricle (Ventura *et al.*, 2008; Wang *et al.*, 2010). This morphological defect correlates with an increase in Islet1-positive cardiac progenitor cells during heart formation (Wang *et al.*, 2010), suggesting that *miR-17–92* miRNAs are involved in the differentiation of the outflow tract myocardium. Evidence that the *miR-17–92* locus is a direct target of the bone morphogenetic protein (BMP) signaling pathway and genetically interacts with *bmp4* locus during outflow tract formation (Wang *et al.*, 2010) suggests that BMP signaling regulates *miR-17–92* to promote myocardial differentiation.

4. miRNAs in Muscle Stem Cells

miRNAs are known to play important roles in regulating stem cell self-renewal and differentiation in a variety of tissues (Gangaraju and Lin, 2009), and a series of papers have addressed whether muscle miRNAs may be involved in controlling satellite cell behavior (Chen *et al.*, 2010; Crist *et al.*, 2009; Hirai *et al.*, 2010). Satellite cells are small, mononucleate cells that reside under the basal lamina of vertebrate muscle fibers. They are quiescent under normal physiological conditions, but in response to environmental factors can restore damaged muscle fibers or form completely new muscle fibers. During this process, satellite cells display two hallmarks of stem cells: they generate differentiated muscle cells and reconstitute the satellite cell population through self-renewal. Two related transcription factors, Pax3 and Pax7, are satellite cell markers and required for the maintenance of embryonic myoblast populations, suggesting that they may also be required for adult satellite cell maintenance (Buckingham, 2006). Recently, Pax3 and Pax7 have been identified as the targets of a series of miRNAs, including *miR-1/206*, *miR-27*, and *miR-486* (Chen *et al.*, 2010; Crist *et al.*, 2009; Dey *et al.*, 2011; Hirai *et al.*, 2010). Since these miRNAs are robustly induced during myogenesis, they may silence Pax3 and Pax7 to repress stem cell identity in differentiating daughter cells. Indeed, a role for miRNAs in satellite cell survival and terminal differentiation is supported by the consequence of Dicer depletion in satellite cells purified from adult mouse muscles (Chen *et al.*, 2010), which display increased apoptosis and defective myogenesis upon differentiation induction. The identification of the specific miRNAs involved as well as their relevant targets will be of great interest.

5. Concluding Remarks

miRNAs collectively display a number of features indicating that they are important for muscle biology: some are located within the introns of dedicated muscle genes like α-MHC and β-MHC (van Rooij *et al.*, 2007, 2009), others are under the direct control of myogenic and cardiogenic transcription factors like MyoD, SRF, and MEF2 (Liu *et al.*, 2007; Xin *et al.*, 2009; Zhao *et al.*, 2005), and many are expressed specifically and at very high level throughout the lifetime of muscle cells from birth in the mesoderm of the zygote to death in the aging adult (Rao *et al.*, 2009; Tang *et al.*, 2007). Unlike myogenic transcription factors, however, muscle miRNAs are not absolutely required for the specification or differentiation of muscle cells during development: partially or completely functional muscle systems form in mutant animals that are depleted of specific muscle miRNAs, including *miR-1* (Simon *et al.*, 2008; Sokol and Ambros, 2005; Zhao *et al.*, 2007), *miR-17–92* (Ventura *et al.*, 2008), *miR-133a-1/miR-133a-2* (Liu *et al.*, 2008), *miR-143/miR-145* (Boettger *et al.*, 2009; Elia *et al.*, 2009; Xin *et al.*, 2009), *miR-206* (Williams *et al.*, 2009), *miR-208a* (Callis *et al.*, 2009; van Rooij *et al.*, 2007), and *miR-208b/miR-499* (van Rooij *et al.*, 2009). Many of these mutants die either during embryogenesis or soon thereafter, however, displaying significant defects in muscle tissue morphology and/or size. These phenotypes, along with data that miRNAs like *miR-1*, *miR-133*, *miR-143*, and *miR-145* predominantly regulate the expression of cytoskeletal components (Mishima *et al.*, 2009; Xin *et al.*,

2009), indicate that miRNAs tailor gene expression required for muscle cell shape and behavior. In addition to this role in the terminal differentiation of muscle cells, though, experiments involving either the overexpression or antisense-mediated abrogation of miRNAs in animals or cell culture indicate that some miRNAs can have profound effects on directing mesoderm and muscle cell fate as well (Cordes *et al.*, 2009; Ivey *et al.*, 2008). A key question, then, is why such roles for muscle miRNAs have not been gleaned from genetic knockouts. One answer could simply be redundancy, and the construction of compound miRNA mutants should address that. Another possibility is that miRNA mutants display fate changes only under certain conditions, and identifying those conditions would provide great insight into muscle miRNA function. A third possibility is that the early embryonic expression of muscle miRNAs simply reflects regulation predominantly required for postdevelopmental functions. Indeed, many muscle miRNAs regulate adult muscle physiology and behavior in response to stress (see, e.g., van Rooij *et al.*, 2007). In any case, the continued analysis of miRNA function in muscles is sure to produce many exciting discoveries.

REFERENCES

Aboobaker, A. A., Tomancak, P., Patel, N., Rubin, G. M., and Lai, E. C. (2005). *Proc. Natl. Acad. Sci. USA* **102,** 18017–18022.

Albinsson, S., Suarez, Y., Skoura, A., Offermanns, S., Miano, J. M., and Sessa, W. C. (2010). *Arterioscler. Thromb. Vasc. Biol.* **30,** 1118–1126.

Alvarez-Saavedra, E., and Horvitz, H. R. (2010). *Curr. Biol.* **20,** 367–373.

Ambros, V. (2004). *Nature* **431,** 350–355.

Bartel, D. P. (2009). *Cell* **136,** 215–233.

Bernstein, E., Kim, S. Y., Carmell, M. A., Murchison, E. P., Alcorn, H., Li, M. Z., Mills, A. A., Elledge, S. J., Anderson, K. V., and Hannon, G. J. (2003). *Nat. Genet.* **35,** 215–217.

Biemar, F., Zinzen, R., Ronshaugen, M., Sementchenko, V., Manak, J. R., and Levine, M. S. (2005). *Proc. Natl. Acad. Sci. USA* **102,** 15907–15911.

Boettger, T., Beetz, N., Kostin, S., Schneider, J., Kruger, M., Hein, L., and Braun, T. (2009). *J. Clin. Invest.* **119,** 2634–2647.

Boutz, P. L., Chawla, G., Stoilov, P., and Black, D. L. (2007). *Genes Dev.* **21,** 71–84.

Brenner, J. L., Jasiewicz, K. L., Fahley, A. F., Kemp, B. J., and Abbott, A. L. (2010). *Curr. Biol.* **20,** 1321–1325.

Buckingham, M. (2006). *Curr. Opin. Genet. Dev.* **16,** 525–532.

Callis, T. E., Pandya, K., Seok, H. Y., Tang, R. H., Tatsuguchi, M., Huang, Z. P., Chen, J. F., Deng, Z., Gunn, B., Shumate, J., Willis, M. S., Selzman, C. H., *et al.* (2009). *J. Clin. Invest.* **119,** 2772–2786.

Caygill, E. E., and Johnston, L. A. (2008). *Curr. Biol.* **18,** 943–950.

Cheloufi, S., Dos Santos, C. O., Chong, M. M., and Hannon, G. J. (2010). *Nature* **465,** 584–589.

Chen, J. F., Mandel, E. M., Thomson, J. M., Wu, Q., Callis, T. E., Hammond, S. M., Conlon, F. L., and Wang, D. Z. (2006). *Nat. Genet.* **38,** 228–233.

Chen, J. F., Murchison, E. P., Tang, R., Callis, T. E., Tatsuguchi, M., Deng, Z., Rojas, M., Hammond, S. M., Schneider, M. D., Selzman, C. H., Meissner, G., Patterson, C., *et al.* (2008). *Proc. Natl. Acad. Sci. USA* **105,** 2111–2116.

Chen, J. F., Tao, Y., Li, J., Deng, Z., Yan, Z., Xiao, X., and Wang, D. Z. (2010). *J. Cell Biol.* **190,** 867–879.

Cifuentes, D., Xue, H., Taylor, D. W., Patnode, H., Mishima, Y., Cheloufi, S., Ma, E., Mane, S., Hannon, G. J., Lawson, N. D., Wolfe, S. A., and Giraldez, A. J. (2010). *Science* **328,** 1694–1698.

Cordes, K. R., Sheehy, N. T., White, M. P., Berry, E. C., Morton, S. U., Muth, A. N., Lee, T. H., Miano, J. M., Ivey, K. N., and Srivastava, D. (2009). *Nature* **460,** 705–710.

Crist, C. G., Montarras, D., Pallafacchina, G., Rocancourt, D., Cumano, A., Conway, S. J., and Buckingham, M. (2009). *Proc. Natl. Acad. Sci. USA* **106,** 13383–13387.

Davis, B. N., Hilyard, A. C., Lagna, G., and Hata, A. (2008). *Nature* **454,** 56–61.

Dey, B. K., Gagan, J., and Dutta, A. (2011). *Mol. Cell. Biol.* **31,** 203–214.

Elia, L., Quintavalle, M., Zhang, J., Contu, R., Cossu, L., Latronico, M. V., Peterson, K. L., Indolfi, C., Catalucci, D., Chen, J., Courtneidge, S. A., and Condorelli, G. (2009). *Cell Death Differ.* **16,** 1590–1598.

Gangaraju, V. K., and Lin, H. (2009). *Nat. Rev. Mol. Cell Biol.* **10,** 116–125.

Ge, Y., Sun, Y., and Chen, J. (2011). *J. Cell Biol.* **192,** 69–81.

Giraldez, A. J., Cinalli, R. M., Glasner, M. E., Enright, A. J., Thomson, J. M., Baskerville, S., Hammond, S. M., Bartel, D. P., and Schier, A. F. (2005). *Science* **308,** 833–838.

Graveley, B. R., Brooks, A. N., Carlson, J. W., Duff, M. O., Landolin, J. M., Yang, L., Artieri, C. G., van Baren, M. J., Boley, N., Booth, B. W., Brown, J. B., Cherbas, L., *et al.* (2011). *Nature* **471,** 473–479.

Grimson, A., Srivastava, M., Fahey, B., Woodcroft, B. J., Chiang, H. R., King, N., Degnan, B. M., Rokhsar, D. S., and Bartel, D. P. (2008). *Nature* **455,** 1193–1197.

Grishok, A., Pasquinelli, A. E., Conte, D., Li, N., Parrish, S., Ha, I., Baillie, D. L., Fire, A., Ruvkun, G., and Mello, C. C. (2001). *Cell* **106,** 23–34.

Hatfield, S. D., Shcherbata, H. R., Fischer, K. A., Nakahara, K., Carthew, R. W., and Ruohola-Baker, H. (2005). *Nature* **435,** 974–978.

Hirai, H., Verma, M., Watanabe, S., Tastad, C., Asakura, Y., and Asakura, A. (2010). *J. Cell Biol.* **191,** 347–365.

Huang, T. H., Zhu, M. J., Li, X. Y., and Zhao, S. H. (2008). *PLoS One* **3,** e3225.

Ikeda, S., He, A., Kong, S. W., Lu, J., Bejar, R., Bodyak, N., Lee, K. H., Ma, Q., Kang, P. M., Golub, T. R., and Pu, W. T. (2009). *Mol. Cell. Biol.* **29,** 2193–2204.

Ivey, K. N., Muth, A., Arnold, J., King, F. W., Yeh, R. F., Fish, J. E., Hsiao, E. C., Schwartz, R. J., Conklin, B. R., Bernstein, H. S., and Srivastava, D. (2008). *Cell Stem Cell* **2,** 219–229.

Kanellopoulou, C., Muljo, S. A., Kung, A. L., Ganesan, S., Drapkin, R., Jenuwein, T., Livingston, D. M., and Rajewsky, K. (2005). *Genes Dev.* **19,** 489–501.

Kloosterman, W. P., Steiner, F. A., Berezikov, E., de Bruijn, E., van de Belt, J., Verheul, M., Cuppen, E., and Plasterk, R. H. (2006). *Nucleic Acids Res.* **34,** 2558–2569.

Kwon, C., Han, Z., Olson, E. N., and Srivastava, D. (2005). *Proc. Natl. Acad. Sci. USA* **102,** 18986–18991.

Lee, R. C., and Ambros, V. (2001). *Science* **294,** 862–864.

Li, Q., Lin, X., Yang, X., and Chang, J. (2010). *Am. J. Physiol. Heart Circ. Physiol.* **298,** H1340–H1347.

Li, T., Wu, R., Zhang, Y., and Zhu, D. (2011). *BMC Genomics* **12,** 186.

Liu, N., Williams, A. H., Kim, Y., McAnally, J., Bezprozvannaya, S., Sutherland, L. B., Richardson, J. A., Bassel-Duby, R., and Olson, E. N. (2007). *Proc. Natl. Acad. Sci. USA* **104,** 20844–20849.

Liu, N., Bezprozvannaya, S., Williams, A. H., Qi, X., Richardson, J. A., Bassel-Duby, R., and Olson, E. N. (2008). *Genes Dev.* **22,** 3242–3254.

Martinez, N. J., Ow, M. C., Reece-Hoyes, J. S., Barrasa, M. I., Ambros, V. R., and Walhout, A. J. (2008). *Genome Res.* **18,** 2005–2015.

McCarthy, J. J. (2008). *Biochim. Biophys. Acta* **1779,** 682–691.

Mishima, Y., Abreu-Goodger, C., Staton, A. A., Stahlhut, C., Shou, C., Cheng, C., Gerstein, M., Enright, A. J., and Giraldez, A. J. (2009). *Genes Dev.* **23,** 619–632.

Monteys, A. M., Spengler, R. M., Wan, J., Tecedor, L., Lennox, K. A., Xing, Y., and Davidson, B. L. (2010). *RNA* **16,** 495–505.

Murchison, E. P., Partridge, J. F., Tam, O. H., Cheloufi, S., and Hannon, G. J. (2005). *Proc. Natl. Acad. Sci. USA* **102,** 12135–12140.

Naguibneva, I., Ameyar-Zazoua, M., Polesskaya, A., Ait-Si-Ali, S., Groisman, R., Souidi, M., Cuvellier, S., and Harel-Bellan, A. (2006). *Nat. Cell Biol.* **8,** 278–284.

Niu, Z., Li, A., Zhang, S. X., and Schwartz, R. J. (2007). Serum response factor micromanaging cardiogenesis. *Curr. Opin. Cell Biol.* **19**(6), 618–627.

O'Rourke, J. R., Georges, S. A., Seay, H. R., Tapscott, S. J., McManus, M. T., Goldhamer, D. J., Swanson, M. S., and Harfe, B. D. (2007). *Dev. Biol.* **311,** 359–368.

Pan, Y., Balazs, L., Tigyi, G., and Yue, J. (2011). *Biochem. Biophys. Res. Commun.* **408,** 369–374.

Polesskaya, A., Cuvellier, S., Naguibneva, I., Duquet, A., Moss, E. G., and Harel-Bellan, A. (2007). *Genes Dev.* **21,** 1125–1138.

Rao, P. K., Kumar, R. M., Farkhondeh, M., Baskerville, S., and Lodish, H. F. (2006). *Proc. Natl. Acad. Sci. USA* **103,** 8721–8726.

Rao, P. K., Toyama, Y., Chiang, H. R., Gupta, S., Bauer, M., Medvid, R., Reinhardt, F., Liao, R., Krieger, M., Jaenisch, R., Lodish, H. F., and Blelloch, R. (2009). *Circ. Res.* **105,** 585–594.

Rau, F., Freyermuth, F., Fugier, C., Villemin, J. P., Fischer, M. C., Jost, B., Dembele, D., Gourdon, G., Nicole, A., Duboc, D., Wahbi, K., Day, J. W., *et al.* (2011). *Nat. Struct. Mol. Biol.* **18,** 840–845.

Saxena, A., and Tabin, C. J. (2010). *Proc. Natl. Acad. Sci. USA* **107,** 87–91.

Shkumatava, A., Stark, A., Sive, H., and Bartel, D. P. (2009). *Genes Dev.* **23,** 466–481.

Simon, D. J., Madison, J. M., Conery, A. L., Thompson-Peer, K. L., Soskis, M., Ruvkun, G. B., Kaplan, J. M., and Kim, J. K. (2008). *Cell* **133,** 903–915.

Sinkkonen, L., Hugenschmidt, T., Berninger, P., Gaidatzis, D., Mohn, F., Artus-Revel, C. G., Zavolan, M., Svoboda, P., and Filipowicz, W. (2008). *Nat. Struct. Mol. Biol.* **15,** 259–267.

Sokol, N. S., and Ambros, V. (2005). *Genes Dev.* **19,** 2343–2354.

Sokol, N. S., Xu, P., Jan, Y. N., and Ambros, V. (2008). *Genes Dev.* **22,** 1591–1596.

Spruce, T., Pernaute, B., Di-Gregorio, A., Cobb, B. S., Merkenschlager, M., Manzanares, M., and Rodriguez, T. A. (2010). *Dev. Cell* **19,** 207–219.

Suh, N., and Blelloch, R. (2011). *Development* **138,** 1653–1661.

Suzuki, H. I., Yamagata, K., Sugimoto, K., Iwamoto, T., Kato, S., and Miyazono, K. (2009). *Nature* **460,** 529–533.

Tang, X., Gal, J., Zhuang, X., Wang, W., Zhu, H., and Tang, G. (2007). *RNA* **13,** 1803–1822.

Trabucchi, M., Briata, P., Garcia-Mayoral, M., Haase, A. D., Filipowicz, W., Ramos, A., Gherzi, R., and Rosenfeld, M. G. (2009). *Nature* **459,** 1010–1014.

van Rooij, E., Sutherland, L. B., Qi, X., Richardson, J. A., Hill, J., and Olson, E. N. (2007). *Science* **316,** 575–579.

van Rooij, E., Quiat, D., Johnson, B. A., Sutherland, L. B., Qi, X., Richardson, J. A., Kelm, R. J., Jr., and Olson, E. N. (2009). *Dev. Cell* **17,** 662–673.

Ventura, A., Young, A. G., Winslow, M. M., Lintault, L., Meissner, A., Erkeland, S. J., Newman, J., Bronson, R. T., Crowley, D., Stone, J. R., Jaenisch, R., Sharp, P. A., *et al.* (2008). *Cell* **132,** 875–886.
Viswanathan, S. R., and Daley, G. Q. (2010). *Cell* **140,** 445–449.
Viswanathan, S. R., Daley, G. Q., and Gregory, R. I. (2008). *Science* **320,** 97–100.
Vo, N. K., Dalton, R. P., Liu, N., Olson, E. N., and Goodman, R. H. (2010). *Proc. Natl. Acad. Sci. USA* **107,** 19231–19236.
Wang, J., Greene, S. B., Bonilla-Claudio, M., Tao, Y., Zhang, J., Bai, Y., Huang, Z., Black, B. L., Wang, F., and Martin, J. F. (2010). *Dev. Cell* **19,** 903–912.
Wienholds, E., Kloosterman, W. P., Miska, E., Alvarez-Saavedra, E., Berezikov, E., de Bruijn, E., Horvitz, H. R., Kauppinen, S., and Plasterk, R. H. (2005). *Science* **309,** 310–311.
Williams, A. H., Valdez, G., Moresi, V., Qi, X., McAnally, J., Elliott, J. L., Bassel-Duby, R., Sanes, J. R., and Olson, E. N. (2009). *Science* **326,** 1549–1554.
Xin, M., Small, E. M., Sutherland, L. B., Qi, X., McAnally, J., Plato, C. F., Richardson, J. A., Bassel-Duby, R., and Olson, E. N. (2009). *Genes Dev.* **23,** 2166–2178.
Zhang, L., Stokes, N., Polak, L., and Fuchs, E. (2011). *Cell Stem Cell* **8,** 294–308.
Zhao, Y., Samal, E., and Srivastava, D. (2005). *Nature* **436,** 214–220.
Zhao, Y., Ransom, J. F., Li, A., Vedantham, V., von Drehle, M., Muth, A. N., Tsuchihashi, T., McManus, M. T., Schwartz, R. J., and Srivastava, D. (2007). *Cell* **129,** 303–317.

CHAPTER FOUR

Small RNAs in Germ Cell Development

Torsten U. Banisch, Mehdi Goudarzi, *and* Erez Raz

Contents

Abstract

The identification of small RNA molecules and the elucidation of their functions in the cell enhanced our understanding of the mechanisms controlling gene expression at the posttranscriptional level. Here, we review findings concerning the role small RNA molecules play in the development of the germline in various organisms. In this context, microRNAs (miRNAs) function predominantly in regulating mRNA expression and turnover and direct the expression of specific proteins to founders of the germline, the primordial germ cells (PGCs). miRNA

Center for Molecular Biology of Inflammation, Institute of Cell Biology, University of Münster, Münster, Germany

Current Topics in Developmental Biology, Volume 99
ISSN 0070-2153, DOI: 10.1016/B978-0-12-387038-4.00004-5

function is then important for the maintenance of germline stem cells (GSCs) in their niche and for coordinating GSC differentiation into the gametes. The second family of small RNA molecules, the short endo-siRNAs, regulate gene expression posttranscriptionally as well but are also important for transposable elements regulation and thus for genome stability. Last, the Piwi-interacting small RNAs (piRNAs) are specifically expressed in the germline, where they primarily control transposon activity, guarding the genome from potential damage associated with excessive transposition. Together, the function of these three groups of small RNAs ensures proper development and continuation of the germline through the generations.

1. Introduction: The Germ Cell Lineage and Small RNA Molecules

The germline passes the genetic information from one generation to the next. The first cells that exclusively give rise to the germline by mitotic divisions are the primordial germ cells (PGCs). The PGCs then migrate toward the region where the gonad will form, where they settle in their niche and are maintained as germline stem cells (GSCs) that give rise to differentiated progeny. A characteristic of PGCs and GSCs is the possession of a unique structure, the germ plasm, also termed "nuage" (Extavour and Akam, 2003), which is an electron-dense structure associated with mitochondria and contains germline specific mRNAs and proteins. Relevant for this review, small noncoding RNAs have been shown to reside in the germ plasm (Findley *et al.*, 2003). These small regulatory molecules are crucial for the regulation of various processes in the life of germ cells including the specification of germ cell progenitors, maintenance of GSC fate, protection of the genome, and allowance of proper differentiation into the gametes. Three major classes of small RNAs have been characterized, and while they are produced and function in different pathways, related proteins are required for their biogenesis and activity. The first to be discovered and perhaps the best-characterized small RNA molecules are the microRNAs (miRNAs), followed by the later identification of endogenous small interfering RNAs (endo-siRNAs) and Piwi-interacting RNAs (piRNAs) (Aravin *et al.*, 2001; Hamilton *et al.*, 2002; Lee *et al.*, 1993).

In this review, we highlight and discuss the roles of small RNAs belonging to these three groups in various aspects of germline development in animals—from specification of the PGC progenitors to the final differentiation steps producing the gametes. The function of small RNAs in the plant germline has been reviewed elsewhere (Borges *et al.*, 2011; Grant-Downton, 2010; Jones-Rhoades *et al.*, 2006; Voinnet, 2009).

1.1. PGC specification: Preformation/epigenesis

The establishment of the germline is achieved by two different strategies. Specification of germ cells is directed either by localized maternally provided determinants in a process called "preformation" or epigenetically by inductive signals emanating from surrounding tissues.

Many animals such as *Caenorhabditis elegans*, *Drosophila melanogaster*, *Xenopus laevis*, and *Danio rerio* employ germ plasm for specifying their germline. The maternally provided germ plasm is composed of specific RNAs and proteins that when incorporated into cells directs them to develop as germ cells (Fig. 4.1A(a and b) and B(a and b)). In contrast to this strategy, in other organisms (e.g., mammalians and Urodele amphibians) germ cell progenitors are specified in an inductive process. Here, signaling cues from surrounding tissues, involving proteins of the BMP family, specify the germ cells (Fig. 4.1C(a); Lawson *et al.*, 1999; Ying and Zhao, 2001; Ying *et al.*, 2001; for review, see Extavour and Akam, 2003).

A more detailed description of germ cell development and the factors involved is provided in the following section.

1.2. Germline development in different model organisms

1.2.1. Germline development in the fly and the nematode

Cells forming at the posterior pole of the early *D. melanogaster* embryo (pole cells) harbor the maternally provided pole plasm which directs them to become progenitors of the germline (Fig. 4.1A(a and b); Technau and Campos-Ortega, 1986). During embryonic development, a combination of active and passive cell movements bring the PGCs to the region where the gonad develops and where they associate with somatic gonad precursors, reviewed in Richardson and Lehmann (2010). During later stages of gonad formation, GSCs are located in the "niche," a position where they are maintained as pluripotent cells by different signaling events that control self-renewal and differentiation (Kiger *et al.*, 2000; Tran *et al.*, 2000; Xie and Spradling, 1998). One of the better-characterized signals controlling the balance between maintenance of pluripotency and differentiation is the BMP homologue Decapentaplegic (Dpp), constituting a key signal for female GSC maintenance (Xie and Spradling, 1998, 2000). The niche is called "nest" in female, consisting of two GSCs associated with five or six cap cells, which are part of the terminal filament (Fig.4.1A(c); Sahut-Barnola *et al.*, 1995). In male gonads, the niche is referred to as "hub" and consists of 5–9 GSCs and 10–18 somatic cyst progenitor stem cells (Hardy *et al.*, 1979). The early processes of gametogenesis in both the ovaries and testes are very similar: a GSC divides, giving rise to two daughter cells, one that remains in the niche and maintains a GSC fate, while the other becomes located away from the niche and from the "stemness" signals and thus starts

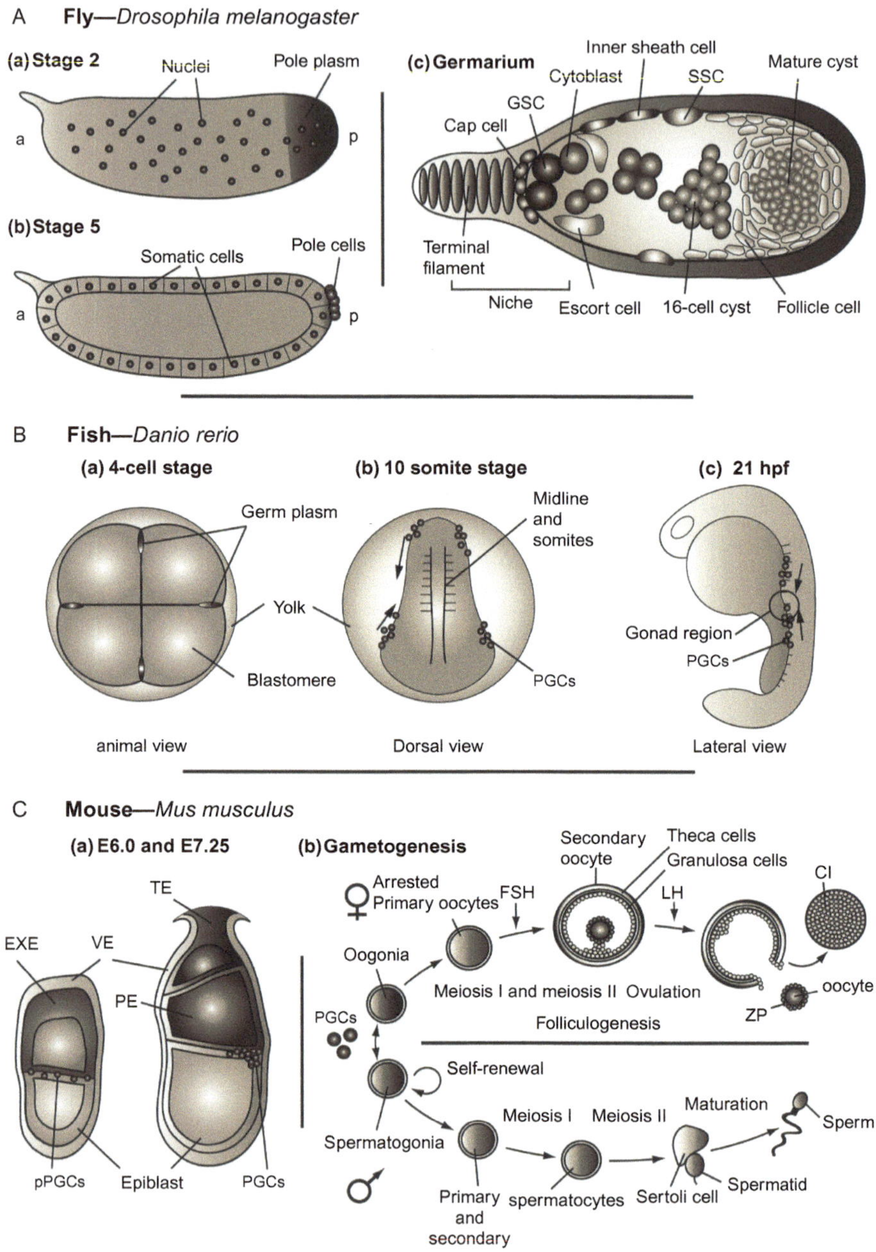

Figure 4.1 (A) Schematic representation of germline development in *D. melanogaster*. (A(a)) The maternally provided pole plasm is localized to the posterior (p) of the developing embryo; a (anterior) (A(b)) Following cellularization, cells located at the posterior of the embryo that inherit the pole plasm become the pole cells, or PGCs. (A(c)) The GSC are maintained by signals provided by cells in the niche, while germline

differentiating (Fig. 4.1A(c); reviewed in Davies and Fuller, 2008 and Xie *et al.*, 2008).

In the nematode *C. elegans*, the germ plasm is aggregated into the "P-granules," which are evenly distributed throughout the oocyte cytoplasm before and shortly after fertilization. During the first division, the granules become localized to the posterior pole of the zygote and are then asymmetrically segregated and stabilized during the following divisions, such that the germ plasm material is inherited by the P4 blastomere, defining the single germline precursor (Deppe *et al.*, 1978; Hird *et al.*, 1996; Seydoux and Fire, 1994; Strome and Wood, 1982). The P4 blastomere gives rise to the two PGCs, Z2 and Z3, that together with two somatic gonadal cells (Z1 and Z4) give rise to the gonad (reviewed in Kemphues and Strome, 1997; Seydoux and Braun, 2006).

1.2.2. Germline development in zebrafish

The mechanisms responsible for PGC specification in zebrafish were revealed when the first early molecular marker for these cells, the RNA encoding for the Vasa protein, was identified (Braat *et al.*, 1999; Yoon *et al.*, 1997). *vasa* mRNA, which resides within an electron-dense germ plasm, localizes to the first cleavage planes in the early embryo (Fig. 4.1B(a)) and is later incorporated into four cells. Cells that maintain the germ plasm in them during subsequent cell divisions develop into PGCs (Knaut *et al.*, 2000). While the topology of germ plasm distribution differs, this scenario is similar to that described for the invertebrate species discussed above, as well as for

cells located away from it further divide and differentiate. The somatic stem cells (SSCs) give rise to the follicle cells that envelope the developing cyst. (B) PGC development in the zebrafish. (B(a)) Maternally inherited germ plasm accumulates at the edges of the first two cleavage furrows and is later incorporated into the four cells that become the PGCs. (B(b and c)) After specification, the PGCs actively migrate toward the region where the gonad develops, eventually forming two bilateral clusters on either sides of the body, positions where they interact with somatic cells and differentiate into gametes. (C) Germ cell development in mouse. (C(a)) Epiblast cells of the early mouse embryo are induced by signals from the extraembryonic ectoderm (EXE) to become progenitors of the PGCs (pPGCs) at E6.0, persistent signaling specifies the PGCs at E7.25; TE (trophoectoderm), VE (visceral ectoderm), PE (primitive endoderm). (C(b)) In the females the oogonia give rise to the primary oocytes that are arrested at meiosis I. Postnatally, the oocyte completes the meiosis, and dependent on FSH signal gives rise to the secondary oocyte, which undergoes folliculogenesis and becomes surrounded by the Zona pellucida (ZP), granulosa- and theca cells. Upon LH stimulus, ovulation occurs and the somatic support cells differentiate to form the Corpus luteum (Cl). In males, the spermatogonia divide postnatally into one daughter cell retaining the spermatogonia cell fate, and another that differentiates into the spermatocyte. Spermatocytes undergo meiosis and give rise to spermatids that interact with the supporting sertoli cells facilitating their development into mature sperm. (See Color Insert.)

X. laevis. The identification of additional zebrafish germ plasm components, many of which are RNA-binding proteins, or encode for such proteins (Hashimoto *et al.*, 2004, 2006; Köprunner *et al.*, 2001; Kosaka *et al.*, 2007; Maegawa *et al.*, 1999; Olsen *et al.*, 1997; Weidinger *et al.*, 2003; Yoon *et al.*, 1997), suggested that similar to its role in other organisms, the germ plasm plays a role in controlling RNA stability, localization, and function (reviewed in Ikenishi, 1998). Whereas many zebrafish germ plasm components such as *dazl*, *nanos*, and *vasa* are conserved between zebrafish and other organisms whose germ cells are specified by germ plasm, some appear to be vertebrate specific, notably the RNA-binding protein Dead end (Dnd) (Hay *et al.*, 1988; Houston *et al.*, 1998; Komiya *et al.*, 1994; Köprunner *et al.*, 2001; Lehmann and Nusslein-Volhard, 1991; Maegawa *et al.*, 1999; Weidinger *et al.*, 2003; Yoon *et al.*, 1997; Zhou and King, 1996).

Following their specification, the PGCs proliferate and migrate toward the region where the gonad develops, where they associate with somatic cells (Fig. 4.1B(b and c)) and differentiate into gametes (Richardson and Lehmann, 2010).

1.2.3. Germline development in the mouse

Mouse PGCs are specified around embryonic day E5.5–E6.0 in response to signals encoded by members of the transforming growth factor β (TGFβ) superfamily and the family of bone morphogenetic proteins (BMPs) originating from cells of the extraembryonic ectoderm and the visceral endoderm cells (Fig. 4.1C(a); Lawson *et al.*, 1999; Ying and Zhao, 2001; Ying *et al.*, 2001). These signals establish "competence" of the cells to assume the fate of PGC precursors as marked by BMP-induced expression of IFITM3 (interferon-induced transmembrane protein3, also referred to as "Fragilis") (de Sousa Lopes *et al.*, 2007; Saitou *et al.*, 2002). At E6.25, some of the responding cells start expressing PRDM1 (PRDI-BF1-RIZ domain-containing 1; BLIMP1) and the closely related PRDM14, becoming the first cells in the mammalian embryo to be committed to a specific fate. These BMP-dependent events ultimately lead to suppression of somatic gene expression (e.g., *Hox* genes) and concomitant induction of germline characteristic gene expression and *de novo* expression of pluripotency genes such as *Sox2*, *Oct4*, and *Nanog* (Kurimoto *et al.*, 2008; Ohinata *et al.*, 2005; Saitou *et al.*, 2002; Vincent *et al.*, 2005; Yamaji *et al.*, 2008). Similar to many other organisms, following their specification, the specified PGCs migrate toward the region where the gonad develops (Richardson and Lehmann, 2010). Several gene products have been shown to be required for the survival of mouse PGC during their migration. Relevant for this review, the function of RNA-binding proteins that are associated with the germ plasm of different organisms such as Nanos and Dead end homolog 1 (DND1, Ter) is essential at this stage (Tsuda *et al.*, 2003; Youngren *et al.*, 2005). The PGCs then loose their migratory behavior and eventually differentiate into gametes, sperm, and egg.

During gametogenesis, the male germ cells undergo mitotic arrest and resume division postnatally as spermatogonia cells. In contrast, the germ cells in the female continue to divide, arrest in the diplotene stage of meiosis I, and resume meiotic division postnatally during the ovarian folliculogenesis (Fig. 4.1C(b); reviewed in Edson *et al.*, 2009).

1.3. Small RNAs in germ cell development

The first evidence for small noncoding RNAs (sRNAs) function in development was the discovery that introduction of double-stranded RNA (dsRNA) fragments triggered effective inhibition of RNA function, namely RNA interference (RNAi) in *C. elegans* (Fire *et al.*, 1998) following the discovery of genes encoding for small RNAs (Lee *et al.*, 1993; Wightman *et al.*, 1993). Prominent species of small RNAs are the miRNAs, the endo-siRNAs, both playing more common roles in all tissues, and the piRNAs whose function is mostly required in the germline (Ambros *et al.*, 2003; Aravin *et al.*, 2007a; Lagos-Quintana *et al.*, 2001; Lau *et al.*, 2001; Lee and Ambros, 2001; Reinhart and Bartel, 2002).

Below, we discuss the function these three small RNA families play in germline development by regulating gene expression and transposon activity.

2. miRNAs in Germ Cell Development

2.1. miRNA biogenesis and function

miRNAs comprise a large family of 21–25-nucleotide (nt)-long noncoding RNAs that function by regulating gene expression posttranscriptionally. The biogenesis of miRNAs was recently reviewed in great detail by Winter *et al.* (2009) and Krol *et al.* (2010). Briefly, miRNAs are processed from precursor molecules (pri-miRNAs) transcribed by RNA polymerase II (Fig. 4.2A). The pri-miRNAs fold into a characteristic hairpin structure that is processed by the RNase III enzyme Drosha and the accessory protein DGCR8 into the ~70-nt-long pre-miRNA (Lund *et al.*, 2004). Following export into the cytoplasm mediated by Exportin 5, the pre-miRNA is further processed into the 21–25-nt-long miRNA/miRNA* duplex by another RNase III, Dicer, supported by the RNA-binding protein TRBP (Carmell and Hannon, 2004; Hutvagner *et al.*, 2001; Yi *et al.*, 2003). To allow targeting of specific mRNAs, the guiding strand of the now mature miRNA is loaded onto an Argonaut (AGO) protein, one of the main components in the miRNA-induced silencing complex, miRISC. The miRNA incorporated into the miRISC will guide the complex to the specific target mRNAs.

miRNAs interact with their target mRNAs by recognizing short sequence signatures often located within their 3′ untranslated region

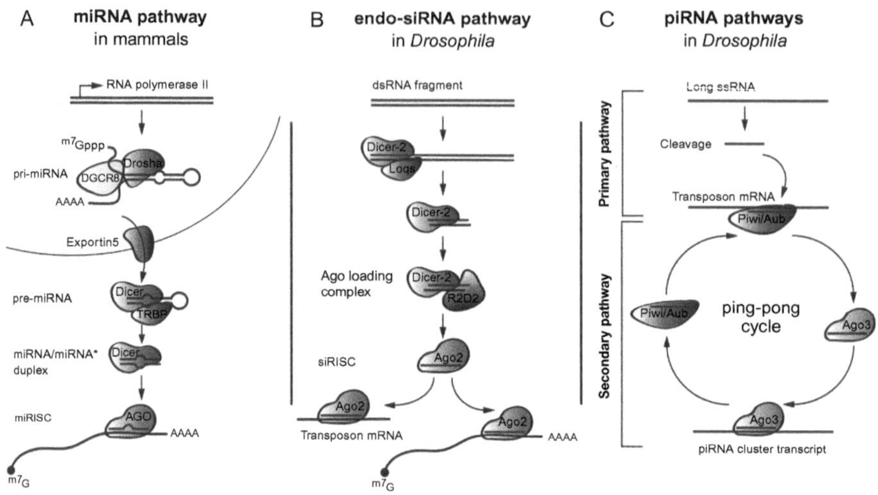

Figure 4.2 (A) Schematic representation of miRNA biogenesis and function in mammals. miRNAs are transcribed by RNA polymerase II, and Drosha and DGCR8 process the stem loop-containing pri-miRNAs. After export into the cytoplasm by Exportin 5, Dicer and the accessory protein TRBP/Loqs (mammals/fly, respectively) further process the pre-miRNA into the miRNA/miRNA duplex. The guide strand is then loaded onto the effector protein AGO forming the miRISC complex that is guided toward the 3′UTR of the target mRNAs, resulting in deadenylation or translational repression. (B) endo-siRNA biogenesis and function in *D. melanogaster*. Precursor double-stranded RNAs (dsRNAs) fragments are processed by Dicer-2 and the dsRNA-binding protein Loqs. The processed RNA bound to the R2D2 and Dicer-2 proteins is loaded onto Ago2 forming the siRISC that targets the slicer activity of Ago2 toward target mRNAs and RNA of transposable elements. (C) piRNA biogenesis and function *D. melanogaster*. The single-strand RNA which is the precursor for piRNAs is cleaved and the resulting antisense piRNA is loaded onto Piwi/Aub protein and binds transposon RNA thereby initiating the ping-pong cycle. Cleavage of the transposon RNA generates a small sense RNA molecule that is loaded onto Ago3 protein and the RNA-protein complex then binds the piRNA transcript. Cleavage by piRNA/Ago3 and interaction of the resulting small RNA with PIWI/Aub protein completes the cycle, leading to an increase in the number of piRNA molecules and inhibition of transposon function. (See Color Insert.)

(3′UTR). Association of miRNAs with their target commonly leads to deadenylation and subsequent degradation of the target and can also result in translational repression (Carrington and Ambros, 2003; Lai, 2002; reviewed in Bartel, 2009; Kim *et al.*, 2009; Krol *et al.*, 2010).

2.2. miRNA function during PGC specification

As mentioned above, PGC specification in the invertebrates *C. elegans*, *D. melanogaster* and in some vertebrates like *X. laevis* and *Danio rerio* follows the preformation strategy that depends on maternally provided factors.

These factors include RNA molecules and proteins that together constitute the germ plasm. Cells inheriting the germ plasm are determined to assume the germ cell fate.

The first indications for a possible role of miRNAs in germline specification were observations in *D. melanogaster* demonstrating that components of the miRNA machinery like Dicer-1 and dFMRP (fragile X mental retardation protein) could influence the expression of Oskar, Vasa, and Nanos, which are crucial pole plasm components (Megosh *et al.*, 2006). Further analysis of the defects in pole cell formation upon loss of the miRNA machinery revealed that the miRNAs themselves, namely miR-6 and miR-9, are key regulators of this process (Leaman *et al.*, 2005).

As in *Drosophila*, the germ plasm of zebrafish is enriched at specific locations of the developing embryo (Braat *et al.*, 1999; Olsen *et al.*, 1997; Yoon *et al.*, 1997). Importantly, whereas germ plasm and mRNAs that are part of it (e.g., *nanos*, *dead end*, and *tdrd7*) are highly concentrated in certain locations in the early embryo, a significant amount of the material is actually not localized and is found in other cells as well. This point is exemplified by the *nanos* mRNA, whose translation is strongly inhibited in cells destined to give rise to somatic cells (Blaser *et al.*, 2005; Köprunner *et al.*, 2001). As development progresses, these mRNA molecules are preferentially cleared from the soma, establishing PGC specific expression of the transcripts.

Two important studies deciphered the mechanisms underlying the somatic silencing and degradation of such germline-enriched mRNAs, demonstrating the role of miRNA 430 (miR-430) in controlling *nanos* and *tdrd7* function (Giraldez *et al.*, 2006; Mishima *et al.*, 2006). miRNAs of the miR-430 cluster are highly abundant in early zebrafish embryos and are key regulators of mRNA turnover during maternal-to-zygotic transition. The role of miRNAs in regulating maternal mRNA turnover was demonstrated in embryos lacking Dicer function (MZ*dicer* mutant embryos) that exhibit grossly abnormal embryonic development due to misexpression of a large set of mRNAs (Giraldez *et al.*, 2005, 2006). Further, mRNAs that are normally silenced and degraded in somatic cells, while being expressed in germ cells, retain their somatic expression in MZ*dicer* mutant embryos. Whereas miRNAs are responsible for the somatic silencing of "germ cell" mRNAs, they are also present in the germ cells (Mishima *et al.*, 2006), but their function is counteracted in these cells. One factor known to inhibit miRNA activity in zebrafish germ cells is the Dead end protein. Dead end binds poly-U stretches within the 3′UTR of the target mRNA (e.g., *nanos*, *tdrd7*, and *hub*) to inhibit miRNA function by a yet unknown mechanism (Kedde *et al.*, 2007; Mickoleit *et al.*, 2011; Slanchev *et al.*, 2009). In addition, the DAZ-like protein (Dazl) was found to increase translation efficiency of mRNAs by promoting poly(A) tail elongation in the germ line and can thereby counteract deadenylation mediated by miR-430 (Takeda *et al.*,

2009). Last, Hu proteins were recently shown to relieve "germline mRNAs" from miRNA repression in *Xenopus* (Koebernick *et al.*, 2010). These results suggest that multiple proteins cooperate to ensure the expression of such mRNAs in the germline despite the presence of miRNAs in the same cells.

Considering that similar mechanisms control this process in *C. elegans*, *D. melanogaster*, and *X. laevis*, the role for miRNAs in promoting clearance of maternally provided transcripts in somatic cells appears to be conserved (Bushati *et al.*, 2008; Lund *et al.*, 2009; Wu *et al.*, 2010). Interestingly, as discussed below, mRNA decay during maternal-to-zygotic transition can also be directed by another group of small RNA molecules, the piRNAs, as shown for *nanos* mRNA (Rouget *et al.*, 2010).

In contrast to germline specification by maternally provided germ plasm described above, in other organisms such as mouse and Urodele amphibians, the germline is specified by inductive signals produced by the embryo itself (Boterenbrood and Nieuwkoop, 1973; Nieuwkoop, 1947; Tam and Zhou, 1996; Tsang *et al.*, 2001).

Two key factors involved in mouse PGC specification are the transcription regulators PRDM1 (also known as BLIMP1) and PRDM14 that are expressed at embryonic day E6.25 in the region where the PGCs arise (Fig. 4.1C(a)). The expression of PRDM1 and PRDM14 is induced by BMPs, and loss-of-function experiments revealed the critical role they play in PGC specification (Ohinata *et al.*, 2005; Robertson *et al.*, 2007; Vincent *et al.*, 2005; Yamaji *et al.*, 2008). Persistent expression of both proteins in the germline allows PGC development by repression of genes characteristic of somatic differentiation (e.g., *Hox* genes). Additionally, PRDM14 functions as a key regulator for the reacquisition of pluripotency and genome-wide epigenetic reprogramming (Kurimoto *et al.*, 2008; Vincent *et al.*, 2005). Whereas the function of PRDM1 is essential for additional processes in embryogenesis, embryos lacking PRDM14 function are viable but infertile (Kurimoto *et al.*, 2008; Vincent *et al.*, 2005), suggesting a specific role in germline development for the latter.

Following the induction by BMP, PRDM14 expression is maintained by PRDM1 (Yamaji *et al.*, 2008), and continued regulation over the extent of PRDM1 expression is important. Interestingly, as originally identified in the context of Hodgkin's lymphoma, PRDM1 is suppressed by the miRNA *let-7*, which targets a miRNA seed within the 3′UTR of the *prdm1* mRNA (Nie *et al.*, 2008). This regulation calls for a mechanism relieving *let-7*-mediated inhibition in the developing germline. The mechanism involves LIN-28, a translational regulator first identified in *C. elegans* (Ambros, 1989), that was shown to modulate *let-7* activity by direct binding to the precursor of *let-7* miRNA (pre-*let-7*) (Newman *et al.*, 2008; Piskounova *et al.*, 2008; Rybak *et al.*, 2008; Viswanathan *et al.*, 2008). In this case, LIN-28 induces uridylation at the 3′ end of the pre-miRNA by recruiting the noncanonical poly(A)polymerase TUTase4 (terminal uridylyltransferase 4), thus inhibiting further processing of *pre-let-7* by Dicer (Heo *et al.*, 2008). Consistently, LIN-28 was found to be a crucial component in mouse PGC development, as inhibition of LIN-28 function in PGCs resulted in loss of PRDM1 expression and failure to establish the germline. Together, LIN-28 can relieve *Prdm1* from miRNA-mediated repression by abrogating *let-7* maturation steps, allowing sustained expression of PRDM1 and proper germline development (West *et al.*, 2009). These results show that control over the biogenesis of miRNAs can serve in regulating germline development.

2.3. miRNAs in germ cell maintenance and development

Following their arrival at the gonad, the germ cells associate with somatic cells that support the maintenance of GSCs and regulate their differentiation into the gametes (Fig. 4.1A(c) and C(b)). Different factors regulating GSC self-renewal and maturation and the role of miRNA function in these processes are discussed below.

2.3.1. miRNA function in maintenance and maturation steps of GSCs

As miRNAs are essential for embryonic development (Bernstein *et al.*, 2003; Giraldez *et al.*, 2005; Grishok *et al.*, 2001), studying their role in the GSC requires analysis of animals in which the function of Dicer is eliminated specifically in these cells. Such experiments, conducted in *Drosophila*, revealed a role for miRNAs in controlling GSC division. Here, miRNAs were found to be crucial for the GSCs to pass the G1/S checkpoint, by repressing the cell cycle inhibitor Dacapo (Dap). Similar to the defects in *dicer-1* females, GSCs of male flies show delay in the cell cycle at G1/S transition (Hatfield *et al.*, 2005). A more recent study suggests that Dap repression mediated by the miRNAs miR-7, miR-278, and miR-309 depends on Insulin receptor signaling, thus providing an example for cooperation between extrinsic and intrinsic signals in regulating GSC division (Yu *et al.*, 2009).

In addition to the role in controlling GSC division, the function of Dicer-1 was found to be critical for maintaining the population of self-renewing, nondifferentiated GSCs (Jin and Xie, 2007). Similar defects in GSC maintenance were observed in flies carrying mutations in genes encoding for different components of the miRNA pathway including the dsRNA-binding protein Loquacious (Loqs), Ago1, and dFMRP (fragile X mental retardation protein) (Forstemann *et al.*, 2005; Neumuller *et al.*, 2008; Park *et al.*, 2007; Pek *et al.*, 2009; Yang *et al.*, 2007a,b). For example, the function of the RNA-binding protein dFMRP, a component of miRISC that specifically associates with the *bantam* miRNA, is important for the ability of somatic stem cells (SSCs) of the ovary to support self-renewal of the GSCs

(Yang *et al.*, 2007b, 2009). Conversely, differentiation of GSCs can be achieved by inhibiting miRNA function as is the case for Maelstrom (Mael), a protein that was originally identified as a PIWI-associated nuage component in mouse (Findley *et al.*, 2003; Soper *et al.*, 2008). Mael represses miR-7 transcription, thereby allowing proper translation level of the Bag-of-Marbles (Bam) protein, a crucial initiator of gametogenesis (Pek *et al.*, 2009). Similarly, the Vasa protein, that is essential for proper germline development, was shown to act by inhibiting miRNA function through its binding to *mei-P26* mRNA. In this case, the positive effect of Vasa on Mei-P26 protein levels results in negative regulation of Ago1 function thereby repressing the miRNA pathway (Liu *et al.*, 2009; Neumuller *et al.*, 2008).

Together, in this context, miRNAs are important for promoting GSC divisions and the maintenance of their pluripotent state, while repressing premature differentiation of the cells into gametes. Therefore, the proper differentiation of germline cells depends on mechanisms that would inhibit miRNA function.

2.3.2. miRNA function in germ cell differentiation—Oogenesis

In contrast to the role of miRNAs in repressing GSC differentiation, further studies in *D. melanogaster* revealed that the highly conserved miRNA-184 is expressed in the female germline where it promotes multiple differentiation steps during oogenesis and early embryogenesis. miR-184 regulates GSC differentiation by tuning the levels of the Saxophone (Sax) receptor, which responds to Dpp signals secreted from the niche cells, leading to suppression of Bam (Iovino *et al.*, 2009; Xie and Spradling, 1998). Thus, miR-184-mediated repression of Sax leads to dampening of the Dpp signal resulting in higher levels of Bam, thereby promoting GSC differentiation. Additionally, miR-184 has an impact on oocyte and embryo dorsoventral and anteroposterior polarity regulating the transport of the ventralizing signal Gurken and the transcriptional repressor Tramtrack69, respectively (Iovino *et al.*, 2009).

Whereas Dicer function appears to be crucial for germline development in some species, it is dispensable in the case of zebrafish where germline cells lacking *dicer* give rise to eggs and sperm (Giraldez *et al.*, 2005). The apparent dispensability of Dicer-dependent small RNAs for zebrafish germline development, however, still lacks an explanation.

2.3.3. miRNA function in germ cell differentiation—Spermatogenesis

Genome-wide miRNA profiling revealed the presence of miRNAs in mouse testis (Barad *et al.*, 2004; Liu *et al.*, 2004a), and that a number of these are differentially expressed during spermatogenesis. Indeed, crucial components of the miRNA machinery such as Dicer, AGO1, Drosha, and testis-expressed miRNAs were found to be localized to the chromatoid body, a germ cell specific structure where RNA-processing takes place (Gonzalez-Gonzalez *et al.*, 2008; Kotaja *et al.*, 2006). Consistently, reducing the function of Dicer in mouse PGCs resulted in germ cell proliferation defects and aberrant differentiation of spermatogonia manifested by abnormal spermatids morphology and sperms with impaired motility (Hayashi *et al.*, 2008; Maatouk *et al.*, 2008). Whereas these phenotypes could also reflect roles for endo-siRNAs in spermatogenesis since both rely on Dicer, the function of miRNAs was demonstrated in this context. An interesting observation was that despite the global meiotic transcriptional silencing of genes located on the X chromosome, many X-linked miRNAs escape this inhibition and are transcribed in spermatocytes during meiosis (Song *et al.*, 2009). Specific roles for miRNAs in spermatogenesis were assigned, for example, to the miR-17-92 cluster that was shown to repress the translation of the transcription factor E2F1, thereby protecting meiotic cells from apoptosis (Novotny *et al.*, 2007). Similarly, Mirn122a was suggested to be predominantly expressed in late stage male germ cells, where it represses specific RNAs expressed in spermatogenesis (Yu *et al.*, 2005).

2.3.4. miRNA function in sex differentiation

The involvement of miRNAs in sex-specific PGC differentiation was suggested by analysis of expression profiles of mouse PGCs at E9.5–E13.5. In these studies, differential expression of miRNAs was detected when comparing male and female PGCs. While the functional importance of the differences in expression should be further investigated, differential expression was detected for *let-*7, miR-125a, miR-9, and miR-29b (Hayashi *et al.*, 2008; Takada *et al.*, 2009). Indeed, repression of *dnmt3a* and *dnmt3b* (*de novo* methyl transferases 3a and b, respectively) by the female-biased expressed miR-29b allows the GSCs to escape the genome-wide methylation observed in males at E15 (Lees-Murdock *et al.*, 2005). Together, these findings provide initial hints concerning the possible function miRNAs play in sex determination and differentiation.

2.4. miRNA function in the somatic cells of the gonad

Proper development of GSC depends on signals provided by surrounding gonadal somatic cells that in turn regulate cell division, self-renewal, and differentiation. Regulation of gene expression in those somatic cells by miRNAs thus influences GSC maintenance and differentiation processes.

2.4.1. miRNAs in gonadal somatic cells support germ cell development

As described above, depletion of *dicer-1* in *Drosophila* GSC leads to failure of GSC maintenance as a result of defects in cell cycle and self-renewal (Hatfield *et al.*, 2005; Jin and Xie, 2007). In addition, removing the function of Dicer-1 in SSCs of the ovary revealed a function in this cell population as well. Similar to the GSC, SSCs lacking Dicer-1 failed to self-renew leading to their depletion in the niche and premature differentiation, resulting in defects in follicle cell proliferation and growth (Jin and Xie, 2007). Consistently, mutants for *loqs* and *ago1* affecting the miRNA pathway displayed abnormally developed germaria (Forstemann *et al.*, 2005; Park *et al.*, 2007; Yang *et al.*, 2007a). An example for a gene regulated by miRNAs in somatic cells of the ovary is *notch* (Deng *et al.*, 2001; Lopez-Schier and St Johnston, 2001). Here, Belle, a component of the miRNA pathway, was found to regulate the timing of Notch activity in follicle cells. Loss of Belle delays Notch activation thus delaying follicle differentiation that in turn affects GSC cell cycle and maturation (Poulton *et al.*, 2011).

A role for miRNAs in somatic gonadal cells was further demonstrated in male mice, where the development of sertoli cells that tightly control spermatogenesis depends on this level of regulation. Selective ablation of *dicer* in these somatic cells results in impaired sertoli cell maturation that leads to a complete absence of spermatozoa and progressive testicular degeneration. Similarly, somatic cell development in the female mouse also relies on miRNA-mediated regulation. In this case, somatic granulosa cells that surround the developing oocyte and support its development appear to require Dicer function. Specific depletion of *dicer* in granulosa cells leads to degeneration of follicles in the mutant ovaries. This effect results from premature onset of folliculogenesis caused by lack of mmu-miR-503, which normally fine tunes the levels of Cyclin D2 (CCND2), an important regulator of cell proliferation during folliculogenesis (Lei *et al.*, 2010). Moreover, *dicer* depletion in somatic cells of the gonad also affects the development of somatic reproductive tissues, resulting in defects in ovulation, oocyte migration from oviduct to uterus, and subsequent implantation (Gonzalez and Behringer, 2009; Hong *et al.*, 2008; Nagaraja *et al.*, 2008).

Together, these results suggest that Dicer function is critical for spermatogenesis and for proper development of oviducts and uterus and plays a role in regulating folliculogenesis by maintaining the communication between germline and somatic cells in the gonad.

2.4.2. miRNAs and hormonal regulation of gametogenesis

During oogenesis and spermatogenesis, factors like the follicle stimulating hormone (FSH), luteinizing hormone (LH) and steroid hormones (e.g., estrogen, progesterone, and androgen) are key regulatory factors (Combelles *et al.*, 2004; Dong *et al.*, 1996; reviewed in Edson *et al.*, 2009) and miR-132 and miR-212 were found to be important for the response to those hormone signals (Fiedler *et al.*, 2008). For example, LH triggers downstream cAMP signaling events, which in turn can activate CREB (cAMP response element binding), a repressor of miR-132 and miR-212 transcription in granulosa cells (Fiedler *et al.*, 2008). Interestingly, the miRNA expression levels can in turn influence the secretion of steroid hormones in human primary granulosa cells (Sirotkin *et al.*, 2010, 2009). Similarly, spermatogenesis is under tight control of FSH and androgens, which regulate miRNA transcription level in sertoli cells (Nicholls *et al.*, 2011). Thus, hormones like LH and FSH can control the transcription of specific miRNAs, which in turn are able to modulate output levels of steroid hormones, thereby controlling gametogenesis.

3. Endogenous-siRNAs in Germ Cell Development

RNAi was first demonstrated when *C. elegans* embryos were subjected to exogenous dsRNAs (Bernstein *et al.*, 2003; Fire *et al.*, 1998). The introduced dsRNAs were cleaved by Dicer and loaded onto the siRISC complex leading to degradation of a targeted gene. It was postulated that this process constitutes a protective mechanism to counteract viral infections in flies and reduce transposition activity in both *C. elegans* and *D. melanogaster* (Bernstein *et al.*, 2003; Fire *et al.*, 1998). endo-siRNAs were first described in plant and *C. elegans* (Ambros *et al.*, 2003; Hamilton *et al.*, 2002) followed by their identification in *D. melanogaster* and vertebrates (Chung *et al.*, 2008; Czech *et al.*, 2008; Ghildiyal *et al.*, 2008; Kawamura *et al.*, 2008; Okamura *et al.*, 2008a,b; Zilberman *et al.*, 2003). The first mammalian endo-siRNA was later characterized in cultured human cells where it regulates the L1 retrotransposon (Yang and Kazazian, 2006). endo-siRNA expression in the germline was detected by deep sequencing analysis performed on *D. melanogaster* and mouse ovaries (Tam *et al.*, 2008; Watanabe *et al.*, 2008).

3.1. endo-siRNA biogenesis and function

The endo-siRNAs are a class of 21-nt-short RNAs, present in both sense and antisense orientations (Czech *et al.*, 2008; Ghildiyal *et al.*, 2008; Okamura *et al.*, 2008b). The biogenesis of endo-siRNAs is reviewed in detail in Ghildiyal and Zamore (2009), Ketting (2011), Kim *et al.* (2009) and Okamura and Lai (2008) and is briefly described below.

endo-siRNA is derived from different templates including transposon transcripts, heterochromatic sequences, sense–antisense transcript duplexes, long stem loops, mRNAs, and from pseudogenes (Chung *et al.*, 2008; Czech *et al.*, 2008; Ghildiyal *et al.*, 2008; Sasidharan and Gerstein, 2008; Tam *et al.*, 2008), which underlies their abundant expression. Considering that transposons as well as endogenous gene sequences serve as templates for endo-siRNA precursor synthesis, it became clear that their function is not restricted to targeting and silencing transposons but rather extends to regulation of gene expression (Ghildiyal *et al.*, 2008; Okamura *et al.*,

2008a,b). In contrast to the miRNAs, whose processing relies on Dicer and on Drosha/DGCR8, the endo-siRNA precursors are processed by Dicer and subsequently are loaded onto Argonaute proteins and following specific RNA recognition results in cleavage of the target (Claycomb *et al.*, 2009; Grishok *et al.*, 2001; Hutvagner and Zamore, 2002; Ketting *et al.*, 2001; Knight and Bass, 2001). In addition to the proteins listed above, the biogenesis and function of endo-siRNAs require the dsRNA-binding proteins Loqs-PD and R2D2 in *Drosophila* (Hartig *et al.*, 2009; Liu *et al.*, 2003; Pham *et al.*, 2004; Tomari *et al.*, 2004; Zhou *et al.*, 2009) (Fig. 4.2B).

In addition to the mechanism described above for *Drosophila* and mouse, dsRNAs in *C. elegans* are typically synthesized by RNA-dependent RNA polymerases (RdRPs). Dicer further processes the dsRNA fragments to generate primary endo-siRNAs, including the 26G RNAs, which are loaded onto a "primary Argonaute" (Gent *et al.*, 2010, 2009; Pavelec *et al.*, 2009). It is thought that primary endo-siRNAs recruit the RdRPs EGO-1 or RRF-1 to the cleaved target mRNA, leading to *de novo* synthesis of the secondary endo-siRNAs, the 22G short antisense RNAs that can directly bind to specialized secondary AGO proteins (SAGOs) (Liu *et al.*, 2004b; Meister *et al.*, 2004; Okamura *et al.*, 2004; Pak and Fire, 2007; Ruby *et al.*, 2006; Sijen *et al.*, 2007; Yigit *et al.*, 2006).

3.2. endo-siRNA function in germ cell differentiation—Oogenesis

Investigating the possible role of endo-siRNAs in early germline development of *Drosophila*, the effect of lack of *dicer-2* and *ago2* proteins whose functions are required for endo-siRNAs function was analyzed. Interestingly, whereas a profound effect on dsRNA processing was observed, the function of both proteins appeared to be dispensable for germline specification and maintenance in this organism (Chung *et al.*, 2008; Lee *et al.*, 2004; Okamura *et al.*, 2008a).

In *C. elegans*, mutants lacking the homologue of the plant RdRP EGO1 or other RdRPs that are required for endo-siRNA precursor synthesis display severe defects in oogenesis and spermatogenesis leading to infertility (Smardon *et al.*, 2000). The RNAi response of the mutants in the germline is reduced, correlating with defects in mitosis, premature meiosis, and failure in meiotic recombination, culminating in the production of defective oocytes.

Similarly, examining the role of endo-siRNAs in mouse oogenesis by analyzing knockouts in the *dicer* or *ago2* genes revealed a critical role the corresponding proteins play in spindle organization in meiosis (Murchison *et al.*, 2007; Tang *et al.*, 2007; Watanabe *et al.*, 2008). In principle, this phenotype could result from lack of endo-siRNAs or reflect a requirement for miRNAs, both of which depend on Dicer function. Since the lack of DGCR8, whose function is required specifically for miRNA biosynthesis (Wang *et al.*, 2007), was nonconsequential for oogenesis, the defects observed in oogenesis in *dicer* mutants reflect a role for endo-siRNAs in the process (Ma *et al.*, 2010; Suh *et al.*, 2010). Consistently, expression profiles of Dicer mutant oocytes revealed strong changes in the level of endo-siRNA-regulated transcripts (Murchison *et al.*, 2007; Tang *et al.*, 2007).

3.3. endo-siRNA function in germ cell differentiation—Spermatogenesis

The role of RNAi in spermatogenesis was first described in *C. elegans* where mutants in the endo-siRNA pathway exhibited defects in this process (Gent *et al.*, 2009; Han *et al.*, 2009; Ketting *et al.*, 1999; Pavelec *et al.*, 2009; Rocheleau *et al.*, 2008; She *et al.*, 2009; Smardon *et al.*, 2000; Tabara *et al.*, 1999). Additionally, mutations affecting the biogenesis of the germline specific 22G and 26G RNAs result in a range of defects in spermatogenesis due to failure in chromosome segregation during mitosis and meiosis (Claycomb *et al.*, 2009; Conine *et al.*, 2010; Gu *et al.*, 2009; Han *et al.*, 2009; van Wolfswinkel *et al.*, 2009).

The function of the 22G RNA in controlling mitosis could be related to the finding that it recruits chromatin-modifying factors that promote local histone modifications and establish perio-centromeric chromatin domains. The subsequent binding and organization of factors crucial for proper mitotic divisions such as Condensins or Cohesins allow proper chromosome segregation (Claycomb *et al.*, 2009; Gu *et al.*, 2009; van Wolfswinkel *et al.*, 2009). The function of a subset of 26G endo-siRNAs loaded on the AGO proteins ALG3 and ALG4 is essential for spermatogenesis by regulating gene expression during this process (Conine *et al.*, 2010; Han *et al.*, 2009).

Similar to the defects observed in oogenesis, male *dicer* mutant mice display severe defects in germ cell proliferation and, subsequently, in spermatogenesis, but the relative contribution of miRNA and endo-siRNA pathways is largely unknown (Hayashi *et al.*, 2008; Maatouk *et al.*, 2008).

4. piRNAs in Germ Cell Development

The Piwi protein was first identified in *D. melanogaster* where its function is important for fertility. In the male fly, loss of Piwi function is associated with severe defects in spermatogenesis giving the gene its name, *piwi*, for "P-element induced wimpy testis" (Lin and Spradling, 1997). Three mammalian Piwi homologues containing conserved PAZ and PIWI domains were subsequently identified and shown to be required for spermatogenesis as well (Carmell *et al.*, 2007; Deng and Lin, 2002; Kuramochi-Miyagawa *et al.*, 2001).

The small RNAs that interact and function in combination with the Piwi proteins called piRNAs were shown to be expressed and function mainly in the germline (Aravin *et al.*, 2006; Girard *et al.*, 2006; Grivna *et al.*, 2006).

4.1. piRNA biogenesis and function

piRNAs are slightly larger than miRNAs (24–30nt as compared to 21–24nt) and they are mapped to distinct, nonoverlapping clusters in the genome that are often mapped to repeat sequences (Aravin *et al.*, 2006; Houwing *et al.*, 2007). The mammalian piRNAs can be divided into two differentially regulated subclasses of different sizes referred to as pachytene (29–31bp) and pre-pachytene (26–28bp) piRNAs (Aravin *et al.*, 2007b). Pre-pachytene piRNAs are expressed in spermatogonia prior to meiosis and become depleted starting at mid-pachytene, while pachytene piRNAs are present in spermatogonia around the pachytene stage of meiosis and are decreased when the sperm reaches the haploid round spermatid stage. Whereas pachytene piRNAs associate with the MIWI protein, pre-pachytene piRNAs preferentially associate with MILI or MIWI-2. Functionally, pre-pachytene piRNAs were suggested to control transposable elements as they are enriched with genome repeat elements (Aravin *et al.*, 2008, 2007b; Brennecke *et al.*, 2007), while the biological role of pachytene piRNAs is not yet clear.

Two biogenesis pathways are responsible for piRNA biosynthesis: the primary processing pathway and the secondary ping-pong amplification loop (Fig. 4.2C). First, the primary piRNA biogenesis pathway provides an initial pool of piRNAs that target multiple transposable elements. Next, the ping-pong cycle further shapes the piRNA population by amplifying sequences that specifically target active transposons. These two pathways are conserved in many animal species, including zebrafish (*Danio rerio*), frogs (*X. laevis*), fly (*D. melanogaster*), and mouse (Brennecke *et al.*, 2007; Gunawardane *et al.*, 2007; Houwing *et al.*, 2007; Robine *et al.*, 2009; Vagin *et al.*, 2006). In general, piRNAs are processed in a Dicer-independent manner (Saito *et al.*, 2006; Vagin *et al.*, 2006); however, the mechanism of primary piRNA production is not well understood. Nevertheless, studies in *Drosophila* revealed that the RNA helicase Armitage, the putative nuclease Zucchini, and Yb protein are involved in the primary pathway generating the piRNAs (Saito and Siomi, 2010). This pathway is thought to give rise to piRNAs that are loaded onto Piwi/Aub (Aubergine, a *Drosophila* Piwi homologue) and is the source for the pachytene piRNAs expressed in mammalian testis, which are subsequently bound to MIWI and MILI (Aravin *et al.*, 2006; Girard *et al.*, 2006; Grivna *et al.*, 2006; Lau *et al.*, 2006; Li *et al.*, 2009; Malone *et al.*, 2009; Saito *et al.*, 2006; Watanabe *et al.*, 2006). The class I piRNAs produced by the primary pathway are loaded onto Piwi/Aub and guide those proteins to their transposon RNA targets. The Piwi-protein-mediated cleavage of the transposon RNA target produces the class II piRNAs (Brennecke *et al.*, 2007;

Gunawardane *et al.*, 2007), which are loaded onto Ago3. The Ago3/piRNA complex then targets the transcripts of piRNA cluster to generate more antisense piRNAs that are loaded onto PIWI/Aub, thus amplifying the pool of piRNAs targeting transposons (Fig. 4.2C).

The two biosynthetic pathways described above have been identified in all model organisms, while in *C. elegans*, a third group of Piwi-binding small RNAs was identified, the 21U-RNAs (Ruby *et al.*, 2006). These RNAs are smaller than other piRNAs and are synthesized by a different pathway and function in transposon silencing as well (Batista *et al.*, 2008; Ruby *et al.*, 2006; Wang and Reinke, 2008). For detailed reviews on piRNA biogenesis, see Ghildiyal and Zamore (2009), Ketting (2011), Lau, 2010 and Siomi *et al.* (2011).

As many steps in piRNA biogenesis still remain elusive, further investigations of the details of their maturation and function would help to better understand how piRNAs can effectively guard the germline genome against transposable elements.

4.1.1. piRNAs defend against transposable element activity

As mentioned above, piRNAs can recognize and destroy transposable element transcripts derived from multiple sites that are spread throughout the genome. The importance of the piRNA pathway and specifically the amplification loop in defending the genome against excessive transposon mobilization was demonstrated in studies involving crosses between different *Drosophila* strains (Brennecke *et al.*, 2008). Such crosses can produce sterile progeny through the phenomenon called "hybrid dysgenesis". The basis for this phenomenon is the lack of maternally provided piRNAs that target transposons carried by the paternal genome (Bourc'his and Voinnet, 2010; Brennecke *et al.*, 2008). In other crosses between the same strains, maternal deposition of piRNAs into the egg provides the embryo with an initial piRNA pool to launch the amplification cycle for effective defense against DNA damage caused by transposon mobilization. Similarly, knockout of two members of the mouse PIWI-protein family, *mili* and *miwi-2*, but not *miwi*, leads to the activation of retrotransposons in the male mouse germline, gametogenesis arrest, and consequently to complete sterility in males (Aravin *et al.*, 2007b; Carmell *et al.*, 2007; Kuramochi-Miyagawa *et al.*, 2008). Similarly, Zebrafish bearing mutations in the Piwi proteins, Ziwi and Zili, show derepression of transposable elements and male sterility.

Interestingly, CpG DNA methylation, which functions in transcriptional silencing of LINE and LTR retrotransposons in the genome, is decreased in the male germline of *mili* and *miwi-2* mutants (Aravin *et al.*, 2007b; Kuramochi-Miyagawa *et al.*, 2008). It was thus hypothesized that the piRISC (piRNA induced silencing complex) can also guide the *de novo* methylation machinery to loci of transposable elements, and consistently MIWI-2 protein loaded with antitransposon piRNAs is present in the embryo at the time when *de novo* DNA methylation takes place (Aravin *et al.*, 2007b; Kuramochi-Miyagawa *et al.*, 2008). In the same direction, a *piwi* mutant defective in slicing activity is capable of rescuing the transposable element silencing thus suggesting a slicer-activity independent PIWI function in this context (Saito *et al.*, 2010). This activity could be related to the fact that PIWI proteins are capable of recruiting Heterochromatin protein 1A (HP1A) to specific genomic loci, potentially leading to gene silencing (Brower-Toland *et al.*, 2007; Pal-Bhadra *et al.*, 2004).

4.1.2. piRNA function in controlling protein-encoding genes in germline

In addition to their function in transposon silencing, piRNAs were found to target repetitive sequences of the protein-coding *Stellate* gene (Aravin *et al.*, 2001) and suggested to control *Fasciclin 3* (*Fas3*) levels in the follicle cells of *D. melanogaster* ovaries (Saito *et al.*, 2009). In the fly testes, piRNAs might target the *vasa* gene, which is a helicase necessary for pole plasm formation (Nishida *et al.*, 2007). Interestingly, it has been recently shown that piRNAs may induce degradation of maternally deposited mRNAs, such as *nanos*, during maternal-to-zygotic transition (Rouget *et al.*, 2010). In this case, piRNA function is required for recruiting the CCR4 deadenylation complex to the specific RNA.

4.2. piRNAs function in germ cell specification

The role of Piwi proteins in germ cell specification in *Drosophila* is well studied. Here, Piwi is localized to the polar granules and is required for pole cell formation in a dose-dependent manner (Megosh *et al.*, 2006) such that an increase in maternal *piwi* dosage leads to a proportional increase in PGC formation. Noteworthy, as Piwi protein was shown to associate with a small number of miRNAs in addition to its association with piRNAs (Megosh *et al.*, 2006), it is not clear if Piwi affects germ cell specification through associated piRNAs or miRNAs. An additional piece of evidence linking piRNAs to the process of germ cell specification relates to the phenotype of embryos laid by female flies mutated for the *aubergine* gene that encodes for a Piwi homolog that is localized to polar granules as well (Harris and Macdonald, 2001; Thomson *et al.*, 2008). Such embryos fail to form germ cells reflecting defects in controlling the translation of *oskar* RNA, but thus far the involvement of piRNAs in the process has not been demonstrated.

4.3. piRNA function in germline stem cell maintenance and gametogenesis

Piwi proteins in *D. melanogaster* are expressed in somatic and germline cells of both male and female (Cox *et al.*, 1998), and flies mutated for *piwi* fail to maintain GSCs in both sexes, while overexpressing Piwi in the somatic cells of the germarium results in an increase in rate of GSC division and number (Cox *et al.*, 1998; Szakmary *et al.*, 2005). Indeed, Piwi proteins are important for gonadal and germline development as the function of both Piwi and Aub is required for male and female fertility (Cox *et al.*, 1998; Lin and Spradling, 1997; Schmidt *et al.*, 1999); consistently Ago3 has been shown to be required for female fertility as well, but only partially for male fertility (Li *et al.*, 2009).

The three mouse PIWI proteins, MIWI, MILI, and MIWI-2, are expressed during spermatogenesis in mitotically arrested prenatal GSCs (Aravin *et al.*, 2008), and consistent with the idea that they function in these stages, both *mili* and *miwi-2* mutants exhibit spermatogenic stem cell arrest defects and massive degeneration of spermatogonia observed in *miwi-2* knockout mice (Carmell *et al.*, 2007; Unhavaithaya *et al.*, 2009). Germ cell development in Zebrafish as well requires the Piwi proteins, Ziwi (zebrafish PIWI) and Zili (zebrafish PIWI-like) (Houwing *et al.*, 2008, 2007), as in zebrafish *ziwi* mutants, germ cells undergo apoptosis, resulting in male infertility, while in fish lacking *zili*, loss of germ cell differentiation to mature oocytes or sperm is observed. A role for Piwi proteins in gametogenesis was also demonstrated in mice, where *miwi-2* mutants show predominant arrest at the Leptotene stage of meiosis (Carmell *et al.*, 2007). The mutant spermatogenic cells show defects in dsDNA break repair, which is consistent with a role in proper meiotic recombination (Carmell *et al.*, 2007; Kuramochi-Miyagawa *et al.*, 2004). Similarly, germ cells in mice lacking MILI function are blocked at the zygotene or early pachytene stages of meiotic prophase (Kuramochi-Miyagawa *et al.*, 2004). Defects in meiosis have also been observed in mutant flies for Piwi proteins and proteins involved in repeat-associated siRNA (Chen *et al.*, 2007; Klattenhoff *et al.*, 2007; Lim and Kai, 2007; Pane *et al.*, 2007).

Last, a role for a piRNAs in spermatogenesis was demonstrated in *C. elegans*. In this case, the Piwi-like protein PRG-1 that is localized to P-granules was shown to be required for successful spermatogenesis. Loss of *prg-1* leads to a strong reduction in the expression of 21U-RNAs during spermatogenesis correlated with extensive defects in sperm activation and fertilization (Batista *et al.*, 2008; Wang and Reinke, 2008).

5. Conclusion

This review highlights the function of the different classes of small RNAs in regulating specification and maintenance of the germline, as well as the role in executing the function of this cell lineage, namely the generation of sperm and egg. Given the importance of the germline in propagation of the species, numerous functional redundancies can be identified in the control over germline development, contributing to the robustness of gametogenesis. As germline development depends on interactions with other cell populations, small RNAs control processes important for those interactions, in addition to their role in regulating cellular processes within the germ cells themselves. Of special importance in the germline is the crucial role for small RNAs in regulating transposon activity, which affects the evolution of genetic variability on one hand and the maintenance of genome stability on the other. Further studies of the different small RNA classes in the germline lineage are thus likely to shed light on processes important for the biogenesis of these molecules in different cell types, on the regulation of cellular events such as cell proliferation and differentiation, on the control over signaling events in the developing organism and on genomic events relevant for evolution.

ACKNOWLEDGMENTS

We thank Michal Reichman-Fried for critical review of this chapter. We are supported by funds from the German Research Foundation (DFG), the Max Planck Society (MPG), and the Medical Faculty of the University of Münster.

REFERENCES

Ambros, V. (1989). A hierarchy of regulatory genes controls a larva-to-adult developmental switch in C. elegans. *Cell* **57,** 49–57.

Ambros, V., Lee, R. C., Lavanway, A., Williams, P. T., and Jewell, D. (2003). MicroRNAs and other tiny endogenous RNAs in C. elegans. *Curr. Biol.* **13,** 807–818.

Aravin, A. A., Naumova, N. M., Tulin, A. V., Vagin, V. V., Rozovsky, Y. M., and Gvozdev, V. A. (2001). Double-stranded RNA-mediated silencing of genomic tandem repeats and transposable elements in the D. melanogaster germline. *Curr. Biol.* **11,** 1017–1027.

Aravin, A., Gaidatzis, D., Pfeffer, S., Lagos-Quintana, M., Landgraf, P., Iovino, N., Morris, P., Brownstein, M. J., Kuramochi-Miyagawa, S., Nakano, T., Chien, M., Russo, J. J., *et al.* (2006). A novel class of small RNAs bind to MILI protein in mouse testes. *Nature* **442,** 203–207.

Aravin, A. A., Hannon, G. J., and Brennecke, J. (2007a). The Piwi-piRNA pathway provides an adaptive defense in the transposon arms race. *Science (New York, N.Y.)* **318,** 761–764.

Aravin, A. A., Sachidanandam, R., Girard, A., Fejes-Toth, K., and Hannon, G. J. (2007b). Developmentally regulated piRNA clusters implicate MILI in transposon control. *Science (New York, N.Y.)* **316,** 744–747.

Aravin, A. A., Sachidanandam, R., Bourc'his, D., Schaefer, C., Pezic, D., Toth, K. F., Bestor, T., and Hannon, G. J. (2008). A piRNA pathway primed by individual transposons is linked to *de novo* DNA methylation in mice. *Mol. Cell* **31,** 785–799.

Barad, O., Meiri, E., Avniel, A., Aharonov, R., Barzilai, A., Bentwich, I., Einav, U., Gilad, S., Hurban, P., Karov, Y., Lobenhofer, E. K., Sharon, E., *et al.* (2004). MicroRNA expression detected by oligonucleotide microarrays: System establishment and expression profiling in human tissues. *Genome Res.* **14,** 2486–2494.

Bartel, D. P. (2009). MicroRNAs: Target recognition and regulatory functions. *Cell* **136,** 215–233.

Batista, P. J., Ruby, J. G., Claycomb, J. M., Chiang, R., Fahlgren, N., Kasschau, K. D., Chaves, D. A., Gu, W., Vasale, J. J., Duan, S., Conte, D., Jr., Luo, S., *et al.* (2008). PRG-1 and 21U-RNAs interact to form the piRNA complex required for fertility in C. elegans. *Mol. Cell* **31,** 67–78.

Bernstein, E., Kim, S. Y., Carmell, M. A., Murchison, E. P., Alcorn, H., Li, M. Z., Mills, A. A., Elledge, S. J., Anderson, K. V., and Hannon, G. J. (2003). Dicer is essential for mouse development. *Nat. Genet.* **35,** 215–217.

Blaser, H., Eisenbeiss, S., Neumann, M., Reichman-Fried, M., Thisse, B., Thisse, C., and Raz, E. (2005). Transition from non-motile behaviour to directed migration during early PGC development in zebrafish. *J. Cell Sci.* **118,** 4027–4038.

Borges, F., Pereira, P. A., Slotkin, R. K., Martienssen, R. A., and Becker, J. D. (2011). MicroRNA activity in the Arabidopsis male germline. *J. Exp. Bot.* **62,** 1611–1620.

Boterenbrood, E. C., and Nieuwkoop, P. D. (1973). The formation of the mesoderm in urodelian amphibians. V. Its regional induction by the endoderm. *Wilhelm Roux' Arch. Devl. Biol.* **173,** 319–332.

Bourc'his, D., and Voinnet, O. (2010). A small-RNA perspective on gametogenesis, fertilization, and early zygotic development. *Science (New York, N.Y.)* **330,** 617–622.

Braat, A., Zandbergen, T., van de Water, S., Goos, H., and Zivkovic, D. (1999). Characterization of zebrafish primordial germ cells: Morphology and early distribution of vasa RNA. *Dev. Dyn.* **216,** 153–167.

Brennecke, J., Aravin, A. A., Stark, A., Dus, M., Kellis, M., Sachidanandam, R., and Hannon, G. J. (2007). Discrete small RNA-generating loci as master regulators of transposon activity in Drosophila. *Cell* **128,** 1089–1103.

Brennecke, J., Malone, C. D., Aravin, A. A., Sachidanandam, R., Stark, A., and Hannon, G. J. (2008). An epigenetic role for maternally inherited piRNAs in transposon silencing. *Science (New York, N.Y.)* **322,** 1387–1392.

Brower-Toland, B., Findley, S. D., Jiang, L., Liu, L., Yin, H., Dus, M., Zhou, P., Elgin, S. C., and Lin, H. (2007). Drosophila PIWI associates with chromatin and interacts directly with HP1a. *Genes Dev.* **21,** 2300–2311.

Bushati, N., Stark, A., Brennecke, J., and Cohen, S. M. (2008). Temporal reciprocity of miRNAs and their targets during the maternal-to-zygotic transition in Drosophila. *Curr. Biol.* **18,** 501–506.

Carmell, M. A., and Hannon, G. J. (2004). RNase III enzymes and the initiation of gene silencing. *Nat. Struct. Mol. Biol.* **11,** 214–218.

Carmell, M. A., Girard, A., van de Kant, H. J., Bourc'his, D., Bestor, T. H., de Rooij, D. G., and Hannon, G. J. (2007). MIWI2 is essential for spermatogenesis and repression of transposons in the mouse male germline. *Dev. Cell* **12,** 503–514.

Carrington, J. C., and Ambros, V. (2003). Role of microRNAs in plant and animal development. *Science (New York, N.Y.)* **301,** 336–338.

Chen, Y., Pane, A., and Schupbach, T. (2007). Cutoff and aubergine mutations result in retrotransposon upregulation and checkpoint activation in Drosophila. *Curr. Biol.* **17,** 637–642.

Chung, W. J., Okamura, K., Martin, R., and Lai, E. C. (2008). Endogenous RNA interference provides a somatic defense against Drosophila transposons. *Curr. Biol.* **18,** 795–802.

Claycomb, J. M., Batista, P. J., Pang, K. M., Gu, W., Vasale, J. J., van Wolfswinkel, J. C., Chaves, D. A., Shirayama, M., Mitani, S., Ketting, R. F., Conte, D., Jr., and Mello, C. C. (2009). The Argonaute CSR-1 and its 22G-RNA cofactors are required for holocentric chromosome segregation. *Cell* **139,** 123–134.

Combelles, C. M., Carabatsos, M. J., Kumar, T. R., Matzuk, M. M., and Albertini, D. F. (2004). Hormonal control of somatic cell oocyte interactions during ovarian follicle development. *Mol. Reprod. Dev.* **69,** 347–355.

Conine, C. C., Batista, P. J., Gu, W., Claycomb, J. M., Chaves, D. A., Shirayama, M., and Mello, C. C. (2010). Argonautes ALG-3 and ALG-4 are required for spermatogenesis-specific 26G-RNAs and thermotolerant sperm in Caenorhabditis elegans. *Proc. Natl. Acad. Sci. USA* **107,** 3588–3593.

Cox, D. N., Chao, A., Baker, J., Chang, L., Qiao, D., and Lin, H. (1998). A novel class of evolutionarily conserved genes defined by piwi are essential for stem cell self-renewal. *Genes Dev.* **12,** 3715–3727.

Czech, B., Malone, C. D., Zhou, R., Stark, A., Schlingeheyde, C., Dus, M., Perrimon, N., Kellis, M., Wohlschlegel, J. A., Sachidanandam, R., Hannon, G. J., and Brennecke, J. (2008). An endogenous small interfering RNA pathway in Drosophila. *Nature* **453,** 798–802.

Davies, E. L., and Fuller, M. T. (2008). Regulation of self-renewal and differentiation in adult stem cell lineages: Lessons from the Drosophila male germ line. *Cold Spring Harb. Symp. Quant. Biol.* **73,** 137–145.

de Sousa Lopes, S. M., Hayashi, K., and Surani, M. A. (2007). Proximal visceral endoderm and extraembryonic ectoderm regulate the formation of primordial germ cell precursors. *BMC Dev. Biol.* **7,** 140.

Deng, W., and Lin, H. (2002). Miwi, a murine homolog of piwi, encodes a cytoplasmic protein essential for spermatogenesis. *Dev. Cell* **2,** 819–830.

Deng, W. M., Althauser, C., and Ruohola-Baker, H. (2001). Notch-Delta signaling induces a transition from mitotic cell cycle to endocycle in Drosophila follicle cells. *Development* **128,** 4737–4746.

Deppe, U., Schierenberg, E., Cole, T., Krieg, C., Schmitt, D., Yoder, B., and von Ehrenstein, G. (1978). Cell lineages of the embryo of the nematode Caenorhabditis elegans. *Proc. Natl. Acad. Sci. USA* **75,** 376–380.

Dong, J., Albertini, D. F., Nishimori, K., Kumar, T. R., Lu, N., and Matzuk, M. M. (1996). Growth differentiation factor-9 is required during early ovarian folliculogenesis. *Nature* **383,** 531–535.

Edson, M. A., Nagaraja, A. K., and Matzuk, M. M. (2009). The mammalian ovary from genesis to revelation. *Endocr. Rev.* **30,** 624–712.

Extavour, C. G., and Akam, M. (2003). Mechanisms of germ cell specification across the metazoans: Epigenesis and preformation. *Development* **130,** 5869–5884.

Fiedler, S. D., Carletti, M. Z., Hong, X., and Christenson, L. K. (2008). Hormonal regulation of microRNA expression in periovulatory mouse mural granulosa cells. *Biol. Reprod.* **79,** 1030–1037.

Findley, S. D., Tamanaha, M., Clegg, N. J., and Ruohola-Baker, H. (2003). Maelstrom, a Drosophila spindle-class gene, encodes a protein that colocalizes with Vasa and RDE1/AGO1 homolog, Aubergine, in nuage. *Development* **130,** 859–871.

Fire, A., Xu, S., Montgomery, M. K., Kostas, S. A., Driver, S. E., and Mello, C. C. (1998). Potent and specific genetic interference by double-stranded RNA in Caenorhabditis elegans. *Nature* **391,** 806–811.

Forstemann, K., Tomari, Y., Du, T., Vagin, V. V., Denli, A. M., Bratu, D. P., Klattenhoff, C., Theurkauf, W. E., and Zamore, P. D. (2005). Normal microRNA maturation and germ-line stem cell maintenance requires Loquacious, a double-stranded RNA-binding domain protein. *PLoS Biol.* **3,** e236.

Gent, J. I., Schvarzstein, M., Villeneuve, A. M., Gu, S. G., Jantsch, V., Fire, A. Z., and Baudrimont, A. (2009). A Caenorhabditis elegans RNA-directed RNA polymerase in sperm development and endogenous RNA interference. *Genetics* **183,** 1297–1314.

Gent, J. I., Lamm, A. T., Pavelec, D. M., Maniar, J. M., Parameswaran, P., Tao, L., Kennedy, S., and Fire, A. Z. (2010). Distinct phases of siRNA synthesis in an endogenous RNAi pathway in C. elegans soma. *Mol. Cell* **37,** 679–689.

Ghildiyal, M., and Zamore, P. D. (2009). Small silencing RNAs: An expanding universe. *Nat. Rev. Genet.* **10,** 94–108.

Ghildiyal, M., Seitz, H., Horwich, M. D., Li, C., Du, T., Lee, S., Xu, J., Kittler, E. L., Zapp, M. L., Weng, Z., and Zamore, P. D. (2008). Endogenous siRNAs derived from transposons and mRNAs in Drosophila somatic cells. *Science (New York, N.Y.)* **320,** 1077–1081.

Giraldez, A. J., Cinalli, R. M., Glasner, M. E., Enright, A. J., Thomson, J. M., Baskerville, S., Hammond, S. M., Bartel, D. P., and Schier, A. F. (2005). MicroRNAs regulate brain morphogenesis in zebrafish. *Science (New York, N.Y.)* **308,** 833–838.

Giraldez, A. J., Mishima, Y., Rihel, J., Grocock, R. J., Van Dongen, S., Inoue, K., Enright, A. J., and Schier, A. F. (2006). Zebrafish MiR-430 promotes deadenylation and clearance of maternal mRNAs. *Science (New York, N.Y.)* **312,** 75–79.

Girard, A., Sachidanandam, R., Hannon, G. J., and Carmell, M. A. (2006). A germline-specific class of small RNAs binds mammalian Piwi proteins. *Nature* **442,** 199–202.

Gonzalez, G., and Behringer, R. R. (2009). Dicer is required for female reproductive tract development and fertility in the mouse. *Mol. Reprod. Dev.* **76,** 678–688.

Gonzalez-Gonzalez, E., Lopez-Casas, P. P., and del Mazo, J. (2008). The expression patterns of genes involved in the RNAi pathways are tissue-dependent and differ in the germ and somatic cells of mouse testis. *Biochim. Biophys. Acta* **1779,** 306–311.

Grant-Downton, R. T. (2010). Through a generation darkly: Small RNAs in the gametophyte. *Biochem. Soc. Trans.* **38,** 617–621.

Grishok, A., Pasquinelli, A. E., Conte, D., Li, N., Parrish, S., Ha, I., Baillie, D. L., Fire, A., Ruvkun, G., and Mello, C. C. (2001). Genes and mechanisms related to RNA interference regulate expression of the small temporal RNAs that control C. elegans developmental timing. *Cell* **106,** 23–34.

Grivna, S. T., Beyret, E., Wang, Z., and Lin, H. (2006). A novel class of small RNAs in mouse spermatogenic cells. *Genes Dev.* **20,** 1709–1714.

Gu, W., Shirayama, M., Conte, D., Jr., Vasale, J., Batista, P. J., Claycomb, J. M., Moresco, J. J., Youngman, E. M., Keys, J., Stoltz, M. J., Chen, C. C., Chaves, D. A., *et al.* (2009). Distinct argonaute-mediated 22G-RNA pathways direct genome surveillance in the C. elegans germline. *Mol. Cell* **36,** 231–244.

Gunawardane, L. S., Saito, K., Nishida, K. M., Miyoshi, K., Kawamura, Y., Nagami, T., Siomi, H., and Siomi, M. C. (2007). A slicer-mediated mechanism for repeat-associated siRNA 5′ end formation in Drosophila. *Science (New York, N.Y.)* **315,** 1587–1590.

Hamilton, A., Voinnet, O., Chappell, L., and Baulcombe, D. (2002). Two classes of short interfering RNA in RNA silencing. *EMBO J.* **21,** 4671–4679.

Han, T., Manoharan, A. P., Harkins, T. T., Bouffard, P., Fitzpatrick, C., Chu, D. S., Thierry-Mieg, D., Thierry-Mieg, J., and Kim, J. K. (2009). 26G endo-siRNAs regulate spermatogenic and zygotic gene expression in Caenorhabditis elegans. *Proc. Natl. Acad. Sci. USA* **106,** 18674–18679.

Hardy, R. W., Tokuyasu, K. T., Lindsley, D. L., and Garavito, M. (1979). The germinal proliferation center in the testis of Drosophila melanogaster. *J. Ultrastruct. Res.* **69,** 180–190.

Harris, A. N., and Macdonald, P. M. (2001). Aubergine encodes a Drosophila polar granule component required for pole cell formation and related to eIF2C. *Development* **128,** 2823–2832.

Hartig, J. V., Esslinger, S., Bottcher, R., Saito, K., and Forstemann, K. (2009). Endo-siRNAs depend on a new isoform of loquacious and target artificially introduced, high-copy sequences. *EMBO J.* **28,** 2932–2944.

Hashimoto, Y., Maegawa, S., Nagai, T., Yamaha, E., Suzuki, H., Yasuda, K., and Inoue, K. (2004). Localized maternal factors are required for zebrafish germ cell formation. *Dev. Biol.* **268,** 152–161.

Hashimoto, Y., Suzuki, H., Kageyama, Y., Yasuda, K., and Inoue, K. (2006). Bruno-like protein is localized to zebrafish germ plasm during the early cleavage stages. *Gene Expr. Patterns* **6,** 201–205.

Hatfield, S. D., Shcherbata, H. R., Fischer, K. A., Nakahara, K., Carthew, R. W., and Ruohola-Baker, H. (2005). Stem cell division is regulated by the microRNA pathway. *Nature* **435,** 974–978.

Hay, B., Jan, L., and Jan, Y. (1988). A protein component of Drosophila polar granules is encoded by vasa and has extensive sequence similarity to ATP-dependent helicases. *Cell* **55,** 577–587.

Hayashi, K., de Sousa, Chuva, Lopes, S. M., Kaneda, M., Tang, F., Hajkova, P., Lao, K., O'Carroll, D., Das, P. P., Tarakhovsky, A., Miska, E. A., and Surani, M. A. (2008). MicroRNA biogenesis is required for mouse primordial germ cell development and spermatogenesis. *PLoS One* **3,** e1738.

Heo, I., Joo, C., Cho, J., Ha, M., Han, J., and Kim, V. N. (2008). Lin28 mediates the terminal uridylation of let-7 precursor microRNA. *Mol. Cell* **32,** 276–284.

Hird, S. N., Paulsen, J. E., and Strome, S. (1996). Segregation of germ granules in living Caenorhabditis elegans embryos: Cell-type-specific mechanisms for cytoplasmic localisation. *Development* **122,** 1303–1312.

Hong, X., Luense, L. J., McGinnis, L. K., Nothnick, W. B., and Christenson, L. K. (2008). Dicer1 is essential for female fertility and normal development of the female reproductive system. *Endocrinology* **149,** 6207–6212.

Houston, D. W., Zhang, J., Maines, J. Z., Wasserman, S. A., and King, M. L. (1998). A Xenopus DAZ-like gene encodes an RNA component of germ plasm and is a functional homologue of Drosophila boule. *Development* **125,** 171–180.

Houwing, S., Kamminga, L. M., Berezikov, E., Cronembold, D., Girard, A., van den Elst, H., Filippov, D. V., Blaser, H., Raz, E., Moens, C. B., Plasterk, R. H., Hannon, G. J., *et al.* (2007). A role for Piwi and piRNAs in germ cell maintenance and transposon silencing in Zebrafish. *Cell* **129,** 69–82.

Houwing, S., Berezikov, E., and Ketting, R. F. (2008). Zili is required for germ cell differentiation and meiosis in zebrafish. *EMBO J.* **27,** 2702–2711.

Hutvagner, G., and Zamore, P. D. (2002). A microRNA in a multiple-turnover RNAi enzyme complex. *Science (New York, N.Y.)* **297,** 2056–2060.

Hutvagner, G., McLachlan, J., Pasquinelli, A. E., Balint, E., Tuschl, T., and Zamore, P. D. (2001). A cellular function for the RNA-interference enzyme Dicer in the maturation of the let-7 small temporal RNA. *Science (New York, N.Y.)* **293,** 834–838.

Ikenishi, K. (1998). Germ plasm in Caenorhabditis elegans, Drosophila, and Xenopus. *Dev. Growth Differ.* **40,** 1–10.

Iovino, N., Pane, A., and Gaul, U. (2009). miR-184 has multiple roles in Drosophila female germline development. *Dev. Cell* **17,** 123–133.

Jin, Z., and Xie, T. (2007). Dcr-1 maintains Drosophila ovarian stem cells. *Curr. Biol.* **17,** 539–544.

Jones-Rhoades, M. W., Bartel, D. P., and Bartel, B. (2006). MicroRNAS and their regulatory roles in plants. *Annu. Rev. Plant Biol.* **57,** 19–53.

Kawamura, Y., Saito, K., Kin, T., Ono, Y., Asai, K., Sunohara, T., Okada, T. N., Siomi, M. C., and Siomi, H. (2008). Drosophila endogenous small RNAs bind to Argonaute 2 in somatic cells. *Nature* **453,** 793–797.

Kedde, M., Strasser, M. J., Boldajipour, B., Oude Vrielink, J. A., Slanchev, K., le Sage, C., Nagel, R., Voorhoeve, P. M., van Duijse, J., Orom, U. A., Lund, A. H., Perrakis, A., *et al.* (2007). RNA-binding protein Dnd1 inhibits microRNA access to target mRNA. *Cell* **131,** 1273–1286.

Kemphues, K. J., and Strome, S. (1997). Fertilization and establishment of polarity in the embryo. *In* "C. elegans II," (D. L. Riddle, T. Blumenthal, B. J. Meyer, and J. R. Priess, eds.), .2nd edn. Cold Spring Harbor Laboratory Press, Cold Spring Harbor (NY).

Ketting, R. F. (2011). The many faces of RNAi. *Dev. Cell* **20,** 148–161.

Ketting, R. F., Haverkamp, T. H., van Luenen, H. G., and Plasterk, R. H. (1999). Mut-7 of C. elegans, required for transposon silencing and RNA interference, is a homolog of Werner syndrome helicase and RNaseD. *Cell* **99,** 133–141.

Ketting, R. F., Fischer, S. E., Bernstein, E., Sijen, T., Hannon, G. J., and Plasterk, R. H. (2001). Dicer functions in RNA interference and in synthesis of small RNA involved in developmental timing in C. elegans. *Genes Dev.* **15,** 2654–2659.

Kiger, A. A., White-Cooper, H., and Fuller, M. T. (2000). Somatic support cells restrict germline stem cell self-renewal and promote differentiation. *Nature* **407,** 750–754.

Kim, V. N., Han, J., and Siomi, M. C. (2009). Biogenesis of small RNAs in animals. *Nat. Rev. Mol. Cell Biol.* **10,** 126–139.

Klattenhoff, C., Bratu, D. P., McGinnis-Schultz, N., Koppetsch, B. S., Cook, H. A., and Theurkauf, W. E. (2007). Drosophila rasiRNA pathway mutations disrupt embryonic axis specification through activation of an ATR/Chk2 DNA damage response. *Dev. Cell* **12,** 45–55.

Knaut, H., Pelegri, F., Bohmann, K., Schwarz, H., and Nusslein-Volhard, C. (2000). Zebrafish vasa RNA but not its protein is a component of the germ plasm and segregates asymmetrically before germline specification. *J. Cell Biol.* **149,** 875–888.

Knight, S. W., and Bass, B. L. (2001). A role for the RNase III enzyme DCR-1 in RNA interference and germ line development in Caenorhabditis elegans. *Science (New York, N.Y.)* **293,** 2269–2271.

Koebernick, K., Loeber, J., Arthur, P. K., Tarbashevich, K., and Pieler, T. (2010). Elr-type proteins protect Xenopus Dead end mRNA from miR-18-mediated clearance in the soma. *Proc. Natl. Acad. Sci. USA* **107,** 16148–16153.

Komiya, T., Itoh, K., Ikenishi, K., and Furusawa, M. (1994). Isolation and characterization of a novel gene of the DEAD box protein family which is specifically expressed in germ cells of Xenopus laevis. *Dev. Biol.* **162,** 354–363.

Köprunner, M., Thisse, C., Thisse, B., and Raz, E. (2001). A zebrafish nanos-related gene is essential for the development of primordial germ cells. *Genes Dev.* **15,** 2877–2885.

Kosaka, K., Kawakami, K., Sakamoto, H., and Inoue, K. (2007). Spatiotemporal localization of germ plasm RNAs during zebrafish oogenesis. *Mech. Dev.* **124,** 279–289.

Kotaja, N., Bhattacharyya, S. N., Jaskiewicz, L., Kimmins, S., Parvinen, M., Filipowicz, W., and Sassone-Corsi, P. (2006). The chromatoid body of male germ cells: Similarity with processing bodies and presence of Dicer and microRNA pathway components. *Proc. Natl. Acad. Sci. USA* **103,** 2647–2652.

Krol, J., Loedige, I., and Filipowicz, W. (2010). The widespread regulation of microRNA biogenesis, function, and decay. *Nat. Rev. Genet.* **11,** 597–610.

Kuramochi-Miyagawa, S., Kimura, T., Yomogida, K., Kuroiwa, A., Tadokoro, Y., Fujita, Y., Sato, M., Matsuda, Y., and Nakano, T. (2001). Two mouse piwi-related genes: Miwi and mili. *Mech. Dev.* **108,** 121–133.

Kuramochi-Miyagawa, S., Kimura, T., Ijiri, T. W., Isobe, T., Asada, N., Fujita, Y., Ikawa, M., Iwai, N., Okabe, M., Deng, W., Lin, H., Matsuda, Y., *et al.* (2004). Mili, a mammalian member of piwi family gene, is essential for spermatogenesis. *Development* **131,** 839–849.

Kuramochi-Miyagawa, S., Watanabe, T., Gotoh, K., Totoki, Y., Toyoda, A., Ikawa, M., Asada, N., Kojima, K., Yamaguchi, Y., Ijiri, T. W., Hata, K., Li, E., *et al.* (2008). DNA methylation of retrotransposon genes is regulated by Piwi family members MILI and MIWI2 in murine fetal testes. *Genes Dev.* **22,** 908–917.

Kurimoto, K., Yabuta, Y., Ohinata, Y., Shigeta, M., Yamanaka, K., and Saitou, M. (2008). Complex genome-wide transcription dynamics orchestrated by Blimp1 for the specification of the germ cell lineage in mice. *Genes Dev.* **22,** 1617–1635.

Lagos-Quintana, M., Rauhut, R., Lendeckel, W., and Tuschl, T. (2001). Identification of novel genes coding for small expressed RNAs. *Science (New York, N.Y.)* **294,** 853–858.

Lai, E. C. (2002). Micro RNAs are complementary to 3' UTR sequence motifs that mediate negative post-transcriptional regulation. *Nat. Genet.* **30,** 363–364.

Lau, N. C. (2010). Small RNAs in the animal gonad: Guarding genomes and guiding development. *Int. J. Biochem. Cell Biol.* **42,** 1334–1347.

Lau, N. C., Lim, L. P., Weinstein, E. G., and Bartel, D. P. (2001). An abundant class of tiny RNAs with probable regulatory roles in Caenorhabditis elegans. *Science (New York, N.Y.)* **294,** 858–862.

Lau, N. C., Seto, A. G., Kim, J., Kuramochi-Miyagawa, S., Nakano, T., Bartel, D. P., and Kingston, R. E. (2006). Characterization of the piRNA complex from rat testes. *Science (New York, N.Y.)* **313,** 363–367.

Lawson, K. A., Dunn, N. R., Roelen, B. A., Zeinstra, L. M., Davis, A. M., Wright, C. V., Korving, J. P., and Hogan, B. L. (1999). Bmp4 is required for the generation of primordial germ cells in the mouse embryo. *Genes Dev.* **13,** 424–436.

Leaman, D., Chen, P. Y., Fak, J., Yalcin, A., Pearce, M., Unnerstall, U., Marks, D. S., Sander, C., Tuschl, T., and Gaul, U. (2005). Antisense-mediated depletion reveals essential and specific functions of microRNAs in Drosophila development. *Cell* **121,** 1097–1108.

Lee, R. C., and Ambros, V. (2001). An extensive class of small RNAs in Caenorhabditis elegans. *Science (New York, N.Y.)* **294,** 862–864.

Lee, R. C., Feinbaum, R. L., and Ambros, V. (1993). The C. elegans heterochronic gene lin-4 encodes small RNAs with antisense complementarity to lin-14. *Cell* **75,** 843–854.

Lee, Y. S., Nakahara, K., Pham, J. W., Kim, K., He, Z., Sontheimer, E. J., and Carthew, R. W. (2004). Distinct roles for Drosophila Dicer-1 and Dicer-2 in the siRNA/miRNA silencing pathways. *Cell* **117,** 69–81.

Lees-Murdock, D. J., Shovlin, T. C., Gardiner, T., De Felici, M., and Walsh, C. P. (2005). DNA methyltransferase expression in the mouse germ line during periods of *de novo* methylation. *Dev. Dyn.* **232,** 992–1002.

Lehmann, R., and Nusslein-Volhard, C. (1991). The maternal gene nanos has a central role in posterior pattern formation of the Drosophila embryo. *Development* **112,** 679–691.

Lei, L., Jin, S., Gonzalez, G., Behringer, R. R., and Woodruff, T. K. (2010). The regulatory role of Dicer in folliculogenesis in mice. *Mol. Cell. Endocrinol.* **315,** 63–73.

Li, C., Vagin, V. V., Lee, S., Xu, J., Ma, S., Xi, H., Seitz, H., Horwich, M. D., Syrzycka, M., Honda, B. M., Kittler, E. L., Zapp, M. L., *et al.* (2009). Collapse of germline piRNAs in the absence of Argonaute3 reveals somatic piRNAs in flies. *Cell* **137,** 509–521.

Lim, A. K., and Kai, T. (2007). Unique germ-line organelle, nuage, functions to repress selfish genetic elements in Drosophila melanogaster. *Proc. Natl. Acad. Sci. USA* **104,** 6714–6719.

Lin, H., and Spradling, A. C. (1997). A novel group of pumilio mutations affects the asymmetric division of germline stem cells in the Drosophila ovary. *Development* **124,** 2463–2476.

Liu, Q., Rand, T. A., Kalidas, S., Du, F., Kim, H. E., Smith, D. P., and Wang, X. (2003). R2D2, a bridge between the initiation and effector steps of the Drosophila RNAi pathway. *Science (New York, N.Y.)* **301,** 1921–1925.

Liu, C. G., Calin, G. A., Meloon, B., Gamliel, N., Sevignani, C., Ferracin, M., Dumitru, C. D., Shimizu, M., Zupo, S., Dono, M., Alder, H., Bullrich, F., *et al.* (2004a). An oligonucleotide microchip for genome-wide microRNA profiling in human and mouse tissues. *Proc. Natl. Acad. Sci. USA* **101,** 9740–9744.

Liu, J., Carmell, M. A., Rivas, F. V., Marsden, C. G., Thomson, J. M., Song, J. J., Hammond, S. M., Joshua-Tor, L., and Hannon, G. J. (2004b). Argonaute2 is the catalytic engine of mammalian RNAi. *Science (New York, N.Y.)* **305,** 1437–1441.

Liu, N., Han, H., and Lasko, P. (2009). Vasa promotes Drosophila germline stem cell differentiation by activating mei-P26 translation by directly interacting with a (U)-rich motif in its 3' UTR. *Genes Dev.* **23,** 2742–2752.

Lopez-Schier, H., and St Johnston, D. (2001). Delta signaling from the germ line controls the proliferation and differentiation of the somatic follicle cells during Drosophila oogenesis. *Genes Dev.* **15,** 1393–1405.

Lund, E., Guttinger, S., Calado, A., Dahlberg, J. E., and Kutay, U. (2004). Nuclear export of microRNA precursors. *Science (New York, N.Y.)* **303,** 95–98.

Lund, E., Liu, M., Hartley, R. S., Sheets, M. D., and Dahlberg, J. E. (2009). Deadenylation of maternal mRNAs mediated by miR-427 in Xenopus laevis embryos. *RNA* **15,** 2351–2363.

Ma, J., Flemr, M., Stein, P., Berninger, P., Malik, R., Zavolan, M., Svoboda, P., and Schultz, R. M. (2010). MicroRNA activity is suppressed in mouse oocytes. *Curr. Biol.* **20,** 265–270.

Maatouk, D. M., Loveland, K. L., McManus, M. T., Moore, K., and Harfe, B. D. (2008). Dicer1 is required for differentiation of the mouse male germline. *Biol. Reprod.* **79,** 696–703.

Maegawa, S., Yasuda, K., and Inoue, K. (1999). Maternal mRNA localization of zebrafish DAZ-like gene. *Mech. Dev.* **81,** 223–226.

Malone, C. D., Brennecke, J., Dus, M., Stark, A., McCombie, W. R., Sachidanandam, R., and Hannon, G. J. (2009). Specialized piRNA pathways act in germline and somatic tissues of the Drosophila ovary. *Cell* **137,** 522–535.

Megosh, H. B., Cox, D. N., Campbell, C., and Lin, H. (2006). The role of PIWI and the miRNA machinery in Drosophila germline determination. *Curr. Biol.* **16,** 1884–1894.

Meister, G., Landthaler, M., Patkaniowska, A., Dorsett, Y., Teng, G., and Tuschl, T. (2004). Human Argonaute2 mediates RNA cleavage targeted by miRNAs and siRNAs. *Mol. Cell* **15,** 185–197.

Mickoleit, M., Banisch, T. U., and Raz, E. (2011). Regulation of hub mRNA stability and translation by miR430 and the dead end protein promotes preferential expression in zebrafish primordial germ cells. *Dev. Dyn.* **240,** 695–703.

Mishima, Y., Giraldez, A. J., Takeda, Y., Fujiwara, T., Sakamoto, H., Schier, A. F., and Inoue, K. (2006). Differential regulation of germline mRNAs in soma and germ cells by zebrafish miR-430. *Curr. Biol.* **16,** 2135–2142.

Murchison, E. P., Stein, P., Xuan, Z., Pan, H., Zhang, M. Q., Schultz, R. M., and Hannon, G. J. (2007). Critical roles for Dicer in the female germline. *Genes Dev.* **21,** 682–693.

Nagaraja, A. K., Andreu-Vieyra, C., Franco, H. L., Ma, L., Chen, R., Han, D. Y., Zhu, H., Agno, J. E., Gunaratne, P. H., DeMayo, F. J., and Matzuk, M. M. (2008). Deletion of Dicer in somatic cells of the female reproductive tract causes sterility. *Mol. Endocrinol.* **22,** 2336–2352.

Neumuller, R. A., Betschinger, J., Fischer, A., Bushati, N., Poernbacher, I., Mechtler, K., Cohen, S. M., and Knoblich, J. A. (2008). Mei-P26 regulates microRNAs and cell growth in the Drosophila ovarian stem cell lineage. *Nature* **454,** 241–245.

Newman, M. A., Thomson, J. M., and Hammond, S. M. (2008). Lin-28 interaction with the Let-7 precursor loop mediates regulated microRNA processing. *RNA* **14,** 1539–1549.

Nicholls, P. K., Harrison, C. A., Walton, K. L., McLachlan, R. I., O'Donnell, L., and Stanton, P. G. (2011). Hormonal regulation of sertoli cell micro-RNAs at spermiation. *Endocrinology* **152,** 1670–1683.

Nie, K., Gomez, M., Landgraf, P., Garcia, J. F., Liu, Y., Tan, L. H., Chadburn, A., Tuschl, T., Knowles, D. M., and Tam, W. (2008). MicroRNA-mediated down-regulation of PRDM1/Blimp-1 in Hodgkin/Reed-Sternberg cells: A potential pathogenetic lesion in Hodgkin lymphomas. *Am. J. Pathol.* **173,** 242–252.

Nieuwkoop, P. D. (1947). Experimental investigations on the origin and determiation of the germ cells, and on the development of the lateral plates and germ ridges in the urodeles. *Arch. Neerl. Zool.* **8,** 1–205.

Nishida, K. M., Saito, K., Mori, T., Kawamura, Y., Nagami-Okada, T., Inagaki, S., Siomi, H., and Siomi, M. C. (2007). Gene silencing mechanisms mediated by Aubergine piRNA complexes in Drosophila male gonad. *RNA* **13,** 1911–1922.

Novotny, G. W., Sonne, S. B., Nielsen, J. E., Jonstrup, S. P., Hansen, M. A., Skakkebaek, N. E., Rajpert-De Meyts, E., Kjems, J., and Leffers, H. (2007). Translational repression of E2F1 mRNA in carcinoma in situ and normal testis correlates with expression of the miR-17-92 cluster. *Cell Death Differ.* **14,** 879–882.

Ohinata, Y., Payer, B., O'Carroll, D., Ancelin, K., Ono, Y., Sano, M., Barton, S. C., Obukhanych, T., Nussenzweig, M., Tarakhovsky, A., Saitou, M., and Surani, M. A. (2005). Blimp1 is a critical determinant of the germ cell lineage in mice. *Nature* **436,** 207–213.

Okamura, K., and Lai, E. C. (2008). Endogenous small interfering RNAs in animals. *Nat. Rev. Mol. Cell Biol.* **9,** 673–678.

Okamura, K., Ishizuka, A., Siomi, H., and Siomi, M. C. (2004). Distinct roles for Argonaute proteins in small RNA-directed RNA cleavage pathways. *Genes Dev.* **18,** 1655–1666.

Okamura, K., Balla, S., Martin, R., Liu, N., and Lai, E. C. (2008a). Two distinct mechanisms generate endogenous siRNAs from bidirectional transcription in Drosophila melanogaster. *Nat. Struct. Mol. Biol.* **15,** 581–590.

Okamura, K., Chung, W. J., Ruby, J. G., Guo, H., Bartel, D. P., and Lai, E. C. (2008b). The Drosophila hairpin RNA pathway generates endogenous short interfering RNAs. *Nature* **453,** 803–806.

Olsen, L. C., Aasland, R., and Fjose, A. (1997). A vasa-like gene in zebrafish identifies putative primordial germ cells. *Mech. Dev.* **66,** 95–105.

Pak, J., and Fire, A. (2007). Distinct populations of primary and secondary effectors during RNAi in C. elegans. *Science (New York, N.Y.)* **315,** 241–244.

Pal-Bhadra, M., Leibovitch, B. A., Gandhi, S. G., Rao, M., Bhadra, U., Birchler, J. A., and Elgin, S. C. (2004). Heterochromatic silencing and HP1 localization in Drosophila are dependent on the RNAi machinery. *Science (New York, N.Y.)* **303,** 669–672.

Pane, A., Wehr, K., and Schupbach, T. (2007). Zucchini and squash encode two putative nucleases required for rasiRNA production in the Drosophila germline. *Dev. Cell* **12,** 851–862.

Park, J. K., Liu, X., Strauss, T. J., McKearin, D. M., and Liu, Q. (2007). The miRNA pathway intrinsically controls self-renewal of Drosophila germline stem cells. *Curr. Biol.* **17,** 533–538.

Pavelec, D. M., Lachowiec, J., Duchaine, T. F., Smith, H. E., and Kennedy, S. (2009). Requirement for the ERI/DICER complex in endogenous RNA interference and sperm development in Caenorhabditis elegans. *Genetics* **183,** 1283–1295.

Pek, J. W., Lim, A. K., and Kai, T. (2009). Drosophila maelstrom ensures proper germline stem cell lineage differentiation by repressing microRNA-7. *Dev. Cell* **17,** 417–424.

Pham, J. W., Pellino, J. L., Lee, Y. S., Carthew, R. W., and Sontheimer, E. J. (2004). A Dicer-2-dependent 80s complex cleaves targeted mRNAs during RNAi in Drosophila. *Cell* **117,** 83–94.

Piskounova, E., Viswanathan, S. R., Janas, M., LaPierre, R. J., Daley, G. Q., Sliz, P., and Gregory, R. I. (2008). Determinants of microRNA processing inhibition by the developmentally regulated RNA-binding protein Lin28. *J. Biol. Chem.* **283,** 21310–21314.

Poulton, J. S., Huang, Y. C., Smith, L., Sun, J., Leake, N., Schleede, J., Stevens, L. M., and Deng, W. M. (2011). The microRNA pathway regulates the temporal pattern of Notch signaling in Drosophila follicle cells. *Development* **138,** 1737–1745.

Reinhart, B. J., and Bartel, D. P. (2002). Small RNAs correspond to centromere heterochromatic repeats. *Science (New York, N.Y.)* **297,** 1831.

Richardson, B. E., and Lehmann, R. (2010). Mechanisms guiding primordial germ cell migration: Strategies from different organisms. *Nat. Rev. Mol. Cell Biol.* **11,** 37–49.

Robertson, E. J., Charatsi, I., Joyner, C. J., Koonce, C. H., Morgan, M., Islam, A., Paterson, C., Lejsek, E., Arnold, S. J., Kallies, A., Nutt, S. L., and Bikoff, E. K. (2007). Blimp1 regulates development of the posterior forelimb, caudal pharyngeal arches, heart, and sensory vibrissae in mice. *Development* **134,** 4335–4345.

Robine, N., Lau, N. C., Balla, S., Jin, Z., Okamura, K., Kuramochi-Miyagawa, S., Blower, M. D., and Lai, E. C. (2009). A broadly conserved pathway generates 3'UTR-directed primary piRNAs. *Curr. Biol.* **19,** 2066–2076.

Rocheleau, C. E., Cullison, K., Huang, K., Bernstein, Y., Spilker, A. C., and Sundaram, M. V. (2008). The Caenorhabditis elegans ekl (enhancer of ksr-1 lethality) genes include putative components of a germline small RNA pathway. *Genetics* **178,** 1431–1443.

Rouget, C., Papin, C., Boureux, A., Meunier, A. C., Franco, B., Robine, N., Lai, E. C., Pelisson, A., and Simonelig, M. (2010). Maternal mRNA deadenylation and decay by the piRNA pathway in the early Drosophila embryo. *Nature* **467,** 1128–1132.

Ruby, J. G., Jan, C., Player, C., Axtell, M. J., Lee, W., Nusbaum, C., Ge, H., and Bartel, D. P. (2006). Large-scale sequencing reveals 21U-RNAs and additional microRNAs and endogenous siRNAs in C. elegans. *Cell* **127,** 1193–1207.

Rybak, A., Fuchs, H., Smirnova, L., Brandt, C., Pohl, E. E., Nitsch, R., and Wulczyn, F. G. (2008). A feedback loop comprising lin-28 and let-7 controls pre-let-7 maturation during neural stem-cell commitment. *Nat. Cell Biol.* **10,** 987–993.

Sahut-Barnola, I., Godt, D., Laski, F. A., and Couderc, J. L. (1995). Drosophila ovary morphogenesis: Analysis of terminal filament formation and identification of a gene required for this process. *Dev. Biol.* **170,** 127–135.

Saito, K., and Siomi, M. C. (2010). Small RNA-mediated quiescence of transposable elements in animals. *Dev. Cell* **19,** 687–697.

Saito, K., Nishida, K. M., Mori, T., Kawamura, Y., Miyoshi, K., Nagami, T., Siomi, H., and Siomi, M. C. (2006). Specific association of Piwi with rasiRNAs derived from retrotransposon and heterochromatic regions in the Drosophila genome. *Genes Dev.* **20,** 2214–2222.

Saito, K., Inagaki, S., Mituyama, T., Kawamura, Y., Ono, Y., Sakota, E., Kotani, H., Asai, K., Siomi, H., and Siomi, M. C. (2009). A regulatory circuit for piwi by the large Maf gene traffic jam in Drosophila. *Nature* **461,** 1296–1299.

Saito, K., Ishizu, H., Komai, M., Kotani, H., Kawamura, Y., Nishida, K. M., Siomi, H., and Siomi, M. C. (2010). Roles for the Yb body components Armitage and Yb in primary piRNA biogenesis in Drosophila. *Genes Dev.* **24,** 2493–2498.

Saitou, M., Barton, S. C., and Surani, M. A. (2002). A molecular programme for the specification of germ cell fate in mice. *Nature* **418,** 293–300.

Sasidharan, R., and Gerstein, M. (2008). Genomics: Protein fossils live on as RNA. *Nature* **453,** 729–731.

Schmidt, A., Palumbo, G., Bozzetti, M. P., Tritto, P., Pimpinelli, S., and Schafer, U. (1999). Genetic and molecular characterization of sting, a gene involved in crystal formation and meiotic drive in the male germ line of Drosophila melanogaster. *Genetics* **151,** 749–760.

Seydoux, G., and Braun, R. E. (2006). Pathway to totipotency: Lessons from germ cells. *Cell* **127,** 891–904.

Seydoux, G., and Fire, A. (1994). Soma-germline asymmetry in the distributions of embryonic RNAs in Caenorhabditis elegans. *Development* **120,** 2823–2834.

She, X., Xu, X., Fedotov, A., Kelly, W. G., and Maine, E. M. (2009). Regulation of heterochromatin assembly on unpaired chromosomes during Caenorhabditis elegans meiosis by components of a small RNA-mediated pathway. *PLoS Genet.* **5,** e1000624.

Sijen, T., Steiner, F. A., Thijssen, K. L., and Plasterk, R. H. (2007). Secondary siRNAs result from unprimed RNA synthesis and form a distinct class. *Science (New York, N.Y.)* **315,** 244–247.

Siomi, M. C., Sato, K., Pezic, D., and Aravin, A. A. (2011). PIWI-interacting small RNAs: The vanguard of genome defence. *Nat. Rev. Mol. Cell Biol.* **12,** 246–258.
Sirotkin, A. V., Ovcharenko, D., Grossmann, R., Laukova, M., and Mlyncek, M. (2009). Identification of microRNAs controlling human ovarian cell steroidogenesis via a genome-scale screen. *J. Cell. Physiol.* **219,** 415–420.
Sirotkin, A. V., Laukova, M., Ovcharenko, D., Brenaut, P., and Mlyncek, M. (2010). Identification of microRNAs controlling human ovarian cell proliferation and apoptosis. *J. Cell. Physiol.* **223,** 49–56.
Slanchev, K., Stebler, J., Goudarzi, M., Cojocaru, V., Weidinger, G., and Raz, E. (2009). Control of Dead end localization and activity—Implications for the function of the protein in antagonizing miRNA function. *Mech. Dev.* **126,** 270–277.
Smardon, A., Spoerke, J. M., Stacey, S. C., Klein, M. E., Mackin, N., and Maine, E. M. (2000). EGO-1 is related to RNA-directed RNA polymerase and functions in germ-line development and RNA interference in C. elegans. *Curr. Biol.* **10,** 169–178.
Song, R., Ro, S., Michaels, J. D., Park, C., McCarrey, J. R., and Yan, W. (2009). Many X-linked microRNAs escape meiotic sex chromosome inactivation. *Nat. Genet.* **41,** 488–493.
Soper, S. F., van der Heijden, G. W., Hardiman, T. C., Goodheart, M., Martin, S. L., de Boer, P., and Bortvin, A. (2008). Mouse maelstrom, a component of nuage, is essential for spermatogenesis and transposon repression in meiosis. *Dev. Cell* **15,** 285–297.
Strome, S., and Wood, W. B. (1982). Immunofluorescence visualization of germ-line-specific cytoplasmic granules in embryos, larvae, and adults of Caenorhabditis elegans. *Proc. Natl. Acad. Sci. USA* **79,** 1558–1562.
Suh, N., Baehner, L., Moltzahn, F., Melton, C., Shenoy, A., Chen, J., and Blelloch, R. (2010). MicroRNA function is globally suppressed in mouse oocytes and early embryos. *Curr. Biol.* **20,** 271–277.
Szakmary, A., Cox, D. N., Wang, Z., and Lin, H. (2005). Regulatory relationship among piwi, pumilio, and bag-of-marbles in Drosophila germline stem cell self-renewal and differentiation. *Curr. Biol.* **15,** 171–178.
Tabara, H., Sarkissian, M., Kelly, W. G., Fleenor, J., Grishok, A., Timmons, L., Fire, A., and Mello, C. C. (1999). The rde-1 gene, RNA interference, and transposon silencing in C. elegans. *Cell* **99,** 123–132.
Takada, S., Berezikov, E., Choi, Y. L., Yamashita, Y., and Mano, H. (2009). Potential role of miR-29b in modulation of Dnmt3a and Dnmt3b expression in primordial germ cells of female mouse embryos. *RNA* **15,** 1507–1514.
Takeda, Y., Mishima, Y., Fujiwara, T., Sakamoto, H., and Inoue, K. (2009). DAZL relieves miRNA-mediated repression of germline mRNAs by controlling poly(A) tail length in zebrafish. *PLoS One* **4,** e7513.
Tam, P., and Zhou, S. (1996). The allocation of epiblast cells to ectodermal and germ-line lineages is influenced by the position of the cells in the gastrulating mouse embryo. *Dev. Biol.* **178,** 124–132.
Tam, O. H., Aravin, A. A., Stein, P., Girard, A., Murchison, E. P., Cheloufi, S., Hodges, E., Anger, M., Sachidanandam, R., Schultz, R. M., and Hannon, G. J. (2008). Pseudogene-derived small interfering RNAs regulate gene expression in mouse oocytes. *Nature* **453,** 534–538.
Tang, F., Kaneda, M., O'Carroll, D., Hajkova, P., Barton, S. C., Sun, Y. A., Lee, C., Tarakhovsky, A., Lao, K., and Surani, M. A. (2007). Maternal microRNAs are essential for mouse zygotic development. *Genes Dev.* **21,** 644–648.
Technau, G., and Campos-Ortega, J. (1986). Lineage analysis of transplanted individual cells in embryos of *Drosophila melanogaster* III. Commitment and proliferation capabilities of pole cells and midgut progenitors. *Roux's Arch. Dev. Biol.* **195,** 489–498.

Thomson, T., Liu, N., Arkov, A., Lehmann, R., and Lasko, P. (2008). Isolation of new polar granule components in Drosophila reveals P body and ER associated proteins. *Mech. Dev.* **125,** 865–873.

Tomari, Y., Matranga, C., Haley, B., Martinez, N., and Zamore, P. D. (2004). A protein sensor for siRNA asymmetry. *Science (New York, N.Y.)* **306,** 1377–1380.

Tran, J., Brenner, T. J., and DiNardo, S. (2000). Somatic control over the germline stem cell lineage during Drosophila spermatogenesis. *Nature* **407,** 754–757.

Tsang, T. E., Khoo, P. L., Jamieson, R. V., Zhou, S. X., Ang, S. L., Behringer, R., and Tam, P. P. (2001). The allocation and differentiation of mouse primordial germ cells. *Int. J. Dev. Biol.* **45,** 549–555.

Tsuda, M., Sasaoka, Y., Kiso, M., Abe, K., Haraguchi, S., Kobayashi, S., and Saga, Y. (2003). Conserved role of nanos proteins in germ cell development. *Science (New York, N.Y.)* **301,** 1239–1241.

Unhavaithaya, Y., Hao, Y., Beyret, E., Yin, H., Kuramochi-Miyagawa, S., Nakano, T., and Lin, H. (2009). MILI, a PIWI-interacting RNA-binding protein, is required for germ line stem cell self-renewal and appears to positively regulate translation. *J. Biol. Chem.* **284,** 6507–6519.

Vagin, V. V., Sigova, A., Li, C., Seitz, H., Gvozdev, V., and Zamore, P. D. (2006). A distinct small RNA pathway silences selfish genetic elements in the germline. *Science (New York, N.Y.)* **313,** 320–324.

van Wolfswinkel, J. C., Claycomb, J. M., Batista, P. J., Mello, C. C., Berezikov, E., and Ketting, R. F. (2009). CDE-1 affects chromosome segregation through uridylation of CSR-1-bound siRNAs. *Cell* **139,** 135–148.

Vincent, S. D., Dunn, N. R., Sciammas, R., Shapiro-Shalef, M., Davis, M. M., Calame, K., Bikoff, E. K., and Robertson, E. J. (2005). The zinc finger transcriptional repressor Blimp1/Prdm1 is dispensable for early axis formation but is required for specification of primordial germ cells in the mouse. *Development* **132,** 1315–1325.

Viswanathan, S. R., Daley, G. Q., and Gregory, R. I. (2008). Selective blockade of microRNA processing by Lin28. *Science (New York, N.Y.)* **320,** 97–100.

Voinnet, O. (2009). Origin, biogenesis, and activity of plant microRNAs. *Cell* **136,** 669–687.

Wang, G., and Reinke, V. (2008). A C. elegans Piwi, PRG-1, regulates 21U-RNAs during spermatogenesis. *Curr. Biol.* **18,** 861–867.

Wang, Y., Medvid, R., Melton, C., Jaenisch, R., and Blelloch, R. (2007). DGCR8 is essential for microRNA biogenesis and silencing of embryonic stem cell self-renewal. *Nat. Genet.* **39,** 380–385.

Watanabe, T., Takeda, A., Tsukiyama, T., Mise, K., Okuno, T., Sasaki, H., Minami, N., and Imai, H. (2006). Identification and characterization of two novel classes of small RNAs in the mouse germline: Retrotransposon-derived siRNAs in oocytes and germline small RNAs in testes. *Genes Dev.* **20,** 1732–1743.

Watanabe, T., Totoki, Y., Toyoda, A., Kaneda, M., Kuramochi-Miyagawa, S., Obata, Y., Chiba, H., Kohara, Y., Kono, T., Nakano, T., Surani, M. A., Sakaki, Y., *et al.* (2008). Endogenous siRNAs from naturally formed dsRNAs regulate transcripts in mouse oocytes. *Nature* **453,** 539–543.

Weidinger, G., Stebler, J., Slanchev, K., Dumstrei, K., Wise, C., Lovell-Badge, R., Thisse, C., Thisse, B., and Raz, E. (2003). *dead end,* a novel vertebrate germ plasm component, is required for zebrafish primordial germ cell migration and survival. *Curr. Biol.* **13,** 1429–1434.

West, J. A., Viswanathan, S. R., Yabuuchi, A., Cunniff, K., Takeuchi, A., Park, I. H., Sero, J. E., Zhu, H., Perez-Atayde, A., Frazier, A. L., Surani, M. A., and Daley, G. Q. (2009). A role for Lin28 in primordial germ-cell development and germ-cell malignancy. *Nature* **460,** 909–913.

Wightman, B., Ha, I., and Ruvkun, G. (1993). Posttranscriptional regulation of the heterochronic gene lin-14 by lin-4 mediates temporal pattern formation in C. elegans. *Cell* **75,** 855–862.

Winter, J., Jung, S., Keller, S., Gregory, R. I., and Diederichs, S. (2009). Many roads to maturity: MicroRNA biogenesis pathways and their regulation. *Nat. Cell Biol.* **11,** 228–234.

Wu, E., Thivierge, C., Flamand, M., Mathonnet, G., Vashisht, A. A., Wohlschlegel, J., Fabian, M. R., Sonenberg, N., and Duchaine, T. F. (2010). Pervasive and cooperative deadenylation of 3'UTRs by embryonic microRNA families. *Mol. Cell* **40,** 558–570.

Xie, T., and Spradling, A. C. (1998). Decapentaplegic is essential for the maintenance and division of germline stem cells in the Drosophila ovary. *Cell* **94,** 251–260.

Xie, T., and Spradling, A. C. (2000). A niche maintaining germ line stem cells in the Drosophila ovary. *Science (New York, N.Y.)* **290,** 328–330.

Xie, T., Song, X., Jin, Z., Pan, L., Weng, C., Chen, S., and Zhang, N. (2008). Interactions between stem cells and their niche in the Drosophila ovary. *Cold Spring Harb. Symp. Quant. Biol.* **73,** 39–47.

Yamaji, M., Seki, Y., Kurimoto, K., Yabuta, Y., Yuasa, M., Shigeta, M., Yamanaka, K., Ohinata, Y., and Saitou, M. (2008). Critical function of Prdm14 for the establishment of the germ cell lineage in mice. *Nat. Genet.* **40,** 1016–1022.

Yang, N., and Kazazian, H. H., Jr. (2006). L1 retrotransposition is suppressed by endogenously encoded small interfering RNAs in human cultured cells. *Nat. Struct. Mol. Biol.* **13,** 763–771.

Yang, L., Chen, D., Duan, R., Xia, L., Wang, J., Qurashi, A., and Jin, P. (2007a). Argonaute 1 regulates the fate of germline stem cells in Drosophila. *Development* **134,** 4265–4272.

Yang, L., Duan, R., Chen, D., Wang, J., and Jin, P. (2007b). Fragile X mental retardation protein modulates the fate of germline stem cells in Drosophila. *Hum. Mol. Genet.* **16,** 1814–1820.

Yang, Y., Xu, S., Xia, L., Wang, J., Wen, S., Jin, P., and Chen, D. (2009). The bantam microRNA is associated with drosophila fragile X mental retardation protein and regulates the fate of germline stem cells. *PLoS Genet.* **5,** e1000444.

Yi, R., Qin, Y., Macara, I. G., and Cullen, B. R. (2003). Exportin-5 mediates the nuclear export of pre-microRNAs and short hairpin RNAs. *Genes Dev.* **17,** 3011–3016.

Yigit, E., Batista, P. J., Bei, Y., Pang, K. M., Chen, C. C., Tolia, N. H., Joshua-Tor, L., Mitani, S., Simard, M. J., and Mello, C. C. (2006). Analysis of the C. elegans Argonaute family reveals that distinct Argonautes act sequentially during RNAi. *Cell* **127,** 747–757.

Ying, Y., and Zhao, G. Q. (2001). Cooperation of endoderm-derived BMP2 and extraembryonic ectoderm-derived BMP4 in primordial germ cell generation in the mouse. *Dev. Biol.* **232,** 484–492.

Ying, Y., Qi, X., and Zhao, G. Q. (2001). Induction of primordial germ cells from murine epiblasts by synergistic action of BMP4 and BMP8B signaling pathways. *Proc. Natl. Acad. Sci. USA* **98,** 7858–7862.

Yoon, C., Kawakami, K., and Hopkins, N. (1997). Zebrafish vasa homologue RNA is localized to the cleavage planes of 2- and 4-cell-stage embryos and is expressed in the primordial germ cells. *Development* **124,** 3157–3165.

Youngren, K. K., Coveney, D., Peng, X., Bhattacharya, C., Schmidt, L. S., Nickerson, M. L., Lamb, B. T., Deng, J. M., Behringer, R. R., Capel, B., Rubin, E. M., Nadeau, J. H., *et al.* (2005). The Ter mutation in the dead end gene causes germ cell loss and testicular germ cell tumours. *Nature* **435,** 360–364.

Yu, Z., Raabe, T., and Hecht, N. B. (2005). MicroRNA Mirn122a reduces expression of the posttranscriptionally regulated germ cell transition protein 2 (Tnp2) messenger RNA (mRNA) by mRNA cleavage. *Biol. Reprod.* **73,** 427–433.

Yu, J. Y., Reynolds, S. H., Hatfield, S. D., Shcherbata, H. R., Fischer, K. A., Ward, E. J., Long, D., Ding, Y., and Ruohola-Baker, H. (2009). Dicer-1-dependent Dacapo suppression acts downstream of Insulin receptor in regulating cell division of Drosophila germline stem cells. *Development* **136,** 1497–1507.

Zhou, Y., and King, M. (1996). Localization of Xcat-2 RNA, a putative germ plasm component, to the mitochondrial cloud in Xenopus stage I oocytes. *Development* **122,** 2947–2953.

Zhou, R., Czech, B., Brennecke, J., Sachidanandam, R., Wohlschlegel, J. A., Perrimon, N., and Hannon, G. J. (2009). Processing of Drosophila endo-siRNAs depends on a specific Loquacious isoform. *RNA* **15,** 1886–1895.

Zilberman, D., Cao, X., and Jacobsen, S. E. (2003). ARGONAUTE4 control of locus-specific siRNA accumulation and DNA and histone methylation. *Science (New York, N.Y.)* **299,** 716–719.

CHAPTER FIVE

Diverse Functions of MicroRNAs in Nervous System Development

Luisa Cochella *and* Oliver Hobert

Contents

Abstract

MicroRNAs (miRNAs) are integral parts of the gene regulatory networks that control most developmental processes. Through their regulatory action, miRNAs introduce an additional layer of genetic complexity that can translate into increased cellular diversity, something that is extremely relevant to nervous system structure. In addition, miRNAs sharpen the spatial and temporal boundaries between different cellular states during development. Here, we illustrate these roles with a number of specific miRNAs that act during distinct steps of neural development. We further discuss specific aspects of miRNA function that make these regulators particularly suited to provide the robustness and

Department of Biochemistry and Molecular Biophysics, Howard Hughes Medical Institute, Columbia University Medical Center, New York, USA

Current Topics in Developmental Biology, Volume 99
ISSN 0070-2153, DOI: 10.1016/B978-0-12-387038-4.00005-7

complexity that are essential for the dynamic nature of both the development and activity of the nervous system.

1. Introduction

The large diversity of neuronal and nonneuronal cells that form a nervous system arises from the combinatorial function of a large number of gene regulatory factors. Nowadays, it is evident that these regulators include not only transcription factors but also small regulatory RNAs, with micro-RNAs (miRNAs) being the most intensely studied ones. While our understanding of transcription factor involvement in neural development far exceeds that of miRNAs, here we review some of the more recent examples of the roles of miRNAs in nervous system development.

miRNAs are an abundant class of short (21–23-nt long) regulatory RNAs with versatile functions in biology (for review, see Bartel, 2009; Chekulaeva and Filipowicz, 2009). They are derived from longer primary transcripts (pri-miRNA) that form stem-loop structures, through the successive cleavages of two RNAse III-type enzymes. The first one, Drosha, excises the hairpin structure (pre-miRNA) from the longer pri-miRNA in the nucleus, while cleavage from the second one, Dicer, in the cytoplasm results in the production of the mature 21–23-nt long miRNA. Mature miRNAs associate with a protein complex with an Argonaute family member at its core, and it is in the context of this complex—also known as RNA-induced silencing complex (RISC)—that miRNAs carry out their regulatory roles. miRNAs are, in the vast majority of cases examined, negative regulators of their targets. In the context of the RISC, miRNAs can guide binding to partially complementary sequences in their target mRNAs and cause posttranscriptional repression of gene expression. The precise mechanism by which this repression is exerted is still under debate. Both translational repression—most likely at the level of translation initiation—and mRNA destabilization—likely triggered by mRNA deadenylation—have been shown to affect target protein levels. However, their relative contributions and how they are mechanistically related to one another are still poorly understood (for a recent review, see Djuranovic *et al.*, 2011).

Regardless of our incomplete understanding of how miRNAs regulate their targets, their cellular and physiological effects are evident. miRNAs affect every step required for the proper development of a nervous system, from patterning of the nervous system to neuronal differentiation and plasticity. These functions have been described in nervous systems as simple as that of *Caenorhabditis elegans*, where the first miRNA with a role in neuronal differentiation was discovered (Johnston and Hobert, 2003), as well as in the more complex vertebrate systems. Studies in this wide range of model organisms have been extremely fruitful as they have uncovered common themes in the roles of miRNAs in the nervous system and will likely help us understand their functions from an evolutionary perspective.

The first miRNAs shown to have a role in development of the correct architecture of the vertebrate nervous system were those of the mir-430 family in zebrafish (Giraldez *et al.*, 2005). Fish embryos lacking Dicer showed severe brain morphogenesis defects which could be rescued by injection of preprocessed mir-430 miRNAs. This pioneering study prompted numerous others that have identified other miRNAs and their targets that act at these early stages of neural development.

While a lot of the earlier evidence for involvement of miRNAs in all steps of vertebrate neural development was obtained from Dicer knockout experiments, results from Dicer knockouts should be interpreted carefully as Dicer is known to be involved in the processing of other essential noncoding RNAs (rRNAs, snoRNAs, tRNAs) (Cole *et al.*, 2009; Ender *et al.*, 2008; Liang and Crooke, 2011; Saraiya and Wang, 2008). Thus, we have chosen to focus here on some of the cases of nervous system development where specific miRNAs have been implicated.

2. Diverse Roles of miRNAs in Nervous System Development and Function

2.1. miRNAs act at all steps of neuronal development

2.1.1. Patterning

Patterning of the developing vertebrate neuroectoderm that gives rise to all different mature structures of the nervous system is a complex process that relies on multiple secreted signals emanating from specific regions. Combinations of these signals result in regional domains with particular patterns of transcription factor expression that will give rise to different neuronal and glial pools. A number of miRNAs have been implicated in refining the boundaries of these domains, perhaps fulfilling a role that is best suited for this kind of repressor. A few of these are presented below.

Patterning of the midbrain and anterior hindbrain has long been known to be controlled by an organizing center located at their boundary (the midbrain–hindbrain boundary or MHB), primarily through Wnt and Fgf signaling (Wurst and Bally-Cuif, 2001). Bally-Cuif and colleagues have shown that in zebrafish, mir-9 is involved in maintaining the MHB limits during development. mir-9 is expressed extensively within the neural tube except in the MHB (Leucht *et al.*, 2008). Ectopic expression of mir-9 causes a loss of this organizing center due to ectopic repression of genes required for MHB function. In contrast, loss of mir-9 causes an expansion of the

MHB along the antero-posterior axis. As mentioned, Fgf signaling from the MHB to the surrounding area is critical for proper patterning around this region, and it is in fact the target of mir-9 function. mir-9 seems to target directly at least three components of the Fgf pathway, *fgfr-1, fgfr-8,* and *canopy* (Leucht *et al.*, 2008). Interestingly, the patterning defect induced by ectopic mir-9 can be fully rescued by protecting *fgfr-1* from mir-9 targeting, arguing that most of the effect of mir-9 on the MHB is through this pathway. Importantly, Fgf signaling is part of a feedback loop that ensures MHB stability and mir-9 may be a part of such loop as well. Integration of miRNAs into known gene regulatory networks is a key aspect of their biological function and evolution and will be discussed in Section 4. In addition to its effect on patterning, mir-9 also affects neurogenesis at the boundary areas. This is a role for mir-9 that seems to be conserved in other areas of the developing nervous system as well as across organisms and will be discussed in Section 2.1.2.

Another well-characterized case of neural patterning during development is the spatial organization of different neuronal and glial types in the spinal cord. Two important mechanisms are used to define these diverse progenitor domains. For dorso-ventral patterning, sonic hedgehog (SHH) signaling from the ventral side of the neural tube and retinoic acid from the mesoderm define five domains of ventral neural progenitors with characteristic gene expression (for review, see Jessell, 2000). For antero-posterior regionalization of the spinal cord, distinct domains of *Hox* gene expression have been shown to be crucial (for review, see Dasen and Jessell, 2009). Two cases have been described where miRNAs are proposed to ensure the correct domains of expression in both mentioned aspects of spinal cord patterning.

In the first case, mir-17-3p restricts the expression of transcription factor OLIG2 to its known domain of expression (Chen *et al.*, 2011). While *Olig2* is normally expressed in the ventral progenitor pool that gives rise to motor neurons (pMN), it was also found to be transiently expressed in one of the neighboring ventral pools that produce V2 interneurons (p2). However, for p2 identity establishment, *Olig2* should be turned off and p2 cells should express IRX3, another transcription factor that forms a cross-repressive loop with OLIG2. mir-17-3p is expressed in the p2 domain (as well as in other Irx3-positive pools) but not in the pMN domain. mir-17-3p can directly repress *Olig2*, and thus, loss of this miRNA results in a significant loss of V2 interneurons due to ectopic expression of OLIG2 in the p2 pool.

In the second case, mir-196 has been implicated in the antero-posterior patterning of the spinal cord. Although further evidence should be sought, it seems likely that mir-196 delimits the domain of expression of *Hoxb8* in the caudal neural tube to ensure *Hoxb8* is absent in the prospective hindlimb field (Asli and Kessel, 2010). Ectopic HOXB8 expression in this lumbar area causes a decrease in the number of motor neurons. However, the precise contribution of mir-196 to this restriction is difficult to assess. A consistent role for mir-196 in restricting *Hoxb8* expression had previously been reported during limb formation, where *Hoxb8* is expressed in the forelimb buds but not in the hindlimb buds (Hornstein *et al.*, 2005). Loss-of-function experiments knocking down mir-196 with an antagomir in developing chick embryos showed derepression of *Hoxb8* at more anterior somites consistent with a transformation of the last cervical vertebrae to a thoracic identity (McGlinn *et al.*, 2009). Interestingly, mir-196 has also been implicated in proper tail and spinal cord regeneration following amputation in axolotls (Sehm *et al.*, 2009). However, in this case, *Hox* gene expression does not seem to be the major target of mir-196.

Finally, the role of miRNAs in other patterning events, such as early blastoderm organization in *Drosophila* (Leaman *et al.*, 2005), neural tube closure (Maller Schulman *et al.*, 2008), and morphogenesis in the cortex and hippocampus (Davis *et al.*, 2008), have been proposed based on individual miRNA knockdown, target-gene analysis, and conditional loss of Dicer, respectively. These will not be discussed here, but further research in these areas should prove interesting.

2.1.2. Neurogenesis

Neurogenesis is a tightly controlled process by which neuroepithelial progenitors or other types of neural stem cells (NSCs), such as those involved in adult neurogenesis, become progressively committed until becoming postmitotic neurons (for reviews, see Conti and Cattaneo, 2010; Li and Jin, 2010). Along their path to become neurons, NSCs have the choice of proliferating to expand the progenitor pool or becoming more committed neural precursor cells (NPCs). In addition, NPCs can give rise to either glial cells (oligodendrocytes (OLs) and astrocytes) or neurons, and so during neurogenesis, alternative fates must be repressed. Both intrinsic and extrinsic factors are involved in these decisions. Among the intrinsic factors, the role of miRNAs is becoming increasingly evident, with two miRNAs, mir-9 and mir-124, being extensively implicated in neurogenesis.

In NSCs from adult mouse forebrains, gain of function of mir-9 causes a dose-dependent decrease in proliferation, and when these cells are induced to differentiate, mir-9 can cause a marked increase in the percentage of differentiated neurons and glia (Zhao *et al.*, 2009). In addition, introduction of mir-9 in NSCs in developing mouse embryos (at E13.5) reduced their proliferation and induced them to migrate to the cortical plate, as differentiated neurons do. These cells also lost expression of a progenitor marker and instead expressed a neuronal marker. A major target of mir-9 for these effects seems to be the transcription factor TLX/Nr2e1, known to be required for stem-cell renewal. Interestingly, TLX/Nr2e1 is a transcriptional repressor of mir-9-1 and mir-9-2 (there are three mir-9 encoding genes in the mouse genome), thereby establishing a cross-repressive feedback loop.

In another report using hESC-derived hNPCs, loss-of-function analysis of mir-9 revealed a somewhat different role. First, mir-9 was necessary for the proliferation of the hNPCs (Delaloy *et al.*, 2010). In addition, in the hESC-derived hNPCs, mir-9 inhibits neuronal migration by repressing Stathmin, a protein that increases microtubule instability. While the opposing effect of mir-9 in these two experimental setups could well be due to their inherent differences (perhaps the main difference is the origin of the studied progenitors), it is also possible that mir-9 gain- (in Zhao *et al.*, 2009) and loss-of-function (in Delaloy *et al.*, 2010) studies have uncovered different functions for this versatile miRNA, and that its effect on proliferation and differentiation results from the balance of its different targets at different developmental stages or in different neuronal or neuronal progenitor types. Further experiments with emphasis on comparing the function of miRNAs along different precursor or mature states of different types of neurons should clarify these discrepancies.

In vivo studies in vertebrate systems have been more consistent with a role for mir-9 as a promoter of neurogenesis, mostly by repressing inhibitors of neuronal differentiation. In mouse and zebrafish, mir-9 is found in proliferating progenitor cells as well as in mature neurons (Lagos-Quintana *et al.*, 2002; Leucht *et al.*, 2008; Wienholds *et al.*, 2005). As mentioned above, mir-9 is expressed in the NPCs just outside the zebrafish MHB where it represses the antineurogenic Hes bHLH transcription factors *her5* and *her9*, thereby promoting differentiation. This effect of mir-9 also explains why its ectopic expression causes loss of the MHB: not only does it repress MHB required genes, but it also causes premature differentiation and thus depletion of the progenitor pool in the MHB. In *Xenopus*, mir-9 has also been implicated in promoting neurogenesis through the inhibition of *hairy1*, another member of the Hes bHLH transcriptional repressors (Bonev *et al.*, 2011).

In the developing mouse brain, mir-9 function has been most extensively studied in the telencephalon. Using gain- and loss-of-function experiments by injection of mir-9 or a mir-9 antisense oligonucleotide, respectively, and more recently analysis of a mir-9-2/mir-9-3 double mutant mouse, Aizawa and colleagues have shown that mir-9 promotes differentiation of Cajal-Retzius cells in the medial pallium as well as other early-born neurons (E12.5–E13.5) (Shibata *et al.*, 2008, 2011). This early effect seems to be mediated by repression of *Foxg1* by mir-9. FOXG1 is a forkhead transcription factor with a known role in promoting proliferation of progenitor cells, and in the medial pallium, the expression patterns of mir-9 and *Foxg1* form reciprocal gradients. Interestingly, no effect on TLX/Nr2e1 was observed in these studies. In the absence of mir-9, increased proliferation of early-differentiating neural progenitors is observed in the medial pallium and also in the subpallium (in this case, mir-9 seems to target both *Foxg1* and *Gsh2*). Interestingly, in the mir-9-2/3 mutant at later stages (E16.5–E18.5), there is a decrease of proliferation of progenitors in the pallium although it is unclear whether this is due to direct action of mir-9 on these progenitors or a consequence of its earlier effect. This decrease in later progenitor proliferation could be related to the decrease in proliferation of hNPCs described above. It is interesting to note that in the mouse and zebrafish, olfactory epithelia miRNAs of the mir-200 family also seem to regulate neurogenesis at least in part through their action on *Foxg1* (Choi *et al.*, 2008).

In addition to the targets mentioned so far, mir-9 and mir-9* (a miRNA partially complementary to mir-9 derived from the opposite strand of the same precursor hairpin) have been shown to repress the well-known, antineuronal transcriptional repressor, REST/NRSF (RE1 silencing transcription factor/neuron restrictive silencer factor) and its cofactor, CoREST (Conaco *et al.*, 2006; Packer *et al.*, 2008). While these studies were conducted in cell lines and their contribution *in vivo* should be further explored, this result is consistent with the proneural role of mir-9 and its coordinated function with mir-124 to promote neurogenesis as discussed below.

mir-9 has also been shown to have a role in a well-characterized model for neurogenesis, *Drosophila* sensory organ development (Li *et al.*, 2006). However, in this case, despite *mir-9* being highly conserved, its role seems to be different than that in vertebrate systems. First, *mir-9* expression in the nervous systems is not as prominent as in other organisms. *mir-9* is present in embryonic epithelial cells and in the larval wing disc, but not in the sensory organ precursors (SOPs) each of which will give rise to a neuron. In fact, ectopic *mir-9* expression causes a reduction in the number of SOPs mediated by repression of *Senseless*, a known proneural gene. As expected, loss of *mir-9* results in additional SOPs, suggesting that *mir-9* in *Drosophila* suppresses neuronal precursor specification in "nonneuronal" tissues. Another miRNA in *Drosophila*, *mir-7*, complements the function of *mir-9*. While *mir-9* is expressed in the non-SOP cells, *mir-7* is present in the SOP where it promotes expression of proneural genes such as *Atonal* and *Senseless* through its negative effect on *E(spl)* (Li *et al.*, 2009). Interestingly, *mir-7* is also part of a complex gene regulatory network—involving different players and a different target—that controls photoreceptor determination (Li and Carthew, 2005).

In mice, mir-124 is expressed somewhat later than mir-9 during development, but it is also expressed in neuronal progenitors and mature neurons as it continues to be expressed into adulthood and is the most abundant miRNA in the adult mammalian brain (Deo *et al.*, 2006; Lagos-Quintana *et al.*, 2002). mir-124 also seems to both repress the expression of neural progenitor genes as well as induce the expression of neuronal genes to promote neuronal differentiation (Coolen and Bally-Cuif, 2009; Vo *et al.*, 2010). These roles of mir-124 are carried out through a number of targets.

In HeLa cells, transfection with mir-124 can cause a significant change in the transcriptome profile, making it more similar to that of the brain (Lim *et al.*, 2005). Changes in transcriptome specificity also arise from the effect of mir-124 on alternative splicing (Makeyev *et al.*, 2007). mir-124 targets a repressor of alternative splicing called PTBP1 in the nervous system; this, in turn, allows for alternative splicing of another splicing regulator, PTBP2, resulting in correctly spliced and functional PTBP2. Increased levels of PTBP2 in the nervous system correlate with neural specific alternative splicing patterns and are necessary for proper neuronal differentiation.

Other *in vitro* experiments using ESCs or other cell lines have shown that mir-124 induces neuronal like differentiation, although in most cases, additional proneurogenic factors were required to uncover the effect of mir-124 (reviewed in Maiorano and Mallamaci, 2010). More recently, the neurogenic effect of mir-124 has been explored *in vivo*. In mouse models, mir-124 has been shown to promote neurogenesis in the embryonic cerebral cortex (Maiorano and Mallamaci, 2009), as well as to control the timing of progression down the lineage during adult neurogenesis in the stem-cell niche of the subventricular zone (Cheng *et al.*, 2009). In the embryonic cortex, a dramatic upregulation in mir-124 expression in the precursors undergoing direct neurogenesis as well as gain-of-function experiments support the role of mir-124 as an inducer of neurogenesis (Maiorano and Mallamaci, 2009). Although no specific target of mir-124 was shown to be responsible for this effect, a more recent study also analyzing progenitors in the developing cortex identified the Ephrin *EfnB1* as a relevant target of mir-124 (Arvanitis *et al.*, 2010). Interestingly, mir-124 and *EfnB1* form a cross-repressive loop resulting in two states, one with high *EfnB1* and low mir-124 that correlates with maintenance of the progenitor pool, and one with high mir-124 and low *EfnB1* that seems to promote neuronal differentiation.

In the chick spinal cord, the role of mir-124 in neurogenesis has been somewhat controversial. In a first report, no effects of mir-124 overexpression or inhibition were observed (Cao *et al.*, 2007). However, a different study showed subtle yet noticeable increase in neuronal differentiation at the expense of proliferation of neural progenitors upon mir-124 overexpression (Visvanathan *et al.*, 2007), again suggesting that mir-124 plays a role in neurogenesis but may not be strictly sufficient. In the chick spinal cord, mir-124 is able to repress SCP1 (small C-terminal domain phosphatase 1) an antineural factor that is recruited by REST to its target genes. However, the effect of mir-124 in spinal cord neurogenesis is mediated only in part by repression of SCP1, as a mir-124-insensitive SCP1 transcript could only partially counteract the neurogenic function of mir-124. Subsequent work has shown that indeed, mir-124 acts on a number of additional targets to favor neurogenesis. Moreover, it is possible that mir-124 acts in conjunction with mir-9/mir-9* to robustly reduce REST/SCP1-mediated repression of neural genes (Conaco *et al.*, 2006; Packer *et al.*, 2008).

This is not the only target that is shared by these miRNAs. In the mouse developing neural tube, mir-124 and mir-9* have also been shown to cooperate in promoting neurogenesis by repressing a progenitor-specific subunit of a Swi/Snf-like chromatin-remodeling complex termed BAF. Chromatin remodeling by BAF is important for the different steps of neural development and its subunit composition changes accordingly. By targeting a progenitor-specific subunit of BAF, BAF53a, mir-124, and mir-9* promote this subunit switch and among other effects, reduce the proliferation of neural progenitor cells (Yoo *et al.*, 2009).

The overall picture that emerges from these studies is that in vertebrates, mir-9 and mir-124 are integral to the successful transition between progenitors and differentiated neurons. They seem to orchestrate this transition through the regulation of multiple targets, all of which either promote progenitor proliferation or directly inhibit neuronal differentiation, and most of which are transcriptional regulators.

2.1.3. Neuron-class differentiation

In the previous section, we discussed the role of miRNAs in the first, more general aspects of neurogenesis of NSCs that become progressively committed to giving rise to neurons. However, miRNAs are also involved in the subsequent steps of defining what type of neurons those committed precursors will become and in controlling the progression of the differentiation program.

It is widely accepted that distinct cell types are the products of combinatorial "codes" of gene regulatory factors; these include, but are not limited to, transcription factors and miRNAs. This aspect of neural development has been widely studied in the nematode *C. elegans*, where specification of pan-neuronal features can be genetically separated from neuron-class specification and where several genes affecting the latter have been identified (Hobert, 2011). In *C. elegans,* mir-124 does not seem to be involved in neurogenesis as broadly as in other systems, but rather it is expressed in a subset of sensory neurons and it may play a role in shaping their cell-specific transcriptome (Clark *et al.*, 2010). One of the best-studied miRNAs involved in neuron-class specification, a miRNA called *lsy-6,* also comes from *C. elegans*. *lsy-6* is responsible for the subclass diversification of the ASE neurons, a pair of sensory neurons on either side of the head of the worm (Johnston and Hobert, 2003). Despite being bilaterally symmetric with respect to a number of criteria (position in the head, connectivity, morphology of its projections, and shared gene expression), each ASE neuron senses different environmental cues and responds to them in different ways. This is largely due to the presence of *lsy-6* only in the left ASE neuron (ASEL), where it represses the Nkx6-type transcription factor *cog-1*. *lsy-6*, *cog-1*, and another zinc-finger transcription factor, *die-1*, form a bistable feedback loop that can exist in either one of two states: a high *lsy-6* and *die-1* state that results in the expression of an ASEL-specific gene battery and a high *cog-1* state that causes the ASE neuron to adopt an ASER fate (Johnston *et al.*, 2005).

Studies in *Drosophila* have also provided an interesting case: *mir-279* was identified in a screen because its loss resulted in ectopic CO_2-sensing neurons in the maxillary palps (MPs) in addition to the CO_2-sensing neurons normally present in the antenna of the fly (Cayirlioglu *et al.*, 2008). These ectopic sensory neurons in the MP were hybrids, with properties of both CO_2-sensing neurons and one of two specific subclasses of olfactory neurons. One of the targets of *mir-279* involved in this process is a transcription factor called Nerfin-1; however, ectopic expression of Nerfin-1 by itself is not sufficient to generate CO_2-sensing neurons, suggesting the involvement of additional targets. Interestingly, discovery of this role for *mir-279* may have uncovered an evolutionary path in which introduction of a miRNA was involved. In the fly, CO_2 causes an aversive response; however, in blood-feeding insects, CO_2 is attractive and it is sensed by neurons in the MP. It is interesting to speculate that the *mir-279*-mutant phenotype may have uncovered an intermediate hybrid state on which selective pressure could have acted to generate the diversity observed nowadays.

In the chick spinal cord, mir-9 has been implicated in the specification of different motor neuron subtypes (Otaegi *et al.*, 2011). mir-9 is transiently expressed in motor neurons of the lateral motor column (LMC), and its overexpression causes a change in identity of these neurons to that of the median motor column. This effect seems to be caused by the mir-9-mediated repression of *FoxP1*. Interestingly, mir-9 and *FoxP1* are coexpressed in LMC motor neurons where mir-9 has been proposed to tune FoxP1 levels. This is consistent with different *FoxP1* dose requirements to generate different motor neuron classes (Dasen *et al.*, 2008; Rousso *et al.*, 2008).

Another miRNA involved in neuronal differentiation is mir-133. In human and mouse midbrains, mir-133 seems to negatively regulate dopaminergic neuron differentiation through the repression of *Pitx3*, a *bicoid*-related transcription factor well known to promote dopaminergic neuron differentiation and survival. In turn, Pitx3 can transcriptionally activate mir-133 (Kim *et al.*, 2007). Why would mir-133 be expressed and repress a prodopaminergic factor in cells that have to become dopaminergic neurons? While the answer is still unclear, one clue may come from an analysis of *Pitx3* levels in different dopaminergic neuron populations in the brain. This study showed that *Pitx3* level is about six times higher in dopaminergic neurons from the ventral tegmental area than in neurons from the substantia nigra (Korotkova *et al.*, 2005). It will be interesting to know whether mir-133 is equally present in these two areas and is perhaps involved in creating this difference, and how this difference may impinge on the function and susceptibility to degenerate of these two classes of dopaminergic neurons. Alternatively, mir-133 could be part of dynamic regulatory interactions required for the progression of differentiation from progenitors to mature dopaminergic neurons.

Such a role for a miRNA has been proposed for mir-96 in controlling the progression of differentiation in cochlear hair cells in mice. In this case, loss of mir-96 function (through a point mutation in its seed region) puts a brake in the differentiation program that leaves both inner and outer hair cells with an immature morphology, electrophysiology, and innervation pattern (Kuhn *et al.*, 2011). Therefore, mir-96 could be in charge of down-regulating genes required for initial specification but that could later impair the acquisition of the fully differentiated fate. Identifying the targets of mir-96 will be essential to pinning down its function.

2.1.4. Maturation

During neuronal maturation, the appropriate connections between neurons and their targets are established. In addition, in vertebrates, those neurons that fail to integrate into the corresponding circuits die through apoptosis, while those that form appropriate connections must inhibit apoptosis since they have to survive throughout the lifetime of the organism. miRNAs play roles in these processes as well.

A number of groups have reported roles for miRNAs in regulating dendritic spine development and plasticity. mir-132 is induced by synaptic activity and has been shown to increase dendritic length, branching, and spine density in *in vitro* models using primary cortical and hippocampal neurons from embryonic or newborn rats (Vo *et al.*, 2005). In addition, the expression of mir-132 in the developing rat hippocampus correlates with a period of active synaptogenesis (Impey *et al.*, 2010). The effect of mir-132 to promote spine formation seems to be mediated by p250GAP, a Rho-family GTPase that regulates actin dynamics (Vo *et al.*, 2005; Wayman *et al.*, 2008). More recently, the effect of mir-132 was confirmed *in vivo* in an adult neurogenesis model: knockout of mir-132 in newborn hippocampal neurons in adult mice decreased dendritic length and arborization (Magill *et al.*, 2010). mir-132 has also been implicated in regulating synaptic maturation during the neonatal period through its action on MeCP2 (Klein *et al.*, 2007). The interaction with MeCP2 results in a complex regulatory network that will be further discussed in Section 3.2.

The structure and function of synapses can also be regulated by mir-125b. Overexpression of this miRNA in cultured hippocampal neurons from rat embryos induced the formation of long and narrow spines with low mEPSC amplitude (Edbauer *et al.*, 2010). This morphology correlates with the timing of mir-125b expression; endogenous levels are high in younger neurons that typically have filopodia-like projections. A target or targets of mir-125b regulating spine morphology are still unknown.

mir-134 is, like mir-124, a nervous system-specific miRNA. However, unlike mir-124 that is expressed earlier, mir-134 expression in the hippocampus peaks around postnatal day 13 correlating with the time of synaptic maturation (Schratt *et al.*, 2006). mir-134 localizes to dendrites, where it regulates spine morphology. First, in hippocampal neurons in culture, mir-134 overexpression resulted in decreased dendritic spine volume, without affecting the number of spines, through the repression of Lim-domain-containing protein kinase 1 (*Limk1*) (Schratt *et al.*, 2006). LIMK1 has been shown to regulate dendritic spine morphology by affecting dynamics of the actin cytoskeleton. However, upon neuronal activation—which is known to stimulate dendritic growth—the inhibition on *Limk1* is relieved. Surprisingly, in younger neurons, neuronal activity induces mir-134 expression and through downregulation of Pumilio 2, mir-134 promotes dendritic outgrowth (Fiore *et al.*, 2009; Khudayberdiev *et al.*, 2009). These seemingly opposite roles of mir-134 can be reconciled in two different ways. First, it is possible that mir-134 plays different roles at different time points, mediated through distinct targets. Alternatively, both effects could be part of homeostasis or plasticity mechanisms to coordinate global and local responses of a neuron to increased or decreased neuronal activity. For example, an increase in dendritic arborization has been shown to be accompanied by a decrease in the strength of individual synapses to maintain the overall excitability within a certain range (Peng *et al.*, 2009). In addition, localized effects of mir-134 in the synapse allow for differential downscaling or strengthening of individual spines. Whether all three, mir-132, mir-134, and mir-125b, act in the same cell to provide a coordinated, balanced response or whether they act at different times and/or different cell types need to be further explored to obtain a more comprehensive picture of their function during dendritic formation and plasticity.

In addition, three different miRNAs have been shown to affect synaptic strength and plasticity in three different organisms. In *Aplysia californica*, mir-124 is, as in *C. elegans*, restricted to sensory neurons. The sensory-motor synapse of *Aplysia* has been extensively studied, and it can be modulated by serotonin. Serotonin is known to promote long-term facilitation at this synapse through the activation of CREB. While mir-124 seems to target CREB in the sensory neuron, interestingly, serotonin was found to inhibit mir-124 biogenesis and thus relieve the repression this miRNA imposes on CREB (Rajasethupathy *et al.*, 2009). This arrangement forms a coherent feedforward loop (FFL), postulated to increase the specificity of the response.

The other two cases involve miRNAs regulating the strength of the neuromuscular junction (NMJ). The first report was from Kim and colleagues who showed that in *C. elegans*, *mir-1*, a conserved muscle-specific miRNA, affects both pre- and postsynaptic function to couple changes in the activity of the muscle to that of the motor neuron (Simon *et al.*, 2008). In the muscle, *mir-1* decreases sensitivity to acetylcholine (ACh) by targeting two subunits of the nicotinic ACh receptor, while in the motor neuron, it exerts a non-cell-autonomous effect mediated through a retrograde signal from the muscle that results in decreased ACh release. As opposed to promoting a transition, as proposed for mir-124 in the *Aplysia* sensory-motor synapse, *mir-1* is likely to be mediating a homeostatic response to refine the activity of the nematode NMJ. At the *Drosophila* larva NMJ, the *mir-310-313* cluster also plays a role in regulating synaptic strength, likely allowing for the establishment of synaptic homeostasis as well. The *mir-310* cluster negatively regulates synaptic strength at the NMJ during larval development when synaptic growth is fast, likely to promote homeostatic compensation (Tsurudome *et al.*, 2010). In contrast to *mir-1* in the worm, the *mir-310* cluster exerts its function in the motor neuron, not in the muscle. In the neuron, it inhibits expression of a kinesin family member, *Khc-73*, which plays a role in recruiting specific proteins to the presynaptic active zones.

As mentioned, the regulation of apoptosis is essential during the development and maturation of the nervous system. It is well known that during their maturation, neurons that establish successful connections become decreasingly sensitive to apoptosis induced by a number of insults. miRNA profiling of developing sympathetic neurons revealed that mir-29 levels are significantly increased in mature neurons compared to developing ones (Kole *et al.*, 2011). This increase is not limited to sympathetic neurons; it is also observed in cerebellar and cortical neurons during their maturation. In neuronal cultures, mir-29 has been shown to repress the apoptotic pathway by targeting at least four different BH3-only inducers of apoptosis. In the developing retina of *Xenopus*, mir-24a plays a similar role in restricting apoptosis through repression of the proapoptotic factors *caspase 9* and *apaf1* (Walker and Harland, 2009).

A summary of the different stages of neuronal development and the discussed miRNAs acting at each stage is presented in Fig. 5.1.

2.2. miRNAs in glia development

Glial cells account for at least half of the cells in the human brain and their functions range from providing structural support, regulating the balance of water, ions, and nutrients in the brain and maintaining the blood–brain barrier, to providing the myelin sheaths that are so essential for neuronal conduction and even modulating neuronal synaptic transmission (for review, see the Nature Insight on Glia issue, editorial by Chouard and Gray, 2010). Glial cells derive from the same NPCs that give rise to neurons, through a similar series of stages. Perhaps not surprisingly at this

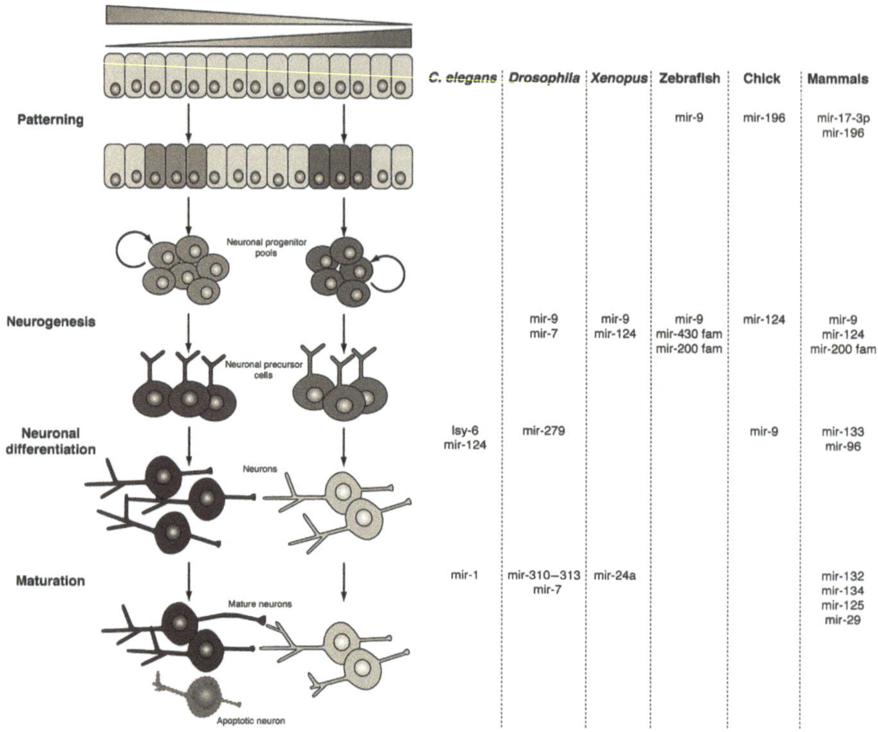

Figure 5.1 Summary of the miRNAs acting during different stages of neuronal development. Only those miRNAs discussed in the text have been included. (See Color Insert.)

point, miRNAs have been implicated in practically all steps of glia development, particularly in the two glial types that produce myelin, OLs (in the CNS), and Schwann cells (in the PNS). Here, we will illustrate the state of our knowledge by discussing the roles of miRNAs in OL development.

The differentiation of oligodendrocyte progenitor cells (OPCs) into OLs occurs in at least two steps. First, a series of intrinsic and external signals promote differentiation into a premyelinating OL that is able to extend processes to contact axons and initiate sheath formation. Subsequently, myelin sheaths start being produced, and at this stage, new processes cannot be extended to contact new axons. Thus, the timing of these events is important for proper myelination. At least five different miRNAs are involved in OL differentiation. mir-19, from the mir-17-92 cluster, first promotes the expansion of the OPC pool, inducing proliferation through the inhibition of its target *PTEN* (Budde *et al.*, 2010). OPCs are then induced to differentiate by a number of factors including mir-219 and mir-338. Together, these two miRNAs target a number of OPC-expressed genes that impair differentiation (mostly transcription factors and signaling pathways) (Dugas *et al.*, 2010; Zhao *et al.*, 2010). mir-23 also represses an OPC-expressed gene, Lamin B1 (*Lmnb1*), that otherwise inhibits the morphological differentiation of OLs (Lin and Fu, 2009). Finally, mir-138 specifically promotes the first stage of OL differentiation, acquisition of the premyelinating OL state, while preventing progression to the later stage, thus extending the intermediate stage that is necessary to establish the number of sheaths that a cell will form (Dugas *et al.*, 2010).

In *C. elegans*, a single miRNA, *mir-228*, has been identified, to date, to be expressed in sheath and/or socket cells that are glia-like support cells for the ciliated sensory neurons in the worm (Pierce *et al.*, 2008). While these cells do not produce myelin, it will be interesting to see whether *mir-228* plays a role in some step during their differentiation.

2.3. miRNAs have highly context-dependent functions

In addition to the commonalities of miRNA functions in diverse systems, it has also become evident that even the most conserved miRNAs can play substantially different roles in different organisms. Even in the same organism, the same miRNA can have distinct roles in different cell types or at different time points in development. This observation that miRNAs can have distinct context-dependent functions has been explored by Gao (2010), and we will expand it here with more recent examples from the literature.

First, even highly conserved miRNAs can have diverse roles in different organisms. For example, as we have seen above, mir-9 promotes neurogenesis in multiple vertebrate systems (Bonev *et al.*, 2011; Conaco *et al.*, 2006; Lagos-Quintana *et al.*, 2002; Leucht *et al.*, 2008; Packer *et al.*, 2008; Wienholds *et al.*, 2005; Zhao *et al.*, 2009). However, in *Drosophila*, it acts in nonneural cells to repress neuronal differentiation (Li *et al.*, 2006). Another case of seemingly distinct functions in different organisms is provided by mir-124. While in vertebrate systems it seems to promote neurogenesis in multiple types of neuronal precursors, in *Aplysia* and in *C. elegans*, mir-124 is only expressed in sensory neurons, and at least in *Aplysia*, while it is not yet known whether it plays a role in neurogenesis, it has been shown to be involved in synaptic plasticity (Rajasethupathy *et al.*, 2009). In *C. elegans*, its precise function is still unknown, but in its absence, the neurons where it is expressed are still present, arguing against a strong role in neurogenesis (Clark *et al.*, 2010).

The second level at which the same miRNA can display segregation of function is at the tissue level within the same organism. A number of miRNAs fit this category, with distinct targets and thus different effects in different tissues. mir-138 was presented above as an important player in OL differentiation; however, it has also been implicated in spine morphogenesis

(Siegel *et al.*, 2009) and cardiac patterning (Morton *et al.*, 2008). As mentioned, mir-338 is involved in OL differentiation but plays a role in axonal function as well (Aschrafi *et al.*, 2008), while mir-132 in addition to its role in dendritic outgrowth also modulates circadian rhythms and has a function in innate immunity (Cheng *et al.*, 2007; Lagos *et al.*, 2010).

A few studies have uncovered that the same miRNA can have different effects in different time points along the development of the same tissue. mir-124 seems to have two different effects during eye development in *Xenopus*. During the optic vesicle stage, mir-124 is necessary and sufficient to repress neurogenesis and promote proliferation of retinal progenitors (Liu *et al.*, 2010). This effect is at least in part through the repression if *NeuroD1*, a known inducer of neuronal differentiation. Interestingly, if after injection with mir-124 for overexpression animals are examined later, at the optic cup stage, a reduction in cell proliferation is observed, more consistent with the role of mir-124 in promoting neurogenesis presented above (Qiu *et al.*, 2009). While further work is necessary to explore whether this later effect of mir-124 is due to a direct effect on proliferation or a consequence of its earlier effect on the progenitor pool, it is interesting to note that careful temporal analysis of miRNA function could address some of the discrepancies mentioned above (e.g., Section 2.1.2).

Another interesting case came from a study in *Xenopus,* where loss of mir-9 caused different effects on neurogenesis in neural progenitors from the forebrain as compared to those in the hindbrain (Bonev *et al.*, 2011). In the hindbrain, knockdown of mir-9 caused an increase in proliferation of neural progenitors, consistent with its role in limiting proliferation and promoting differentiation in other organisms. In contrast, in the forebrain, knockdown of mir-9 reduced the number of progenitors due to increased apoptosis, but when apoptosis was blocked, a similar increase in progenitor proliferation was observed. Interestingly, in both cases, mir-9 seems to exert its function through the repression of *hairy1,* and protection of *hairy1* from mir-9 repression phenocopies both the increased proliferation in the posterior progenitors and the increased apoptosis in the anterior progenitors. This suggests that the specificity of function is downstream of the miRNA target and that context-dependent functions can arise through a variety of mechanisms.

Finally, a given miRNA can cause different effects even within the same cell. This is of particular relevance to neurons, where localization in different cellular compartments can lead to distinct functions. One example presented above is that of mir-134 which localizes to puncta in the dendritic terminals of neurons where it can have a local effect on the strength of individual spines by targeting Limk1 but also seems to be able to cause more global effects on dendritic outgrowth by targeting a more general regulator, Pumilio2 (Khudayberdiev *et al.*, 2009).

The fact that miRNAs have functions that can be so dependent on the cellular context is consistent with a number of observations that suggest that miRNAs evolve in a highly dynamic manner with a high birth rate and presumably also a high rate of incorporation into preexisting gene regulatory networks (Grimson *et al.*, 2008; Heimberg *et al.*, 2008; Hobert, 2008; Liu *et al.*, 2008; Lu *et al.*, 2008). The rapid evolvability of miRNAs could result in the acquisition of diverse targets in different organisms or different tissues. Interestingly, in those cases where the miRNAs are 100% conserved and still have distinct targets, some other kind of constraint, either structural or that imposed by one shared target, must also be in place.

In addition to evolutionary implications, the fact that cellular context can have such an impact on miRNA function has a very practical consequence: it is definitely an important point to keep in mind when analyzing potential targets of a miRNA, especially since so many approaches to validate miRNA targets rely on artificial expression of both miRNA and target in heterologous systems.

3. Neuronal-Specific Aspects of miRNA Function

3.1. Subcellular localization and spatial regulation

As most posttranscriptional regulators, miRNAs carry out their function in the cytosol. In neurons, the cytosol is divided in distinct subcellular domains, the soma, the dendrites, and the axon, and miRNAs have been detected in all these compartments. Translation regulation of synaptic-specific mRNAs in the neurites has been shown to provide a fast response, at the site where it is required, to allow for synaptic plasticity, and miRNAs are able to participate in these localized regulatory responses. Importantly, RISC is also present at synaptic terminals, and some of its components are regulated by neuronal activity, something we will discuss in Section 3.2.

While a few miRNAs have been observed to be present or even enriched in neuronal processes by *in situ* hybridization, four studies undertook an unbiased approach to identify dendrite- and axon-enriched miRNAs. Using different methods, all four managed to isolate miRNAs from either the soma or the projections and identified a number of miRNAs that are enriched in the different compartments.

First, Kosik and colleagues isolated miRNAs from the soma or the dendrites from hippocampal neurons in culture and identified mir-26 and mir-292-5p as being highly enriched in dendrites (Kye *et al.*, 2007). Smalheiser and colleagues fractionated adult mouse brains and isolated miRNAs from different synaptic fractions (synaptoneurosomes, enriched in dendritic spines and synaptosomes, enriched in synaptic membranes); this led to the identification of subsets of miRNAs enriched in these synaptic fractions as compared to whole brain homogenates (Lugli *et al.*, 2008). Further analysis of these miRNAs will reveal whether they have specific functions at the synapse. Work from Schratt and colleagues identified a number of miRNAs enriched in synaptosomes prepared from hippocampal neurons in culture. Among these, mir-218 and mir-138 were the most significant (Siegel *et al.*, 2009). The authors further showed that mir-138 is a negative regulator of dendritic spine size, acting through the local repression of the synthesis of the depalmitoylation enzyme Lypla1/APT1 which, in turn, affects the membrane localization of a G protein involved in Rho signaling and actomyosin contraction. Finally, Kaplan and colleagues characterized the pool of miRNAs present in distal axons and compared them to those in the soma of sympathetic neurons in culture (Natera-Naranjo *et al.*, 2010). Kaplan and colleagues had previously shown that mir-338, which was not detected as being significantly enriched in axons in their later work, is present in axons as shown by *in situ* hybridization and is able to regulate cytochrome *c* oxidase IV, ATP levels, and thus the rate of neurotransmitter uptake in axons (Aschrafi *et al.*, 2008). Therefore, even those miRNAs which are not significantly enriched in these compartments can potentially have localized roles, for example, if their targets are spatially restricted.

While there is little overlap between the miRNAs identified in all these different studies, which could be due to the differences in samples and methodologies, these will likely prove to be useful resources for further studies. In addition, some conclusions are shared by different studies. For example, mir-124, one of the most abundant neuronal miRNAs, has been shown by two of these studies to be enriched in the soma (this is in contrast to mir-124 in *Aplysia* where it has been seen in the projections of sensory neurons), suggesting some kind of exclusion mechanism from the processes.

Having miRNA-mediated regulation in the synaptic terminals in addition to the regulation in the soma has two obvious advantages. First, a miRNA present both in the processes and in the soma can cause two types of effects as illustrated by mir-134 and mir-132, a rapid, localized one at the former (typically by targeting synaptic proteins), and perhaps a slower, more sustained one at the latter (typically by targeting transcriptional or posttranscriptional regulators). And second, it can allow for uncoupled global and local responses carried out by the same miRNA, as discussed for mir-134 in Section 2.1.4. Further characterization of how miRNAs are transported through the different compartments and how this transport is regulated will be critical to our understanding of how miRNAs contribute to neuronal development and function.

3.2. Activity-dependent regulation of miRNA biogenesis and activity

Neuronal activity is an important player during the maturation phase of neuronal development, as it modulates the establishment and refinement of neuronal connections, mainly through its effects on dendrite morphology and synaptic plasticity. Upon neuronal activation with brain-derived neurotrophic factor (BDNF) or KCl, a number of signaling events, most notably activation of the CaMKII (calcium/calmodulin-dependent kinase II) phosphorylation cascade, result in local synaptic changes as well as in the activation of transcription factors in the nucleus. Two of these transcription factors, CREB and MeF2, have been shown to activate the transcription of mir-132 and the cluster containing mir-134, respectively (Fiore *et al.*, 2009; Wayman *et al.*, 2008).

Increased mir-132 levels upon CREB phosphorylation have two described consequences. As mentioned above, mir-132 promotes dendritic growth and branching, through its effect on the actin cytoskeleton. In addition, mir-132 downregulates MeCP2 (methyl CpG-binding protein 2), a broad transcriptional regulator with a strong implication in the neurodevelopmental disorder, Rett syndrome (Klein *et al.*, 2007). Among MeCP2's targets is BDNF itself; thus downregulation of MeCP2 by mir-132 results in a decrease in BDNF transcription, suggesting that mir-132 plays a role in neuronal homeostasis. Interestingly, CREB-mediated activation of mir-132 also occurs in the suprachiasmatic nucleus, where it plays a role in modulation of the circadian clock by light (Cheng *et al.*, 2007), again illustrating how network modules involving miRNAs can adopt different functions in different contexts. Notably, the miRNA *bantam* in *Drosophila* also plays a role in circadian rhythm modulation (Kadener *et al.*, 2009). The molecular oscillations that underlie the circadian rhythms are sustained by interconnected feedforward and feedback loops; miRNAs were likely an advantageous addition to these networks to enhance not only their robustness but also their flexibility (O'Neill and Hastings, 2007).

Just as miRNA, biogenesis can be stimulated by neuronal activity, so can miRNA catabolism (Krol *et al.*, 2010). Filipowicz and colleagues found that many miRNAs decay with much faster rates in neurons than in nonneuronal cells and that miRNA turnover in neurons is regulated by neuronal activity. For example, blocking glutamate receptors in hippocampal neurons slowed the decay of mir-124, -128, -134, and -138, while adding glutamate made it faster (Krol *et al.*, 2010). A rapid turnover of miRNAs (given by fast rates of degradation but also fast rates of biogenesis) likely allows neurons to adjust their repertoire of miRISC to the changing environment, in order to respond accordingly by changing its morphology or adjusting the strength of its synapses.

Finally, not only are the levels of miRNAs themselves affected by neuronal activity but so is the composition of the RISC. Work in both *Drosophila* olfactory interneurons (Ashraf *et al.*, 2006) and in rat hippocampal neurons (Banerjee *et al.*, 2009) has shown that upon neuronal activation, the DExD-box protein Armitage/MOV10, which is found at the synapses, is degraded via the ubiquitin–proteasome pathway. Reduction in the level of this key component of the silencing complex results in the release of a number of synaptic mRNAs from the repressed state and into the polysome fraction. In both systems, CaMKII protein synthesis was increased, but interestingly, so were Limk1 and Lypla1/APT1 in the rat neurons. These two mRNAs had previously been shown to be targets of mir-134 (Schratt *et al.*, 2006) and mir-138 (Siegel *et al.*, 2009), respectively, and their repression had been shown to be relieved at the synapse by neuronal activity.

Overall, miRNA-mediated translational repression seems to be deeply integrated in the dynamic responses that neurons elicit during activity and to maintain homeostasis. Their subcellular localization, the rapid kinetics, and reversibility of their function and their target diversity make them very suitable to fulfill this role.

3.3. A role for miRNAs in generating neuronal diversity

The cellular diversity in the nervous system is unparalleled by that of any other organ system. Even an organism with a simple nervous system, such as *C. elegans*, has at least 118 different classes of neurons. In general, the properties of each cell class are a consequence of the genes each class expresses, and this is, in turn, defined by the gene regulatory factors present in each cell type. Transcription factors (TFs) and miRNAs are the two most abundant and diverse classes of gene expression regulators, and it has been proposed that combinatorial "codes" of TFs and miRNAs can define all different cell types (Hobert, 2004).

The magnitude of the contribution of miRNAs to generating this diversity is beginning to be grasped. While a number of examples have been provided so far, many more are likely to be uncovered. Given their numbers, high evolvability, their diverse spatial and temporal expression patterns (Kapsimali *et al.*, 2007), and their ability to modify preexisting genetic networks to produce stable, heritable phenotypes, miRNAs are very good candidates to introduce an additional level of complexity.

We have already discussed the role of *lsy-6* in diversifying two sensory neurons in *C. elegans* that would otherwise be practically identical. In this case, it seems likely that the incorporation of a single regulatory factor (*lsy*-6) into preexisting regulatory networks during evolution could be responsible for this diversification. *lsy-6* is a nematode-specific miRNA; however, it is not present in all nematodes. While species such as *Caenorhabditis briggsae*, *Caenorhabditis remanei*, and *Caenorhabditis brenneri* have homologs of *lsy-6* and its target *cog-1, Pristionchus pacificus,* a more distant relative, does not seem to have a *lsy-6* homolog, and interestingly, while the *cog-1* ORF is conserved with that of *C. elegans,* their 3′UTRs are not. Further analysis of the properties of the ASE sensory neurons in *Pristionchus* will likely provide new insight into the incorporation of miRNAs into gene regulatory networks during evolution.

Other miRNAs seem to distinguish different neuronal—and glial—types during development, as discussed throughout this chapter. For example, mir-9 distinguishes cells outside the zebrafish MHB from the progenitors in the actual MHB; mir-17-3p is important to robustly diversify two pools of progenitors in the mouse spinal cord, and the presence of *mir-279* allows for the generation of distinct classes of sensory neurons in *Drosophila.*

Notably, miRNAs may also play a role in maintaining the identity of distinct neuronal classes postdevelopmentally. mir-16 seems to be present at higher levels in noradrenergic than in serotoninergic neurons, and it is able to repress serotonin-metabolic enzymes in noradrenergic neurons (Baudry *et al.*, 2010). Both *in vivo* and *in vitro*—in a bipotent neuroectodermal cell line that can differentiate into either serotoninergic or noradrenergic neurons—mir-16 inhibits the expression of the serotonin transporter (directly) as well as of tryptophan hydroxylase (likely indirectly). Loss of mir-16 in noradrenergic neurons gives them the ability to synthesize, store, and degrade serotonin without affecting the noradrenaline metabolism. These findings suggest that miRNAs are likely involved in maintaining well-defined neuronal classes.

Overall, it seems likely that miRNAs with their versatile repressive abilities have been able, during evolution, to segregate functions contained in common precursors into distinct neuronal subpopulations. Further studies analyzing this role of miRNAs as diversifiers, to generate the vast neuronal complexity, are necessary and will undoubtedly prove extremely interesting.

4. Integration of miRNA Function into Gene Regulatory Networks

4.1. Switches or modulators with a few or "hundreds" of targets?

To fully understand the contribution of miRNA regulation to a biological phenomenon, it is important to answer questions such as does a given miRNA act by regulating one or a few major targets, or is it really able to target dozens or hundreds of them, as prediction algorithms and genome-wide transcriptome and proteome analyses propose? And, do miRNAs act as molecular switches, turning off the expression of their targets, or do they modulate their expression to lower but still detectable levels? These questions have been raised in multiple occasions since the birth of the field, and while here we discuss them briefly, for further reading we suggest, among others, Flynt and Lai (2008).

The proposal that a miRNA can target numerous mRNAs is based on bioinformatic studies that use combinations of RNA sequence, secondary structure, and conservation information. These predict that each miRNA could have dozens, or even "hundreds," of target mRNAs. In addition, the finding that many miRNAs cause not only translation repression but also a decrease in the levels of their target mRNA has led to the use of microarray analysis to identify potential miRNA targets. In these studies, as well as in proteomic analyses, typically the level of numerous mRNAs and proteins change, in general quite subtly, when a given miRNA is knocked out or overexpressed, and the subsequent statistical analyses in general show enrichment among these for potential targets of the miRNAs under study (based on the same prediction algorithms). This has been taken as proof that each miRNA indeed targets a large number of mRNAs. Unfortunately, it seems to be disregarded that our prediction algorithms still suffer from very high false-positive rates and that in these genome-wide analyses the majority of the changes observed could be indirectly caused by miRNA misexpression, even those that have a putative miRNA binding site. In addition, methods to validate miRNA targets usually rely on overexpression of the miRNA and a sensor in a heterologous system, which also likely produces false positives. Moreover, as mentioned above, a miRNA can have distinct targets in different cell types or even in the same cell type at distinct time points. It is therefore crucial to conduct appropriate experiments to interpret the target-prediction data correctly. Therefore, while it is a valid possibility that a miRNA causes its effects by targeting dozens of mRNAs, it has not been satisfactorily proven yet, and it is in fact very challenging to do so, as it requires systematic testing of all predicted targets in the proper experimental setup with rigorous quantitative approaches.

In contrast, it has been easier to validate those cases where a miRNA has one or a few major targets. In these cases, the strongest evidence has come from studying genetic interactions. For example, reduction in the dose of a major target can fully (or almost fully) suppress the loss-of-function phenotype of the miRNA. Or alternatively, protection of the predicted target from miRNA repression fully (or almost fully) suppresses the miRNA overexpression phenotype. We have discussed some of these examples along the previous sections, but to name a couple, the function of *lsy-6* can be fully accounted for by its effect on *cog-1*; mir-9 has a handful of targets, but specific targets seem to mediate distinct functions in different cell types or at different times. Other examples include *mir-8* and its target *atrophin* in the *Drosophila* nervous system (Karres *et al.*, 2007) and mir-150 and its target c-Myb in the mouse B-cell lineage (Xiao *et al.*, 2007).

Regarding the mode of action of miRNAs, experimental evidence suggests two main types of miRNAs: those that are coexpressed with their targets and thus likely modulate the target concentration and those that are mutually exclusive and are therefore considered to have a switch-like behavior. Whether a miRNA will act as a modulator or as a switch depends on a number of variables. An important one to consider is the cellular concentration of the miRNA, given that the number of turnovers a given miRNA-RISC can perform seems to be rather low. According to a number of studies, miRNAs are present at a range of 10–10,000 copies per cell (Chen *et al.*, 2005; Kye *et al.*, 2007). It seems unlikely that a miRNA that is present at very low abundance will have a switch-like behavior, while those that are present at higher concentrations have a higher probability of acting that way. In addition, in neurons, miRNAs are distributed across the soma and the processes as we previously discussed. So, for example, a miRNA that is present even at 1000 copies per cell might end up at one or less copies per dendrite, and this will have a direct impact on the type of regulatory role that miRNA can execute.

The mode of action will also of course be dependent on the target mRNA concentration, and so it will be important as well how many additional targets that miRNA has. At the same cellular concentration, a miRNA with several targets is less likely to be able to switch all of them off, while a miRNA with one or a few targets has a higher chance of fully repressing its target.

Importantly, the observed effect of a miRNA on its target may not be reflective of the direct consequence of miRNA-mediated repression, but rather it may be the result of additional interactions. In fact, this is probably one of the main contributors to the mode of action of a miRNA given the fact that most miRNAs form complex gene regulatory networks with their targets. Thus, even a modest effect of a miRNA on its target can be amplified by feedback loops to end up in mutually exclusive expression of the two. In such a case, the miRNA could be acting as a genetic switch even without being a strict molecular switch. These types of interactions will be discussed in the next section.

4.2. Network motifs involving miRNAs

It is evident from a number of the examples we have explored so far that many miRNAs regulate the expression of specific transcription factors and that, in turn, these TFs feedback to regulate expression of the miRNAs themselves. More directed analyses have shown that a few network motifs that integrate transcriptional and miRNA-mediated posttranscriptional regulation are overrepresented in regulatory networks in organisms ranging from *C. elegans* to humans (Li *et al.*, 2009; Martinez and Walhout, 2009; Martinez *et al.*, 2008; Osella *et al.*, 2011; Tsang *et al.*, 2007). Such network motifs have been implicated not only in providing robustness and stability to genetic programs, maintaining stable steady-state levels of the regulatory factors, but also in providing precise transitions (temporal and spatial) in response to external stimuli or intrinsic signals (Alon, 2007).

The two most recurring motifs are feedback and feedforward loops. In a typical double-negative feedback loop, a miRNA represses a target, a TF, which, in turn, represses expression of the miRNA itself (Fig. 5.2A).

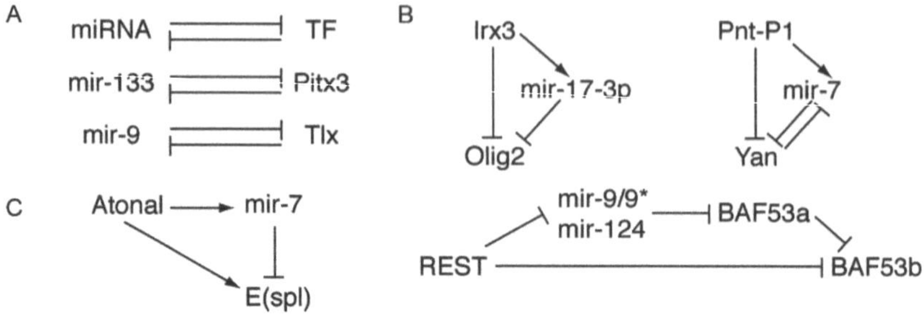

Figure 5.2 Examples of regulatory network motifs including miRNAs. (A) Examples of simple double-negative feedback loops involving one miRNA and one transcription factor. These small motifs are typically embedded in more complex networks. (B) Examples of coherent feedforward loops involving a few of the miRNAs mentioned in the text. (C) An incoherent feedforward loop involving a miRNA that has been implicated in providing robustness to a broader gene regulatory network.

This type of configuration has two possible outcomes. First, if the miRNA is able to fully turn off the TF, and the TF is able to shut down expression of the miRNA, this will result in a bistable switch where only one of the two components can be active at a given time. These relatively simple motifs are typically embedded in more complex networks. Which one of the two remains active will depend on additional input biasing the loop to one or the other side, or alternatively, initial stochastic fluctuations can be amplified to result in one or the other state. In cases where the TF is under an autoregulatory positive feedback, a negative feedback loop with a miRNA can rather act as a noise filter and provide stability against fluctuations in the level of the TF that could trigger an unwanted response, increasing the specificity of a response.

Feedforward motifs also provide a number of advantageous properties to gene regulatory networks. These can belong to two main classes, coherent and incoherent FFLs. Coherent FFLs are those in which an upstream regulator affects a target through two different paths, both of which affect the target levels in the same direction (i.e., both activate or both repress the target). In contrast, in incoherent FFLs, the two regulatory paths cause opposite effects on the target level (i.e., one activates and one represses the target). Coherent FFLs can provide robustness to a biological response and could reinforce a switch-like effect, as is illustrated by the relationship of *Irx3*, *Olig2*, and mir-17-3p in the mammalian spinal cord (Fig. 5.2B). In addition, coherent FFLs will also likely affect the dynamics of activation and repression of the target(s).

One of the functions of incoherent FFLs is to buffer noise in gene expression, defining and maintaining the steady-state level of a network component. This results in more stable states, preventing random switching to the alternative state due to stochastic fluctuations (Fig. 5.2C). Examples of the robustness conferred by both coherent and incoherent FFLs are given by the role of *mir-7* in photoreceptor and SOP differentiation in *Drosophila*. *mir-7* participates in complex interlocking loops and its function is essential to stabilize the expression of different network components in face of environmental fluctuations (Li *et al.*, 2009).

To sum up, addition of a miRNA to an existing gene regulatory network will have distinct consequences depending on the position of its target in the network. In some cases, it will provide a switch-like effect to remove expression of its target and generate a new cellular state, while in others, it will dampen fluctuations and thus provide robustness to a preexisting cellular state.

5. Concluding Remarks

miRNAs are a class of gene regulatory factors with versatile functions and as such they have been adopted during the course of evolution to serve a variety of purposes. During development, miRNAs are a source of robustness and reproducibility; they control spatial and temporal gene expression to allow for proper patterning and specification of different structures. At the same time, they provide heritable variability, by diversifying genetic programs and thus increasing the complexity of any system. Their modes of action and their molecular properties make them particularly suited to play regulatory roles in the dynamic cellular environment of the nervous system. And while their roles are already evidently widespread, we expect to find them involved in many more.

ACKNOWLEDGMENTS

We sincerely apologize to all those colleagues whose work could not be included due to space constraints. We thank Inés Carrera, Xantha Karp, Tao Sun, Hynek Wichterle, Jun-An Chen, and Fiona Doetsch for comments on the chapter. L. C. was funded by the Helen Hay Whitney Foundation and HHMI. O. H. is an HHMI investigator and is also funded by NIH.

REFERENCES

Alon, U. (2007). *Nat. Rev. Genet.* **8,** 450–461.
Arvanitis, D. N., Jungas, T., Behar, A., and Davy, A. (2010). *Mol. Cell. Biol.* **30,** 2508–2517.
Aschrafi, A., Schwechter, A. D., Mameza, M. G., Natera-Naranjo, O., Gioio, A. E., and Kaplan, B. B. (2008). *J. Neurosci.* **28,** 12581–12590.
Ashraf, S. I., McLoon, A. L., Sclarsic, S. M., and Kunes, S. (2006). *Cell* **124,** 191–205.
Asli, N. S., and Kessel, M. (2010). *Dev. Biol.* **344,** 857–868.
Banerjee, S., Neveu, P., and Kosik, K. S. (2009). *Neuron* **64,** 871–884.

Bartel, D. P. (2009). *Cell* **136,** 215–233.

Baudry, A., Mouillet-Richard, S., Schneider, B., Launay, J. M., and Kellermann, O. (2010). *Science* **329,** 1537–1541.

Bonev, B., Pisco, A., and Papalopulu, N. (2011). *Dev. Cell* **20,** 19–32.

Budde, H., Schmitt, S., Fitzner, D., Opitz, L., Salinas-Riester, G., and Simons, M. (2010). *Development* **137,** 2127–2132.

Cao, X., Pfaff, S. L., and Gage, F. H. (2007). *Genes Dev.* **21,** 531–536.

Cayirlioglu, P., Kadow, I. G., Zhan, X., Okamura, K., Suh, G. S., Gunning, D., Lai, E. C., and Zipursky, S. L. (2008). *Science* **319,** 1256–1260.

Chekulaeva, M., and Filipowicz, W. (2009). *Curr. Opin. Cell Biol.* **21,** 452–460.

Chen, C., Ridzon, D. A., Broomer, A. J., Zhou, Z., Lee, D. H., Nguyen, J. T., Barbisin, M., Xu, N. L., Mahuvakar, V. R., Andersen, M. R., Lao, K. Q., Livak, K. J., *et al.* (2005). *Nucleic Acids Res.* **33,** e179.

Chen, J. A., Huang, Y. P., Mazzoni, E. O., Tan, G. C., Zavadil, J., and Wichterle, H. (2011). *Neuron* **69,** 721–735.

Cheng, H. Y., Papp, J. W., Varlamova, O., Dziema, H., Russell, B., Curfman, J. P., Nakazawa, T., Shimizu, K., Okamura, H., Impey, S., and Obrietan, K. (2007). *Neuron* **54,** 813–829.

Cheng, L. C., Pastrana, E., Tavazoie, M., and Doetsch, F. (2009). *Nat. Neurosci.* **12,** 399–408.

Choi, P. S., Zakhary, L., Choi, W. Y., Caron, S., Alvarez-Saavedra, E., Miska, E. A., McManus, M., Harfe, B., Giraldez, A. J., Horvitz, H. R., Schier, A. F., and Dulac, C. (2008). *Neuron* **57,** 41–55.

Chouard, T., and Gray, N. (2010). *Nature* **468,** 213–262.

Clark, A. M., Goldstein, L. D., Tevlin, M., Tavare, S., Shaham, S., and Miska, E. A. (2010). *Nucleic Acids Res.* **38,** 3780–3793.

Cole, C., Sobala, A., Lu, C., Thatcher, S. R., Bowman, A., Brown, J. W., Green, P. J., Barton, G. J., and Hutvagner, G. (2009). *RNA* **15,** 2147–2160.

Conaco, C., Otto, S., Han, J. J., and Mandel, G. (2006). *Proc. Natl. Acad. Sci. USA* **103,** 2422–2427.

Conti, L., and Cattaneo, E. (2010). *Nat. Rev. Neurosci.* **11,** 176–187.

Coolen, M., and Bally-Cuif, L. (2009). *Curr. Opin. Neurobiol.* **19,** 461–470.

Dasen, J. S., and Jessell, T. M. (2009). *Curr. Top. Dev. Biol.* **88,** 169–200.

Dasen, J. S., De Camilli, A., Wang, B., Tucker, P. W., and Jessell, T. M. (2008). *Cell* **123,** 477–491.

Davis, T. H., Cuellar, T. L., Koch, S. M., Barker, A. J., Harfe, B. D., McManus, M. T., and Ullian, E. M. (2008). *J. Neurosci.* **28,** 4322–4330.

Delaloy, C., Liu, L., Lee, J. A., Su, H., Shen, F., Yang, G. Y., Young, W. L., Ivey, K. N., and Gao, F. B. (2010). *Cell Stem Cell* **6,** 323–335.

Deo, M., Yu, J. Y., Chung, K. H., Tippens, M., and Turner, D. L. (2006). *Dev. Dyn.* **235,** 2538–2548.

Djuranovic, S., Nahvi, A., and Green, R. (2011). *Science* **331,** 550–553.

Dugas, J. C., Cuellar, T. L., Scholze, A., Ason, B., Ibrahim, A., Emery, B., Zamanian, J. L., Foo, L. C., McManus, M. T., and Barres, B. A. (2010). *Neuron* **65,** 597–611.

Edbauer, D., Neilson, J. R., Foster, K. A., Wang, C. F., Seeburg, D. P., Batterton, M. N., Tada, T., Dolan, B. M., Sharp, P. A., and Sheng, M. (2010). *Neuron* **65,** 373–384.

Ender, C., Krek, A., Friedlander, M. R., Beitzinger, M., Weinmann, L., Chen, W., Pfeffer, S., Rajewsky, N., and Meister, G. (2008). *Mol. Cell* **32,** 519–528.

Fiore, R., Khudayberdiev, S., Christensen, M., Siegel, G., Flavell, S. W., Kim, T. K., Greenberg, M. E., and Schratt, G. (2009). *EMBO J.* **28,** 697–710.

Flynt, A. S., and Lai, E. C. (2008). *Nat. Rev. Genet.* **9,** 831–842.

Gao, F. B. (2010). *Neural Dev.* **5,** 25.

Giraldez, A. J., Cinalli, R. M., Glasner, M. E., Enright, A. J., Thomson, J. M., Baskerville, S., Hammond, S. M., Bartel, D. P., and Schier, A. F. (2005). *Science* **308,** 833–838.

Grimson, A., Srivastava, M., Fahey, B., Woodcroft, B. J., Chiang, H. R., King, N., Degnan, B. M., Rokhsar, D. S., and Bartel, D. P. (2008). *Nature* **455,** 1193–1197.

Heimberg, A. M., Sempere, L. F., Moy, V. N., Donoghue, P. C., and Peterson, K. J. (2008). *Proc. Natl. Acad. Sci. USA* **105,** 2946–2950.

Hobert, O. (2004). *TIBS.* **29,** 462–468.

Hobert, O. (2008). *Science* **319,** 1785–1786.

Hobert, O. (2011). *Ann. Rev. Cell Dev. Biol.* **27,** 681–696.

Hornstein, E., Mansfield, J. H., Yekta, S., Hu, J. K., Harfe, B. D., McManus, M. T., Baskerville, S., Bartel, D. P., and Tabin, C. J. (2005). *Nature* **438,** 671–674.

Impey, S., Davare, M., Lasiek, A., Fortin, D., Ando, H., Varlamova, O., Obrietan, K., Soderling, T. R., Goodman, R. H., and Wayman, G. A. (2010). *Mol. Cell. Neurosci.* **43,** 146–156.

Jessell, T. M. (2000). *Nat. Rev. Genet.* **1,** 20–29.

Johnston, R. J., and Hobert, O. (2003). *Nature* **426,** 845–849.

Johnston, R. J., Jr., Chang, S., Etchberger, J. F., Ortiz, C. O., and Hobert, O. (2005). *Proc. Natl. Acad. Sci. USA* **102,** 12449–12454.

Kadener, S., Menet, J. S., Sugino, K., Horwich, M. D., Weissbein, U., Nawathean, P., Vagin, V. V., Zamore, P. D., Nelson, S. B., and Rosbash, M. (2009). *Genes Dev.* **23,** 2179–2191.

Kapsimali, M., Kloosterman, W. P., de Bruijn, E., Rosa, F., Plasterk, R. H., and Wilson, S. W. (2007). *Genome Biol.* **8,** R173.

Karres, J. S., Hilgers, V., Carrera, I., Treisman, J., and Cohen, S. M. (2007). *Cell* **131,** 136–145.

Khudayberdiev, S., Fiore, R., and Schratt, G. (2009). *Commun. Integr. Biol.* **2,** 411–413.

Kim, J., Inoue, K., Ishii, J., Vanti, W. B., Voronov, S. V., Murchison, E., Hannon, G., and Abeliovich, A. (2007). *Science* **317,** 1220–1224.

Klein, M. E., Lioy, D. T., Ma, L., Impey, S., Mandel, G., and Goodman, R. H. (2007). *Nat. Neurosci.* **10,** 1513–1514.

Kole, A. J., Swahari, V., Hammond, S. M., and Deshmukh, M. (2011). *Genes Dev.* **25,** 125–130.

Korotkova, T. M., Ponomarenko, A. A., Haas, H. L., and Sergeeva, O. A. (2005). *Eur. J. Neurosci.* **22,** 1287–1293.

Krol, J., Busskamp, V., Markiewicz, I., Stadler, M. B., Ribi, S., Richter, J., Duebel, J., Bicker, S., Fehling, H. J., Schubeler, D., Oertner, T. G., Schratt, G., *et al.* (2010). *Cell* **141,** 618–631.

Kuhn, S., Johnson, S. L., Furness, D. N., Chen, J., Ingham, N., Hilton, J. M., Steffes, G., Lewis, M. A., Zampini, V., Hackney, C. M., Masetto, S., Holley, M. C., *et al.* (2011). *Proc. Natl. Acad. Sci. USA* **108,** 2355–2360.

Kye, M. J., Liu, T., Levy, S. F., Xu, N. L., Groves, B. B., Bonneau, R., Lao, K., and Kosik, K. S. (2007). *RNA* **13,** 1224–1234.

Lagos, D., Pollara, G., Henderson, S., Gratrix, F., Fabani, M., Milne, R. S., Gotch, F., and Boshoff, C. (2010). *Nat. Cell Biol.* **12,** 513–519.

Lagos-Quintana, M., Rauhut, R., Yalcin, A., Meyer, J., Lendeckel, W., and Tuschl, T. (2002). *Curr. Biol.* **12,** 735–739.

Leaman, D., Chen, P. Y., Fak, J., Yalcin, A., Pearce, M., Unnerstall, U., Marks, D. S., Sander, C., Tuschl, T., and Gaul, U. (2005). *Cell* **121,** 1097–1108.

Leucht, C., Stigloher, C., Wizenmann, A., Klafke, R., Folchert, A., and Bally-Cuif, L. (2008). *Nat. Neurosci.* **11,** 641–648.

Li, X., and Carthew, R. W. (2005). *Cell* **123,** 1267–1277.

Li, X., and Jin, P. (2010). *Nat. Rev. Neurosci.* **11,** 329–338.

Li, Y., Wang, F., Lee, J. A., and Gao, F. B. (2006). *Genes Dev.* **20,** 2793–2805.
Li, X., Cassidy, J. J., Reinke, C. A., Fischboeck, S., and Carthew, R. W. (2009). *Cell* **137,** 273–282.
Liang, X. H., and Crooke, S. T. (2011). *Nucleic Acids Res.* **39,** 4875–4889.
Lim, L. P., Lau, N. C., Garrett-Engele, P., Grimson, A., Schelter, J. M., Castle, J., Bartel, D. P., Linsley, P. S., and Johnson, J. M. (2005). *Nature* **433,** 769–773.
Lin, S. T., and Fu, Y. H. (2009). *Dis. Model. Mech.* **2,** 178–188.
Liu, N., Okamura, K., Tyler, D. M., Phillips, M. D., Chung, W. J., and Lai, E. C. (2008). *Cell Res.* **18,** 985–996.
Liu, K., Liu, Y., Mo, W., Qiu, R., Wang, X., Wu, J. Y., and He, R. (2011). *Nucleic Acids Res.* **39,** 2869–2879.
Lu, J., Shen, Y., Wu, Q., Kumar, S., He, B., Shi, S., Carthew, R. W., Wang, S. M., and Wu, C. I. (2008). *Nat. Genet.* **40,** 351–355.
Lugli, G., Torvik, V. I., Larson, J., and Smalheiser, N. R. (2008). *J. Neurochem.* **106,** 650–661.
Magill, S. T., Cambronne, X. A., Luikart, B. W., Lioy, D. T., Leighton, B. H., Westbrook, G. L., Mandel, G., and Goodman, R. H. (2010). *Proc. Natl. Acad. Sci. USA* **107,** 20382–20387.
Maiorano, N. A., and Mallamaci, A. (2009). *Neural Dev.* **4,** 40.
Maiorano, N. A., and Mallamaci, A. (2010). *RNA Biol.* **7,** 528–533.
Makeyev, E. V., Zhang, J., Carrasco, M. A., and Maniatis, T. (2007). *Mol. Cell* **27,** 435–448.
Maller Schulman, B. R., Liang, X., Stahlhut, C., DelConte, C., Stefani, G., and Slack, F. J. (2008). *Cell Cycle* **7,** 3935–3942.
Martinez, N. J., and Walhout, A. J. (2009). *Bioessays* **31,** 435–445.
Martinez, N. J., Ow, M. C., Barrasa, M. I., Hammell, M., Sequerra, R., Doucette-Stamm, L., Roth, F. P., Ambros, V. R., and Walhout, A. J. (2008). *Genes Dev.* **22,** 2535–2549.
McGlinn, E., Yekta, S., Mansfield, J. H., Soutschek, J., Bartel, D. P., and Tabin, C. J. (2009). *Proc. Natl. Acad. Sci. USA* **106,** 18610–18615.
Morton, S. U., Scherz, P. J., Cordes, K. R., Ivey, K. N., Stainier, D. Y., and Srivastava, D. (2008). *Proc. Natl. Acad. Sci. USA* **105,** 17830–17835.
Natera-Naranjo, O., Aschrafi, A., Gioio, A. E., and Kaplan, B. B. (2010). *RNA* **16,** 1516–1529.
O'Neill, J. S., and Hastings, M. H. (2007). *Curr. Biol.* **17,** R760–R762.
Osella, M., Bosia, C., Cora, D., and Caselle, M. (2011). *PLoS Comput. Biol.* **7,** e1001101.
Otaegi, G., Pollock, A., Hong, J., and Sun, T. (2011). *J. Neurosci.* **31,** 809–818.
Packer, A. N., Xing, Y., Harper, S. Q., Jones, L., and Davidson, B. L. (2008). *J. Neurosci.* **28,** 14341–14346.
Peng, Y. R., He, S., Marie, H., Zeng, S. Y., Ma, J., Tan, Z. J., Lee, S. Y., Malenka, R. C., and Yu, X. (2009). *Neuron* **61,** 71–84.
Pierce, M. L., Weston, M. D., Fritzsch, B., Gabel, H. W., Ruvkun, G., and Soukup, G. A. (2008). *Evol. Dev.* **10,** 106–113.
Qiu, R., Liu, K., Liu, Y., Mo, W., Flynt, A. S., Patton, J. G., Kar, A., Wu, J. Y., and He, R. (2009). *Mech. Dev.* **126,** 804–816.
Rajasethupathy, P., Fiumara, F., Sheridan, R., Betel, D., Puthanveettil, S. V., Russo, J. J., Sander, C., Tuschl, T., and Kandel, E. (2009). *Neuron* **63,** 803–817.
Rousso, D. L., Gaber, Z. B., Wellik, D., Morrisey, E. E., and Novitch, B. G. (2008). *Neuron* **59,** 226–240.
Saraiya, A. A., and Wang, C. C. (2008). *PLoS Pathog.* **4,** e1000224.
Schratt, G. M., Tuebing, F., Nigh, E. A., Kane, C. G., Sabatini, M. E., Kiebler, M., and Greenberg, M. E. (2006). *Nature* **439,** 283–289.
Sehm, T., Sachse, C., Frenzel, C., and Echeverri, K. (2009). *Dev. Biol.* **334,** 468–480.

Shibata, M., Kurokawa, D., Nakao, H., Ohmura, T., and Aizawa, S. (2008). *J. Neurosci.* **28,** 10415–10421.

Shibata, M., Nakao, H., Kiyonari, H., Abe, T., and Aizawa, S. (2011). *J. Neurosci.* **31,** 3407–3422.

Siegel, G., Obernosterer, G., Fiore, R., Oehmen, M., Bicker, S., Christensen, M., Khudayberdiev, S., Leuschner, P. F., Busch, C. J., Kane, C., Hubel, K., Dekker, F., *et al.* (2009). *Nat. Cell Biol.* **11,** 705–716.

Simon, D. J., Madison, J. M., Conery, A. L., Thompson-Peer, K. L., Soskis, M., Ruvkun, G. B., Kaplan, J. M., and Kim, J. K. (2008). *Cell* **133,** 903–915.

Tsang, J., Zhu, J., and van Oudenaarden, A. (2007). *Mol. Cell* **26,** 753–767.

Tsurudome, K., Tsang, K., Liao, E. H., Ball, R., Penney, J., Yang, J. S., Elazzouzi, F., He, T., Chishti, A., Lnenicka, G., Lai, E. C., and Haghighi, A. P. (2010). *Neuron* **68,** 879–893.

Visvanathan, J., Lee, S., Lee, B., Lee, J. W., and Lee, S. K. (2007). *Genes Dev.* **21,** 744–749.

Vo, N., Klein, M. E., Varlamova, O., Keller, D. M., Yamamoto, T., Goodman, R. H., and Impey, S. (2005). *Proc. Natl. Acad. Sci. USA* **102,** 16426–16431.

Vo, N. K., Cambronne, X. A., and Goodman, R. H. (2010). *Curr. Opin. Neurobiol.* **20,** 457–465.

Walker, J. C., and Harland, R. M. (2009). *Genes Dev.* **23,** 1046–1051.

Wayman, G. A., Davare, M., Ando, H., Fortin, D., Varlamova, O., Cheng, H. Y., Marks, D., Obrietan, K., Soderling, T. R., Goodman, R. H., and Impey, S. (2008). *Proc. Natl. Acad. Sci. USA* **105,** 9093–9098.

Wienholds, E., Kloosterman, W. P., Miska, E., Alvarez-Saavedra, E., Berezikov, E., de Bruijn, E., Horvitz, H. R., Kauppinen, S., and Plasterk, R. H. (2005). *Science* **309,** 310–311.

Wurst, W., and Bally-Cuif, L. (2001). *Nat. Rev. Neurosci.* **2,** 99–108.

Xiao, C., Calado, D. P., Galler, G., Thai, T. H., Patterson, H. C., Wang, J., Rajewsky, N., Bender, T. P., and Rajewsky, K. (2007). *Cell* **131,** 146–159.

Yoo, A. S., Staahl, B. T., Chen, L., and Crabtree, G. R. (2009). *Nature* **460,** 642–646.

Zhao, C., Sun, G., Li, S., and Shi, Y. (2009). *Nat. Struct. Mol. Biol.* **16,** 365–371.

Zhao, X., He, X., Han, X., Yu, Y., Ye, F., Chen, Y., Hoang, T., Xu, X., Mi, Q. S., Xin, M., Wang, F., Appel, B., *et al.* (2010). *Neuron* **65,** 612–626.

CHAPTER SIX

MicroRNAs and Hematopoietic Cell Development

Ryan M. O'Connell* *and* David Baltimore†

Contents

Abstract

Hematopoiesis is a dynamic and highly complex developmental process that gives rise to a multitude of the cell types that circulate in the blood of multicellular organisms. These cells provide tissues with oxygen, guard against infection,

* Division of Microbiology and Immunology, Department of Pathology, University of Utah, Salt Lake City, Utah, USA
† Division of Biology, California Institute of Technology, Pasadena, California, USA

Current Topics in Developmental Biology, Volume 99
ISSN 0070-2153, DOI: 10.1016/B978-0-12-387038-4.00006-9

prevent bleeding by clotting, and mediate inflammatory reactions. Because the hematopoietic system plays such a central role in human diseases such as infections, cancer, autoimmunity, and anemia, it has been intensely studied for more than a century. This scrutiny has helped to shape many of the developmental paradigms that exist today and has identified specific protein factors that serve as master regulators of blood cell lineage specification. Despite this progress, many aspects of blood cell development remain obscure, suggesting that novel layers of regulation must exist. Consequently, the emergence of regulatory noncoding RNAs, such as the microRNAs (miRNAs), is beginning to provide new insights into the molecular control networks underlying hematopoiesis and diseases that stem from aberrations in this process. This review will discuss how miRNAs fit into our current understanding of hematopoietic development in mammals and how breakdowns in these pathways can trigger disease.

1. Introduction

During embryogenesis, blood cell development takes place in the yolk sac, placenta, and fetal liver. Shortly after birth, the primary site of hematopoiesis shifts to the bone marrow, where it remains throughout adulthood. Production of 10^{11}–10^{12} new blood cells must take place daily to maintain homeostatic levels in adult humans. Mature blood cells are made up of many different lineages that carry out diverse functions such as providing immunity against pathogens, carrying oxygen throughout the body, and mediating the process of clotting. The hematopoietic system is central to mammalian life and is involved either directly or indirectly in most human diseases. Thus, this elegant developmental system warrants the amount of study that has and will continue to go into understanding its intricate processes.

Hematopoietic differentiation is the quintessential stem cell-driven developmental process (Orkin and Zon, 2008). In this, a self-renewing stem cell population spawns offspring that change their properties in a bifurcating process of specification leading to a myriad of ultimate cell products. Thus, the process involves a cascade of cell choices. From the years of effort that have gone into understanding these developmental choices, we have learned that protein-coding genes are critical regulators of these cell fate decisions, as they are throughout mammalian development (Orkin and Zon, 2008; Robb, 2007). Cellular decisions such as self-renewal versus differentiation, proliferation versus quiescence, and survival versus death have all been shown to involve specific sets of proteins. Despite this knowledge, many aspects of hematopoietic development and mature blood cell function remain unclear. Now, with the identification of noncoding (nc) RNAs, and in particular miRNAs, many unanswered questions about hematopoietic development are beginning to be answered.

Most of the genome is transcribed. However, only about 1.5% of transcripts encode proteins; the vast majority of cellular RNAs are noncoding (Mattick, 2007). In recent years, much progress has been made in our understanding of certain classes of noncoding RNAs, including the miRNAs, piwi interacting RNAs, small nucleolar RNAs, long intergenic noncoding RNAs, and RNA transcripts antisense to coding genes (Bartel and Chen, 2004; Guttman *et al.*, 2009; Khanna and Stamm, 2010; Khurana and Theurkauf, 2010; Malecova and Morris, 2010). Many of these RNA species are evolutionarily conserved in sequence, exhibit tissue-specific expression patterns, and are involved in modulating specific cellular pathways and functions. Here, we will focus primarily on the miRNAs.

miRNAs play critical regulatory roles at many stages of hematopoiesis, ranging from stem cells to terminal differentiation. We begin with a discussion of why miRNAs may have evolved to regulate developmental processes like hematopoiesis. The types of regulatory strategies used by miRNAs and the pathways that regulate miRNA function are then discussed, followed by an analysis of how miRNAs have emerged to impact distinct stages of hematopoiesis. Finally, the relevance of miRNAs to hematopoietic diseases that stem from developmental mishaps, as well as translational applications, is addressed. More complete catalogs of specific studies on these topics are presented elsewhere (Lodish *et al.*, 2008; O'Connell *et al.*, 2010c; Xiao and Rajewsky, 2009); this discussion is intended to be more of a conceptual overview.

2. The Emerging Importance of MicroRNAs During Hematopoietic Development

2.1. MicroRNAs and the stabilization of complex phenotypes

As more genomes have been fully sequenced, it has become clear that increased organismal complexity is primarily driven by new forms of gene regulation and not gene number. Aside from the primary event of transcriptional regulation, relevant activities include posttranscriptional mechanisms like alternative splicing or polyadenylation, which can generate many protein derivatives from a single gene. Other types of posttranscriptional regulation include modifications of transcript stability, which are regulated in part by miRNAs. Following their biogenesis, miRNAs enter the cytoplasmic RNA-induced silencing complex (RISC) which helps guide them to interact with mRNA 3′UTRs through the agency of specific base pairing (Filipowicz *et al.*, 2008). This, in turn, triggers reduction of target mRNA levels and protein output, although other mechanisms of action may be relevant. Because there are many hundreds of different miRNAs in humans—with many predicted to target hundreds of mRNAs—miRNAs are thought to have a broad impact on the gene regulatory networks that orchestrate mammalian physiology.

The evolutionarily oldest, most conserved miRNAs are commonly expressed at the highest levels and alterations in their concentrations often result in distinct phenotypes (Johnnidis *et al.*, 2008; Klein *et al.*, 2010; Lu *et al.*, 2008b, 2010; Patrick *et al.*, 2010; Rodriguez *et al.*, 2007; Thai *et al.*, 2007; Ventura *et al.*, 2008; Xiao *et al.*, 2007). Alternatively, younger miRNA families are commonly expressed at much lower levels and have a higher probability of being lost during evolution (Lu *et al.*, 2008b). In these cases, it has been hypothesized that newly evolved miRNAs might be substrates for the formation of new regulatory circuits. If this new interaction is advantageous, a new miRNA and target mRNA connection can become fixed. Because the miRNA "seed" interaction with a cognate mRNA 3′UTR only requires 7–8 nucleotides to become functional, acquisition of such sites due to spontaneous mutations within an mRNA's 3′UTRs should happen more often than changes that require a longer sequence alteration (Shomron *et al.*, 2009).

The hematopoietic system has a fascinating evolutionary history marked by the seminal event of the acquisition of adaptive immunity when the jawed fishes evolved (Cooper and Alder, 2006). Over evolutionary time, one can see a positive correlation between the number of miRNAs expressed by different organisms and their phenotypic complexity (Peterson *et al.*, 2009). It is thought that transcriptional noise, which varies in the expression of specific genes within a given cellular population, must be limited in order to allow for stable, complex phenotypes to take shape (Hornstein and Shomron, 2006). Through their ability to regulate mRNA concentrations posttranscriptionally, miRNAs are well suited for this job. In fact, mRNAs with conserved 3′UTR miRNA binding sites are expressed at more comparable levels between species than mRNAs without miRNA sites (Cui *et al.*, 2007).

As an aside, one might expect that quantitative control of the expression of genes would be regulated by transcriptional modulation. That certainly happens, but it may be that the evolutionary fine-tuning of gene transcription is difficult, perhaps partly because of the inherently cooperative nature of transcription due to the multiple transcription factors that control single genes through established binding site motifs. Thus, miRNAs may have evolved because of their ability to provide finer regulation than is conveniently achieved at the transcriptional event.

2.2. MicroRNAs fine-tune gene expression levels in the hematopoietic system

There are more than 100 different miRNAs expressed in the hematopoietic system (O'Connell *et al.*, 2010a). A recent study investigated the types of protein-coding genes that are predicted to be regulated by immune system miRNAs (Asirvatham *et al.*, 2008). Interestingly, miRNAs preferentially target transcripts encoding transcription factors and the upstream signaling proteins that activate them, while mRNAs encoding cellular receptors and ligands have substantially fewer miRNA binding sites. Certain transcription factors play indispensable roles in determining cell fate during hematopoiesis by acting as "master regulators" of gene expression programs. Thus, their modulation by miRNAs is consistent with the impact that miRNAs have been shown genetically to have on cellular lineage choices during development.

The importance of precise levels of gene expression during hematopoiesis (both for transcription factors and other regulatory proteins) has been clearly demonstrated in recent years. Many papers have described distinct hematopoietic phenotypes in mice lacking only one of the two protein-coding alleles at a particular locus (Chan *et al.*, 2011; Dahl *et al.*, 2003; Egle *et al.*, 2004; Eischen *et al.*, 2004; He, 2010; Joslin *et al.*, 2007; Kamimura *et al.*, 2007; Kinjyo *et al.*, 2002; Le Toriellec *et al.*, 2008; Puebla-Osorio *et al.*, 2011; Schraml *et al.*, 2008; Sernandez *et al.*, 2008; Sportoletti *et al.*, 2008; Sun and Downing, 2004; Wang *et al.*, 2010; Xiang *et al.*, 2010; Xiao *et al.*, 2007; Zandi *et al.*, 2008) (Table 6.1). A 50% reduction in expression of some genes can impact cell lineage decisions, as has been elegantly demonstrated for PU.1 (Dahl *et al.*, 2003). At this locus, loss of a single allele triggers a skewing of the output of granulocytes compared to monocytes. Subtle changes in gene dosage can also impact cancer development, as in the cases of Pten, p53, and cMyc (He, 2010; Salmena *et al.*, 2008), and this can precede or even be distinct from phenotypes caused by loss of heterozygosity. Another example is Socs1, where heterozygous mice display enhanced inflammatory responses to endotoxin (Kinjyo *et al.*, 2002). The twofold effect of heterozygosity in mice is quantitatively similar to the degree of protein expression change mediated by many miRNAs. Thus, it is reasonable that loci displaying haploinsufficient phenotypes are often ones that have functional miRNA binding sites within their 3′UTRs because they should be genes that are susceptible to small, quantitative alterations.

Changes in gene expression dosage can also come about through aneuploidy. One clear example of this is trisomy 21 which causes Down's syndrome. The extra copy of chromosome 21 leads to elevated levels of the genes it carries (Arron *et al.*, 2006). This triggers many phenotypes including an increased likelihood of developing acute leukemia. Thus, it is clear that many lineage-determining signaling proteins and transcription factors must be expressed at accurate levels to ensure proper hematopoietic development, and miRNAs have likely evolved to meet such demands.

2.3. Combinatorial gene regulation by miRNAs

Most miRNAs have a broad repertoire of predicted mRNA targets. However, the number of targets involved in a specific miRNA-dependent phenotype appears to vary. There have been many examples of miRNAs

Table 6.1 Genes with haploinsufficient hematopoietic phenotypes

Gene	Haploinsufficient phenotypes	Conserved miRNA sites (targetscan)	Transcription factor/nuclear protein	Receptor signaling	Apoptosis/cell cycle
Myb	Altered lymphocyte development	13	×		
Tp53	UV-induced B cell lymphoma	3	×		
PU.1	Granulocyte versus monocyte skewing	1	×		
Runx1	Decreased HSCs	19	×		
Aicda	Decreased Ig diversification and translocations	2	×		
Arf	Rescued B cell development	2	×		
Egr1	Myeloid disorders	3	×		
Mcl1	Resistant to cMyc induced AML	10	×		
Prpf19	Defective HSC function	4	×		
EBF1	Defective B cell development	17	×		
Bcl11b	Decreased T cell development	33	×		
Npm1	Increased blood cancers	1		×	
Shp2	Decreased HSC function	7		×	
Apc	Ineffective hematopoiesis	12		×	
Socs1	Enhanced sensitivity to endotoxemia	4		×	
Pten	Myelodysplasia	22		×	
Bim	Increased B cell lymphoma	25			×
Cdkn1b	Increased T cell leukemia	5			×

targeting multiple genes involved in specific pathways. For instance, miR-181a represses several phosphatases in developing T lymphocytes (Li *et al.*, 2007). miR-125 family members in hematopoietic stem cells (HSCs) repress many targets with roles in apoptosis and cell differentiation (Guo *et al.*, 2010; Ooi *et al.*, 2010). miR-146a has been shown to dampen the inflammatory response by targeting Irak1, Traf6, and Stat1 (Lu *et al.*, 2010; Taganov *et al.*, 2006). However, other reports have found that single miRNA targets in particular cells can play dominant roles. For instance, repression of cMyb by miR-150 and FoxP1 by miR-34a are both important for proper B cell development (Rao *et al.*, 2010; Xiao *et al.*, 2007). A significant part of miR-223 function in myeloid cells involves targeting of Mef2c (Johnnidis *et al.*, 2008). Thus, different gene regulatory strategies involving miRNAs and their targets have evolved, with some consisting of multiple targets that impact a specific pathway and others utilizing only a single target.

Some mRNAs possess multiple binding sites for different miRNAs within their 3′UTRs. Such an array of target sites suggests that miRNAs might work collaboratively to repress the expression of certain genes. For instance, the 3′UTRs of Mtpn and p21Cip1 contain many conserved miRNA binding sites, and greater repression of their expression is observed in response to multiple miRNAs compared to any one miRNA alone (Krek *et al.*, 2005; Wu *et al.*, 2010). These studies indicate that in some cases, alterations in certain miRNAs should be studied in the context of changes in other relevant miRNAs in order to fully appreciate the extent of collaborative miRNA control.

2.4. MicroRNAs and aging

Interestingly, some of the phenotypes that have been observed in miRNA-deficient mice take long periods of time to develop. For instance, knockouts of either miR-15a/16-1 or miR-146a in mice suffer from hematopoietically derived tumors, but these are only detectable after at least 1 year of age (Boldin *et al.*, 2011; Klein *et al.*, 2010). miR-155−/− Tregs exhibit a competitive disadvantage versus wild-type Tregs only after more than 3 months of bone marrow reconstitution (Lu *et al.*, 2009). These delayed phenotypes, which are preceded by long periods of unapparent phenotypic effects, suggest that relatively small changes in target gene expression can "build up" over time and eventually result in distinct hematopoietic consequences. These mutational scenarios imply that normally miRNAs ensure that the hematopoietic system operates with high fidelity over long periods of time. It points to an important role for miRNAs in the aging process and suggests that these ncRNAs may be of particular significance in the biology of organisms with longer life spans.

3. Controlling MicroRNA-Mediated Repression of mRNA Targets

In recent years, much has been learned about the mechanistic basis of miRNA biogenesis and subsequent target repression, and this has been reviewed exhaustively elsewhere (Filipowicz *et al.*, 2008). It has become very clear that this is an extensively regulated process that is controlled by a variety of different molecular mechanisms at different steps during the biogenesis process (Fig. 6.1). In this section, many of these points of regulation will be discussed including ways in which the highly dynamic hematopoietic system benefits through control by miRNA levels.

3.1. MicroRNA biogenesis and function

miRNAs are transcribed most commonly by RNA polymerase II, and this process is therefore regulated by the armamentarium of transcription factors available in any one cell type. Thus, like the protein-coding genes of specific cells, miRNAs are first controlled at the level of transcription and can be part of the suite of genes regulated by master regulators to help confer cellular identity. After the primary miRNA transcript is produced, it begins being processed into a pre-miRNA by the enzymes Drosha and DGCR8 (Filipowicz *et al.*, 2008). In addition, other factors such as KSRP, which are expressed in blood cells, also bind to the loop region of the primary transcript and further manage processing of the nascent RNA molecule

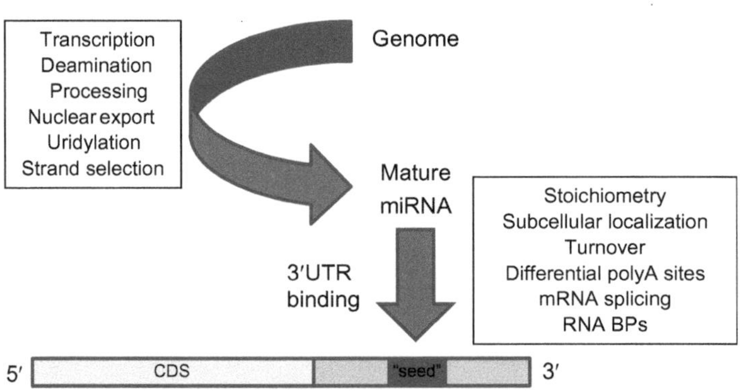

Figure 6.1 The regulation of miRNA function—miRNAs are first controlled by transcription which gives rise to a primary transcript. Following a number of indicated regulatory events, the mature miRNA is loaded into the RISC complex. There are also distinct regulatory steps that control miRNA–RISC complex interactions with target mRNAs. (For color version of this figure, the reader is referred to the Web version of this chapter.)

(Ruggiero *et al.*, 2009; Trabucchi *et al.*, 2009). The pre-miRNA is then relocated to the cytoplasm via Exportin 5 where it is processed further by Dicer. Interestingly, it has been recently shown that certain miRNAs, such as the erythrocyte regulator miR-451, are processed independently of Dicer although via an Ago2-dependent pathway (Cheloufi *et al.*, 2010). Additional forms of posttranscriptional modification, and thus control, have also been reported for certain miRNAs such as let-7, where Tutase mediates uridylation of pre-let-7 in a Lin28-dependent manner (Hagan *et al.*, 2009). This, in turn, causes a block to subsequent let-7 processing. The sequence of the miRNA itself can be subject to variations that include alterations in length at the 3′end (Wu *et al.*, 2007), as well as Adar-dependent deamination that can impact processing and/or "seed" targeting (Yang *et al.*, 2006).

Once a miRNA duplex is created, it is loaded into the RISC complex. The miRNA duplex is unwound and a guide strand is used to direct RISC to cognate 3′UTRs. This is followed by reduced transcript stability through uncapping and deadenylation, and/or a block in translation, ultimately resulting in lower amounts of protein expression (Filipowicz *et al.*, 2008).

3.2. MicroRNA turnover

The relative proportions of miRNAs and their targets needed for adequate repression of gene expression are just beginning to be carefully assessed with the use of next-generation sequencing approaches. Like all chemical reactions, the stoichiometry must be within an optimal range for a miRNA to have a meaningful impact on its targets. It has been shown that the most highly expressed miRNAs make up the bulk of those loaded into the RISC complex (Landthaler *et al.*, 2008). However, because miRNAs can work in concert to enhance target repression (as mentioned above), it is likely that miRNAs expressed at lower levels can work together to repress common targets. Although such details are still being studied, it is clear that the abundance of a mature miRNA is an important determinant of its functionality. Thus, mechanisms that control both its initial production and the turnover of its mature form will have a significant impact on miRNA biology.

While miRNA biogenesis has been well studied, active mechanisms governing miRNA turnover are just beginning to be understood. Perhaps, the best examples of active miRNA turnover have come from experiments using nonmammalian model organisms. Studies in plants and nematodes have identified the exonucleases Sdn1 and Xrn2, respectively, as mediators of miRNA degradation (Chatterjee and Grosshans, 2009; Ramachandran and Chen, 2008). However, recent studies using human cells have found that the exosome 3′–5′ exoribonuclease complex is also involved in miRNA decay (Bail *et al.*, 2010). Further investigation into the mechanisms regulating miRNA turnover remains an important direction for the field.

A hallmark of hematopoiesis is the uncoupled processes of cellular differentiation followed by expansion of the newly formed cell type. This process is continued until terminal differentiation is reached. Such a dynamic process involves the expression of stage-specific genes, including both mRNAs and miRNAs (Kuchen *et al.*, 2010). Based upon these features, one can imagine that at proliferative stages where synthesis of a particular miRNA ceases, cell division and subsequent growth will lead to effective dilution of the miRNA as a mechanism of reducing its concentration. When developing cell populations exit the cell cycle and begin to further differentiate, the miRNAs they produce will be able to efficiently increase in concentration within the quiescent cell. Thus, such a dynamic system can make use of its various cell states to regulate miRNA abundance and therefore gene expression even though the miRNAs themselves may be intrinsically quite stable.

3.3. Regulating miRNA interactions with mRNA 3′UTRs

While many mRNA targets can be predicted for a given miRNA based upon the presence of conserved 3′UTR sites coupled with thermodynamic considerations, it is common to observe only a fraction of predicted target mRNAs being repressed in response to their cognate miRNA within a given cell type. Obviously, an important prerequisite of miRNA targeting of an mRNA is concurrent spatial and temporal expression of both molecules. However, this does not seem a sufficient explanation of why certain targets are not repressed. Thus, additional levels of regulation must exist.

Current studies have begun to expand our understanding of why coexpression of a miRNA and its target mRNA is not always sufficient to trigger repression. One explanation is a loss of the miRNA binding site by the target mRNA under certain conditions. For instance, when T lymphocytes proliferate following their activation, they have been shown to express mRNAs with shortened 3′UTRs as a result of usage of alternative polyadenylation sites (Sandberg *et al.*, 2008). This site shift can result in a loss of miRNA binding sites, rendering these transcripts unresponsive to miRNA-mediated repression. To date, the consequences of this transcript shortening are still under investigation.

On the other hand, some miRNA-targeted mRNAs can undergo alternative splicing and lose terminal exons and adjacent downstream miRNA binding sites in the process. This has been shown for some targets of miR-155, which is a major regulator of hematopoietic development and function (Xu *et al.*, 2010). Yet another regulatory mechanism involves RNA-binding proteins, including Dnd1 and Elavl1, which have the potential to interact with specific 3′UTRs and guard them against or promote miRNA-mediated repression (Kedde *et al.*, 2007; Kim *et al.*, 2009). mRNAs have also been shown to undergo compartmentalization within certain subcellular vesicles, and this could impact repression by miRNAs. Subcellular structures such as P-bodies and stress granules have been identified as sites at which miRNAs accumulate, but their roles in miRNA biology remain obscure (Leung *et al.*, 2006; Pontes and Pikaard, 2008). Taken together, miRNA binding sites within a 3′UTR appear to be subjected to a variety of regulatory mechanisms, and this is reminiscent of the diverse ways in which promoter regions upstream of genes are used to control transcription.

4. MicroRNAs Regulate Different Stages of Hematopoiesis

A plethora of recent reports indicate that miRNAs are essential for proper hematopoietic development and act at various points throughout the process, from HSCs through terminal differentiation (O'Connell *et al.*, 2010c). It is also becoming clear that miRNAs can have both positive and negative impacts on specific aspects of developing blood cells, indicating that their functions must be properly integrated to ensure blood cell homeostasis (Fig. 6.2A).

4.1. Hematopoietic stem cells

Mammalian hematopoiesis is a hierarchical developmental system that depends upon a rare stem cell population to ensure production of blood cells throughout the lifetime of an individual (Orkin and Zon, 2008). Because HSCs are long lived and must strike a careful balance between differentiation and self-renewal, they are good candidates to be regulated by miRNAs. In fact, a number of reports have linked miRNAs to proper HSC function (Guo *et al.*, 2010; O'Connell *et al.*, 2010a; Ooi *et al.*, 2010). Inducible deletion of Dicer or Ars2, involved in miRNA processing, conferred a competitive disadvantage on engrafting HSCs and overall hematopoiesis (Gruber *et al.*, 2009; Guo *et al.*, 2010). This reduction in fitness in the absence of miRNAs could be a consequence of many different aspects of stem cell biology. Stem cell maintenance and function involve many cellular processes including apoptosis, proliferation, differentiation, and cellular trafficking. Specific miRNAs have started to be linked to these events. For instance, the miR-125 family is enriched in HSCs and has been shown to target genes involved in apoptosis (Guo *et al.*, 2010; Ooi *et al.*, 2010). miR-196b is also expressed in HSCs (O'Connell *et al.*, 2010a; Popovic *et al.*, 2009) and regulates specific Hox family members that control differentiation (Yekta *et al.*, 2004). The mechanisms underlying the HSC choice between self-renewal and differentiation remain to be clearly

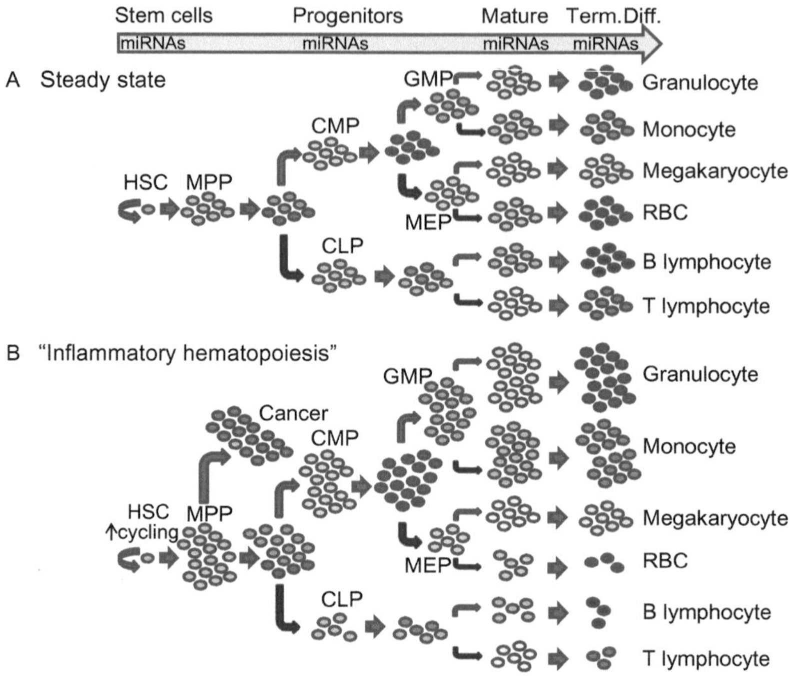

Figure 6.2 Steady-state and "Inflammatory hematopoiesis"—(A) Mammalian hematopoiesis in the bone marrow begins with the hematopoietic stem cell (HSC) which gives rise to progenitors that continue to develop into lineage-restricted progeny. Ultimately, cells reach maturity and become terminally differentiated. miRNAs have been shown to play important regulatory roles at the different stages of blood cell development. (B) The hematopoietic process is sensitive to inflammatory stress caused by microbial infection or elevations in specific types of cytokines. During "inflammatory hematopoiesis," the HSC increases its cell cycle rate and produces more progenitors with a bias toward making granulocyte and monocyte progeny at the expense of other cell lineages. This condition also resembles that observed during preleukemia suggesting a link between "inflammatory hematopoiesis" and cancer. Several miRNAs have also been implicated in the regulation of this process. MPP, multipotent progenitor; CMP, common myeloid progenitor; CLP, common lymphoid progenitor; GMP, granulocyte–monocyte progenitor; MEP, megakaryocyte–erythroid progenitor; RBC, red blood cell. (See Color Insert.)

understood. How miRNAs are integrated into the molecular networks governing these decisions, as they have been shown to be in embryonic stem cells (Melton *et al.*, 2010), is still under investigation.

HSCs first arise during embryonic development and continue to ensure blood production throughout adulthood. However, the HSC does not have an unchanging phenotype. As mice age, they progressively alter the output of their HSCs, favoring progeny cells of the myeloid lineage while exhibiting a reduction in their repopulating capacity despite appearing in increased numbers in the bone marrow (Ergen and Goodell, 2010). Consistent with an ever-changing HSC phenotype, many of the pathways that manage fetal HSCs are distinct from those involved in adult HSCs (Rossi *et al.*, 2008). This suggests that the specific miRNAs relevant to HSC biology might also change depending on the developmental stage of the organism. While we begin to identify miRNAs that are important for HSC function, careful consideration should be given to age-related aspects. This may help to unravel more of the underlying causes of why HSC function varies with age.

4.2. Lymphoid versus myeloid

As progeny hematopoietic cells differentiate away from the HSC, they proceed down one of two general lineages: lymphoid or myeloid. Specific deletion of Dicer in T cells (and thus a majority of miRNAs) results in dramatic defects in the T lymphocyte lineage (Cobb *et al.*, 2005; Muljo *et al.*, 2005). This is characterized by a substantial reduction in the total number of T lymphocytes with the magnitude of this decrease varying depending on the stage of thymic development at which Dicer is removed. Specific miRNAs involved in thymopoiesis have been identified, and these include miRs-17–92 and miR-181a (Li *et al.*, 2007; Xiao *et al.*, 2008). An important aspect of T cell development is selecting clones with the proper antigen specificity to ensure immunity against infection while preventing autoreactive T cells from reaching the periphery. miR-181a functions to modulate the signaling strength of the TCR during development and is therefore important for accurate thymic selection (Ebert *et al.*, 2009).

Dicer is also required for Treg development, and its absence leads to systemic autoimmunity. In this case, miR-155 and miR-146a are implicated in Treg development and function, respectively (Lu *et al.*, 2009, 2010). Evidence points to a critical role for these miRNAs in the regulation of Jak–Stat pathways in Tregs.

A lack of Dicer during early B cell development triggers reduced cell survival and diminished antigen receptor diversity (Koralov *et al.*, 2008). miRs-17-92 and miR-150 appear to be involved in these phenotypes by targeting Bim, Pten, and cMyb (Koralov *et al.*, 2008; Xiao *et al.*, 2007). Although there is a requirement for certain miRNAs for B lymphocytes to progress through their developmental stages, other miRNAs function as inhibitors of B cell maturation and function, limiting B cell output. Among these, enforced overexpression of miR-34a restricts B cells from passing through the proB to preB cell stage of differentiation (Rao *et al.*, 2010). This is mediated through repression of the transcription factor FoxP1.

To date, global disruption of miRNA expression in the myeloid lineage has not been reported to be as deleterious as it is in lymphocytes. However, removal of Dicer using a CD11b-driven Cre mouse strain does lead to diminished osteoclast differentiation and function (Sugatani and Hruska, 2009). Another study found that the development of Langerhans cells in the skin was impaired when Dicer was deleted in CD11c+ cells (Kuipers *et al.*, 2010). Perhaps, this disparity between the importance of Dicer in lymphoid versus myeloid cells is due to the innate immune system being much older and less complex than the adaptive branch, potentially having already evolved before miRNAs themselves had evolved to be able to play critical roles in cellular identity and functionality. This suggests an important general principle that miRNAs play more of a role in recently evolved systems than in more ancient systems. However, it is clear that in myeloid cells, the effects of some miRNAs are balanced by opposing effects of others. By eliminating almost all miRNAs, one might be removing both positive and negative regulators of development leaving the system still able to function, while gain or loss of function of one miRNA does not involve a counterbalance triggering a phenotype. Such a situation is evident for miRNAs 146a and 155, which clearly oppose one another in their actions (see below). This interpretation suggests that the role of these miRNAs might be as buffers against excursions of the transcriptional apparatus as well as providing fine-tuning for a system under the tension of opposing forces.

An impact on the subtypes of myeloid development has been observed when individual miRNAs have been studied. miR-223-deficient mice exhibit an expanded granulocytic compartment and show hyperactivation of granulocytes during fungal infections (Johnnidis *et al.*, 2008). Deletion of miR-146a eventually causes an overproduction of myeloid cells of the granulocyte–monocyte (GM) lineage (Boldin *et al.*, 2011). Conversely, overexpression of miR-155, miR-29a, or miR-125b in the bone marrow all trigger a bias toward GM cell production, suggesting that these miRNAs manage the lymphoid/myeloid balance (Bousquet *et al.*, 2010; Han *et al.*, 2010; O'Connell *et al.*, 2008, 2010a).

Megakaryocytes (which generate platelets) and red blood cells are also part of the myeloid lineage and arise from a common megakaryocyte–erythrocyte progenitor (MEP). Megakaryocytes have been shown to have a distinct miRNA expression "fingerprint" initially suggesting a role for miRNAs in directing megakaryocyte development (Garzon *et al.*, 2006). miR-150 was later shown to promote MEP differentiation into megakaryocytes through repression of cMyb (Lu *et al.*, 2008a). Alternatively, deletion of miR-451 has revealed a role for this miRNA in RBC development (Patrick *et al.*, 2010; Rasmussen *et al.*, 2010). This is a good example of specific miRNAs assisting in lineage decisions made by progenitor populations.

4.3. Reactivation of development

In some cases, naïve cells of the immune system can become activated and this potentiates additional developmental programs. Although considered to have reached maturity, properly activated B or T lymphocytes can continue to differentiate into effector cells or long-lived memory cells. In such cases, a productive immune response can license a continuation in lymphocyte development and lifespan and miRNAs have also been linked to these later stages of development.

B lymphocytes must recognize cognate antigens and receive the necessary costimulatory signals in order to develop into antibody-secreting plasma cells. This process occurs in the germinal center (GC) of lymph nodes and spleen and involves both Ig class switching and affinity maturation of the antigen receptor. A fraction of B cells develop into long-lived memory cells ready to mount a secondary response. miR-155 has been shown to promote GC formation and class switching to IgG and recall responses are diminished in miR-155−/− mice (Rodriguez *et al.*, 2007; Thai *et al.*, 2007; Vigorito *et al.*, 2007). This involves repression of the miR-155 targets Pu.1 and Aicda (Dorsett *et al.*, 2008; Rodriguez *et al.*, 2007; Thai *et al.*, 2007; Vigorito *et al.*, 2007). In distinction to this role for miR-155, there is evidence that miR-125b inhibits the GC response and terminal differentiation of B cells by repression of Blimp1 and IRF4 (Gururajan *et al.*, 2010).

T lymphocytes also continue their developmental programs following antigen encounter and proper costimulation. Naïve helper CD4+ T cells can be skewed into different effector subsets including Th1, Th2, and Th17, which promote distinct types of immune responses. miR-155 impacts these choices and this influence involves repression of the transcription factor cMaf and likely other targets that have yet to be defined (Rodriguez *et al.*, 2007). Loss-of-function studies have found that miR-155 is necessary for adequate differentiation into Th1 and Th17 cell types *in vivo* (O'Connell *et al.*, 2010b), while miR-155−/− CD4+ T cells exhibit a Th2 bias *in vitro* (Rodriguez *et al.*, 2007; Thai *et al.*, 2007). Th17 development is also regulated by miR-326 which represses Ets1, an established negative regulator of the Th17 lineage (Du *et al.*, 2009).

5. MicroRNAs During Hematopoietic Stress and Disease

Breakdowns in the hematopoietic development process or in cellular functioning are underlying causes of hematopoietic disease, and these can be initiated by inflammatory stress, mutations in cells of a specific lineage, or genetic predispositions. In some cases, developmental blocks lead to deficiencies in certain blood cell populations such as RBCs, in the case of anemia, or white blood cells, resulting in immunodeficiency. Alternatively, when excess production of blood cells occurs, it typically involves a single lineage. For instance, some myeloproliferative disorders (MPDs) can be characterized by increases in myeloid cells of the GM lineage. In both of these cases (developmental blocks and enhanced proliferation), hematopoietic malignancies can arise. Alternatively, when inappropriate development of lymphocytes takes place, inflammatory responses against self-tissues can ensue, leading to debilitating autoimmunity. miRNAs have been connected to all of these disorders owing to their critical roles in regulating blood cell development.

5.1. Inflammatory hematopoiesis

A majority of our understanding of mammalian hematopoiesis is based upon animal experiments that have been performed under steady-state, noninflammatory conditions. However, mammals are regularly exposed to a variety of stress conditions that have a clear impact on blood cell development, including microbial infection. The resulting inflammatory response produces cytokines and growth factors with the propensity to impact hematopoietic development resulting in "inflammatory hematopoiesis" (Fig. 6.2B). Mouse models have revealed an enhanced output of GM populations and a concurrent reduction in B lymphocytes and erythroid precursors under such conditions (O'Connell *et al.*, 2008; Ueda *et al.*, 2005). In conjunction with host-produced growth factors, this acute developmental shift also appears to be influenced by direct recognition of pathogen-associated molecular patterns by stem and progenitor populations (Nagai *et al.*, 2006; O'Connell *et al.*, 2008; Takizawa *et al.*, 2011; Ueda *et al.*, 2005). Cells that make up the stem and progenitor compartment have recently been shown to express Toll-like receptors which can directly impact the expression of miRNAs among other genes (O'Connell *et al.*, 2007, 2008; O'Neill *et al.*, 2011).

A number of recent studies have provided evidence that specific miRNAs may be linked to inflammatory hematopoiesis. miR-155 is upregulated in the bone marrow compartment in response to inflammatory stimuli, and its overexpression triggers a myeloid phenotype resembling that caused by treatment with endotoxin (O'Connell *et al.*, 2008). Although transcription of BIC (the precursor to miR-155) is controlled by NF-κB and JNK (O'Connell *et al.*, 2007; Thai *et al.*, 2007), a recent report found that Hoxa9, which is expressed in stem and progenitor cells, also regulates miR-155 levels (Hu *et al.*, 2010). miR-146a is also upregulated by inflammatory signaling pathways and has the opposite impact on hematopoiesis. miR-146a−/− mice develop an MPD characteristic of chronic inflammatory disease owing to hyperactivation of NF-κB (Boldin *et al.*, 2011; Zhao *et al.*, 2011). This appears to be a consequence of the enhanced expression of IRAK1 and TRAF6 (Taganov *et al.*, 2006), and possibly STAT1 (Lu *et al.*, 2010), all proinflammatory signaling proteins that are direct targets of miR-146a.

Many human disorders of hematological origin are phenotypically similar to inflammatory hematopoiesis, suggesting an overlap between the molecular mechanisms that govern these distinctly pathological and physiological processes. Although the studies to date have taken important first steps, much remains to be determined regarding how miRNAs modulate hematopoiesis during inflammation and how these pathways relate to proper immune function versus hematological diseases.

5.2. Cancer

It is quite clear that miRNA expression is dysregulated in a variety of different human leukemias and lymphomas, and that this dysregulation is sufficient to cause cancer in some cases (Bousquet *et al.*, 2010; Calin and Croce, 2006; Costinean *et al.*, 2006; Garzon *et al.*, 2008; O'Connell *et al.*, 2010a). It is not surprising that many of the same miRNAs that regulate hematopoiesis under physiological conditions are involved in disease states. Some miRNAs function as onco-miRs, and their overexpression is sufficient to cause precancerous neoplasms, or in some cases, frank malignancies. Other miRNAs act to suppress cancer development, and loss of these molecular safeguards can result in the onset of malignant disease.

Because miRNAs regulate many different stages of hematopoiesis, malignancies involving perturbations in their levels can arise from cells at unique points within the developmental series. For example, overexpression of miR-155 or miR-29a in HSCs causes an MPD or a frank leukemia, respectively (Han *et al.*, 2010; O'Connell *et al.*, 2008). Alternatively, B cell malignancies are triggered when miR-21 or miR-155 is overexpressed or miR-15a/16-1 is deleted in B lymphocytes (Costinean *et al.*, 2006; Klein *et al.*, 2010; Medina *et al.*, 2010).

As we begin to define the protein targets of specific cancer-related miRNAs, our understanding of how this class of ncRNAs influences cancer phenotypes continues to expand. For example, miR-155 has been shown to repress certain negative regulators of inflammation and cellular activation such as Ship1 and Socs1 (Androulidaki *et al.*, 2009; Costinean *et al.*, 2009; Lu *et al.*, 2009; O'Connell *et al.*, 2009). Upon inappropriate overexpression of miR-155 stemming from chronic inflammation or mutations triggering constant cellular activation, these and other targets will be continually repressed. Consequently, the pathways they normally restrict will become hyperactivated and over time will cause hyperproliferative disease.

Viruses have also been shown to hijack miRNAs and their modes of cellular regulation for their own purposes. For instance, Epstein–Barr virus utilizes miR-155 to transform B cells (Linnstaedt *et al.*, 2011). Kaposi's sarcoma-associated herpesvirus encodes an ortholog of miR-155 with overlapping target specificity, and this is thought to be involved in its pathogenicity (Gottwein *et al.*, 2007). Thus, miR-155's ability to regulate cellular developmental and activation pathways can also be exploited by a variety of oncogenic viruses, underscoring the importance of miRNAs in hematopoietic disease.

Although some miRNAs play oncogenic roles in the hematopoietic system, others have emerged as powerful tumor suppressors. Mice deficient in miR-15a/16-1 develop CLL through a mechanism involving derepression of the survival factor BCL2 (Klein *et al.*, 2010). miR-146a−/− mice develop a range of malignancies, possibly driven by sustained inflammatory stress (Boldin *et al.*, 2011). Of note, expression profiling of many different human cancers indicates that a majority of miRNAs are expressed at reduced levels in malignant cells consistent with a more common tumor suppressor role. Dicer has been shown to function as a haploinsufficient tumor suppressor, with one copy of Dicer leading to a more aggressive malignancy than a complete loss of Dicer (Kumar *et al.*, 2009). Perhaps, reductions in Dicer protein levels lead to lower expression of tumor suppressor miRNAs while maintaining some expression of relevant onco-miRs. This might have a stronger impact on cancer development than complete loss of most miRNAs.

Beyond cell intrinsic roles for miRNAs in the promotion of hematopoietic cancers, the miRNA pathway is also necessary for proper function of the bone marrow "stroma." The bone marrow "stroma" is comprised of fibroblasts, adipocytes, osteoblasts, osteoclasts, and endothelial cells that support stem cell function and hematopoiesis in general. Conditional deletion of Dicer in osteoprogenitors triggers a myelodysplasia that transitions into a secondary leukemia (Raaijmakers *et al.*, 2010). This clearly demonstrates that miRNAs can also impact hematopoiesis and cancer indirectly by perturbing supporting tissues and niches.

5.3. Other hematopoietic disorders

Blood disorders not involving malignant transformation can also be influenced by miRNAs. As mentioned above, an essential function of the hematopoietic system is delivery of oxygen to tissues. Reductions in RBC numbers and/or function causes anemia, and a number of miRNAs have been shown to trigger anemia in mice. A loss of miR-146a or miR-451, or overexpression of miR-155, miR-125b, or miR-29a, all lead to reduced RBC levels (Boldin *et al.*, 2011; Han *et al.*, 2010; O'Connell *et al.*, 2008, 2010a; Patrick *et al.*, 2010; Rasmussen *et al.*, 2010). In most of these cases, the anemia is in conjunction with an MPD, suggesting that myeloid progenitors preferentially produce GM cells at the expense of RBCs in such settings.

Deficiencies in WBCs such as B and T cells are also observed in the absence of Dicer, as described above, and this impairs the immune response. Individual miRNAs that regulate lymphocyte development have the propensity to perturb immune function and create an immunodeficiency. In such cases, an inadequate immune response would lead to greater susceptibility to infection. The interplay between the immune system and solid tumors could also be impacted and might result in enhanced tumor growth.

Conversely, if the immune system develops inappropriately, it can trigger responses against self-tissues. miRNAs have recently been linked to a variety of autoimmune conditions, with their altered expression correlating with disease (Pauley *et al.*, 2009). The importance of miRNAs during autoimmunity is demonstrated by the consequences of a lack of Dicer on Tregs, impairing their development and unleashing a lethal autoimmune condition. Among the specific miRNAs expressed in Tregs, genetic deletion of miR-146a has been shown to impair Treg function resulting in autoimmune inflammation (Lu *et al.*, 2010). Furthermore, two recent studies have found that a lack of miR-155 or inhibition of miR-326 in mice can reduce symptoms of experimental autoimmune encephalomyelitis, a model of human multiple sclerosis (Du *et al.*, 2009; O'Connell *et al.*, 2010b). In both of these cases, a reduction in inflammatory T cell development was observed.

6. Concluding Remarks

The studies described in this review indicate that hematopoietic miRNAs commonly regulate certain types of cellular pathways (Table 6.2). Some miRNAs, such as miR-155, miR-146a, miR-181a, miR-126, miR-451, and miR-21, target proteins involved in cytokine, TLR, or antigen receptor signaling pathways and consequently impact the magnitude of the cellular response to ligands (Li *et al.*, 2007, 2008; Lu *et al.*, 2009; O'Connell *et al.*, 2009; Patrick *et al.*, 2010; Sheedy *et al.*, 2009; Taganov *et al.*, 2006). Other miRNAs repress protein regulators of apoptosis and influence cell survival. These include the miR-125 family, miR-15a/16-1, and miRs-17-92 (Cimmino *et al.*, 2005; Guo *et al.*, 2010; Klein *et al.*, 2010; Ooi *et al.*, 2010; Xiao *et al.*, 2008). Several hematopoietic miRNAs, including miR-150, miR-155, miR-29a, miR-146a, miR-34a, miR-223, miR-196b, and miR-326, directly target transcription factors that control lineage choice (Du *et al.*, 2009; Fabbri *et al.*, 2007; Hu *et al.*, 2010; Johnnidis *et al.*, 2008; Lu *et al.*, 2010; Popovic *et al.*, 2009; Rao *et al.*,

Table 6.2 MicroRNAs that regulate hematopoietic development

miRNA	Knockout/loss-of-function phenotype	Overexpression phenotype	Critical targets	Transcription factor/nuclear protein	Receptor signaling	Apoptosis
miR-150	Defective B cell development	Block in B cell development	Myb	×		
miR-34a	Enhanced B cell output	Block in B cell development	Foxp1	×		
miR-223	Increased granulocyte compartment	nd	Mef2c	×		
miR-196b	nd	Decreased HSC repopulating potential	Hox genes	×		
miR-29a	nd	MPD progressing to AML	Dnmt3a	×		
miR-326	Inhibited Th17 development	Increased Th17 development	Ets1	×		
miR-155	Impaired Ig class switching, decreased inflammatory T cells, and dysfunctional GM cells	MPD or B cell leukemia	PU.1, Ship1, cMaf, Socs1, Aicda	×	×	
miR-146a	Eventual MPD progressing to cancer	Perturbed hematopoiesis	Irak1, Traf6, Stat1	×	×	
miR-21	nd	B cell lymphoma	Pdcd4	×	×	
miR-451	Impaired RBC development	nd	14-3-3zeta		×	
miR-181a	Defective TCR signaling	Increased B cell development	Dusp5, Dusp6, Shp2 and Ptpn22		×	

miR-126	nd	Increased HSC repopulating potential	Plk2	×	
miRs-17-92	Impaired B cell development	Autoimmunity	Bim, Pten	×	×
miR-15a/16-1	CLL	nd	Bcl2		×
miR-125	nd	Increased HSC repopulating potential; MPD and leukemia	Bak1, Bmf		×

2010; Rodriguez *et al.*, 2007; Vigorito *et al.*, 2007; Xiao *et al.*, 2007; Yekta *et al.*, 2004). Although there are clearly exceptions to these types of miRNA regulation in the hematopoietic system, these appear to be the most commonly targeted nodes in the molecular networks that control blood cell development.

Novel and exciting aspects of miRNA biology are continuing to unfold. For instance, recent evidence suggests that miRNAs are involved in intercellular regulatory mechanisms. Microvesicles, or exosomes, found in the serum are loaded with mature miRNAs (Hunter *et al.*, 2008). Many cancer cells, including those of hematopoietic origin, have been shown to release exosomes containing miRNAs (Zhang *et al.*, 2010). While these offer important diagnostic opportunities, their roles during normal physiology are currently being evaluated. Such a system could allow for a cell to downregulate genes in a neighboring or distal cell type, making miRNAs important mediators of communication between cells.

Another emerging role for miRNAs is in the production of induced pluripotent stem (iPS) cells. Specific miRNAs are able to replace the iPS factor cMyc during the conversion of mature cells into iPS cells (Judson *et al.*, 2009). The ability to reverse hematopoietic development using biotechnology has many practical and clinically important applications, and recent evidence indicates that this is also possible (Eminli *et al.*, 2009). As we begin to learn more about which miRNAs are important for hematopoiesis, we may be able to exploit their programming to direct mature immune cells back into iPS cells and/or HSCs. Additionally, cells that make up the immune system appear to have great developmental plasticity, and miRNAs might be used to skew immune cells from one lineage to another.

Taken together, miRNAs represent a promising frontier in the arena of mammalian hematopoiesis and molecular biology in general. Although it is becoming clear that miRNAs are integral components of the signaling networks underlying blood cell development, the field is young and thus the identity and function of many hematopoietically relevant miRNAs and other ncRNAs remain unknown. Identifying and characterizing miRNAs and other ncRNAs involved in these fields of study will be important for the progress of basic science, including our understanding of blood cell development. Additionally, continued advancement in these areas will undoubtedly reveal novel biomarkers and therapeutic targets to be used in the diagnosis and treatment of hematological diseases.

ACKNOWLEDGMENTS

R. M. O. was funded by award number 4R00HL102228-03 from the National Heart, Lung and Blood Institute. This work was also supported by NIH Grant 1R01AI079243-01.

REFERENCES

Androulidaki, A., Iliopoulos, D., Arranz, A., Doxaki, C., Schworer, S., Zacharioudaki, V., Margioris, A. N., Tsichlis, P. N., and Tsatsanis, C. (2009). The kinase Akt1 controls macrophage response to lipopolysaccharide by regulating microRNAs. *Immunity* **31,** 220–231.

Arron, J. R., Winslow, M. M., Polleri, A., Chang, C. P., Wu, H., Gao, X., Neilson, J. R., Chen, L., Heit, J. J., Kim, S. K., Yamasaki, N., Miyakawa, T., *et al.* (2006). NFAT dysregulation by increased dosage of DSCR1 and DYRK1A on chromosome 21. *Nature* **441,** 595–600.

Asirvatham, A. J., Gregorie, C. J., Hu, Z., Magner, W. J., and Tomasi, T. B. (2008). MicroRNA targets in immune genes and the Dicer/Argonaute and ARE machinery components. *Mol. Immunol.* **45,** 1995–2006.

Bail, S., Swerdel, M., Liu, H., Jiao, X., Goff, L. A., Hart, R. P., and Kiledjian, M. (2010). Differential regulation of microRNA stability. *RNA* **16,** 1032–1039.

Bartel, D. P., and Chen, C. Z. (2004). Micromanagers of gene expression: The potentially widespread influence of metazoan microRNAs. *Nat. Rev. Genet.* **5,** 396–400.

Boldin, M. P., Taganov, K. D., Rao, D. S., Yang, L., Zhao, J. L., Kalwani, M., Garcia-Flores, Y., Luong, M., Devrekanli, A., Xu, J., Sun, G., Tay, J., *et al.* (2011). miR-146a is a significant brake on autoimmunity, myeloproliferation, and cancer in mice. *J. Exp. Med.,* **208**(6), 1189–1201.

Bousquet, M., Harris, M. H., Zhou, B., and Lodish, H. F. (2010). MicroRNA miR-125b causes leukemia. *Proc. Natl. Acad. Sci. USA* **107,** 21558–21563.

Calin, G. A., and Croce, C. M. (2006). MicroRNA signatures in human cancers. *Nat. Rev. Cancer* **6,** 857–866.

Chan, G., Cheung, L. S., Yang, W., Milyavsky, M., Sanders, A. D., Gu, S., Hong, W. X., Liu, A. X., Wang, X., Barbara, M., Sharma, T., Gavin, J., *et al.* (2011). Essential role for Ptpn11 in survival of hematopoietic stem and progenitor cells. *Blood* **117,** 4253–4261.

Chatterjee, S., and Grosshans, H. (2009). Active turnover modulates mature microRNA activity in Caenorhabditis elegans. *Nature* **461,** 546–549.

Cheloufi, S., Dos Santos, C. O., Chong, M. M., and Hannon, G. J. (2010). A dicer-independent miRNA biogenesis pathway that requires Ago catalysis. *Nature* **465,** 584–589.

Cimmino, A., Calin, G. A., Fabbri, M., Iorio, M. V., Ferracin, M., Shimizu, M., Wojcik, S. E., Aqeilan, R. I., Zupo, S., Dono, M., Rassenti, L., Alder, H., *et al.* (2005). miR-15 and miR-16 induce apoptosis by targeting BCL2. *Proc. Natl. Acad. Sci. USA* **102,** 13944–13949.

Cobb, B. S., Nesterova, T. B., Thompson, E., Hertweck, A., O'Connor, E., Godwin, J., Wilson, C. B., Brockdorff, N., Fisher, A. G., Smale, S. T., and Merkenschlager, M. (2005). T cell lineage choice and differentiation in the absence of the RNase III enzyme Dicer. *J. Exp. Med.* **201,** 1367–1373.

Cooper, M. D., and Alder, M. N. (2006). The evolution of adaptive immune systems. *Cell* **124,** 815–822.

Costinean, S., Zanesi, N., Pekarsky, Y., Tili, E., Volinia, S., Heerema, N., and Croce, C. M. (2006). Pre-B cell proliferation and lymphoblastic leukemia/high-grade lymphoma in E (mu)-miR155 transgenic mice. *Proc. Natl. Acad. Sci. USA* **103,** 7024–7029.

Costinean, S., Sandhu, S. K., Pedersen, I. M., Tili, E., Trotta, R., Perrotti, D., Ciarlariello, D., Neviani, P., Harb, J., Kauffman, L. R., Shidham, A., and Croce, C. M. (2009). Src homology 2 domain-containing inositol-5-phosphatase and CCAAT enhancer-binding protein beta are targeted by miR-155 in B cells of Emicro-MiR-155 transgenic mice. *Blood* **114,** 1374–1382.

Cui, Q., Yu, Z., Purisima, E. O., and Wang, E. (2007). MicroRNA regulation and interspecific variation of gene expression. *Trends Genet.* **23,** 372–375.

Dahl, R., Walsh, J. C., Lancki, D., Laslo, P., Iyer, S. R., Singh, H., and Simon, M. C. (2003). Regulation of macrophage and neutrophil cell fates by the PU.1:C/EBPalpha ratio and granulocyte colony-stimulating factor. *Nat. Immunol.* **4,** 1029–1036.

Dorsett, Y., McBride, K. M., Jankovic, M., Gazumyan, A., Thai, T. H., Robbiani, D. F., Di Virgilio, M., San-Martin, B. R., Heidkamp, G., Schwickert, T. A., Eisenreich, T., Rajewsky, K., *et al.* (2008). MicroRNA-155 suppresses activation-induced cytidine deaminase-mediated Myc-Igh translocation. *Immunity* **28,** 630–638.

Du, C., Liu, C., Kang, J., Zhao, G., Ye, Z., Huang, S., Li, Z., Wu, Z., and Pei, G. (2009). MicroRNA miR-326 regulates T(H)-17 differentiation and is associated with the pathogenesis of multiple sclerosis. *Nat. Immunol.* **10,** 1252–1259.

Ebert, P. J., Jiang, S., Xie, J., Li, Q. J., and Davis, M. M. (2009). An endogenous positively selecting peptide enhances mature T cell responses and becomes an autoantigen in the absence of microRNA miR-181a. *Nat. Immunol.* **10,** 1162–1169.

Egle, A., Harris, A. W., Bouillet, P., and Cory, S. (2004). Bim is a suppressor of Myc-induced mouse B cell leukemia. *Proc. Natl. Acad. Sci. USA* **101,** 6164–6169.

Eischen, C. M., Alt, J. R., and Wang, P. (2004). Loss of one allele of ARF rescues Mdm2 haploinsufficiency effects on apoptosis and lymphoma development. *Oncogene* **23,** 8931–8940.

Eminli, S., Foudi, A., Stadtfeld, M., Maherali, N., Ahfeldt, T., Mostoslavsky, G., Hock, H., and Hochedlinger, K. (2009). Differentiation stage determines potential of hematopoietic cells for reprogramming into induced pluripotent stem cells. *Nat. Genet.* **41,** 968–976.

Ergen, A. V., and Goodell, M. A. (2010). Mechanisms of hematopoietic stem cell aging. *Exp. Gerontol.* **45,** 286–290.

Fabbri, M., Garzon, R., Cimmino, A., Liu, Z., Zanesi, N., Callegari, E., Liu, S., Alder, H., Costinean, S., Fernandez-Cymering, C., Volinia, S., Guler, G., *et al.* (2007). MicroRNA-29 family reverts aberrant methylation in lung cancer by targeting DNA methyltransferases 3A and 3B. *Proc. Natl. Acad. Sci. USA* **104,** 15805–15810.

Filipowicz, W., Bhattacharyya, S. N., and Sonenberg, N. (2008). Mechanisms of post-transcriptional regulation by microRNAs: Are the answers in sight? *Nat. Rev. Genet.* **9,** 102–114.

Garzon, R., Pichiorri, F., Palumbo, T., Iuliano, R., Cimmino, A., Aqeilan, R., Volinia, S., Bhatt, D., Alder, H., Marcucci, G., Calin, G. A., Liu, C. G., *et al.* (2006). MicroRNA fingerprints during human megakaryocytopoiesis. *Proc. Natl. Acad. Sci. USA* **103,** 5078–5083.

Garzon, R., Garofalo, M., Martelli, M. P., Briesewitz, R., Wang, L., Fernandez-Cymering, C., Volinia, S., Liu, C. G., Schnittger, S., Haferlach, T., Liso, A., Diverio, D., *et al.* (2008). Distinctive microRNA signature of acute myeloid leukemia bearing cytoplasmic mutated nucleophosmin. *Proc. Natl. Acad. Sci. USA* **105,** 3945–3950.

Gottwein, E., Mukherjee, N., Sachse, C., Frenzel, C., Majoros, W. H., Chi, J. T., Braich, R., Manoharan, M., Soutschek, J., Ohler, U., and Cullen, B. R. (2007). A viral microRNA functions as an orthologue of cellular miR-155. *Nature* **450,** 1096–1099.

Gruber, J. J., Zatechka, D. S., Sabin, L. R., Yong, J., Lum, J. J., Kong, M., Zong, W. X., Zhang, Z., Lau, C. K., Rawlings, J., Cherry, S., Ihle, J. N., *et al.* (2009). Ars2 links the nuclear cap-binding complex to RNA interference and cell proliferation. *Cell* **138,** 328–339.

Guo, S., Lu, J., Schlanger, R., Zhang, H., Wang, J. Y., Fox, M. C., Purton, L. E., Fleming, H. H., Cobb, B., Merkenschlager, M., Golub, T. R., and Scadden, D. T.

(2010). MicroRNA miR-125a controls hematopoietic stem cell number. *Proc. Natl. Acad. Sci. USA* **107,** 14229–14234.

Gururajan, M., Haga, C. L., Das, S., Leu, C. M., Hodson, D., Josson, S., Turner, M., and Cooper, M. D. (2010). MicroRNA 125b inhibition of B cell differentiation in germinal centers. *Int. Immunol.* **22,** 583–592.

Guttman, M., Amit, I., Garber, M., French, C., Lin, M. F., Feldser, D., Huarte, M., Zuk, O., Carey, B. W., Cassady, J. P., Cabili, M. N., Jaenisch, R., *et al.* (2009). Chromatin signature reveals over a thousand highly conserved large non-coding RNAs in mammals. *Nature* **458,** 223–227.

Hagan, J. P., Piskounova, E., and Gregory, R. I. (2009). Lin28 recruits the TUTase Zcchc11 to inhibit let-7 maturation in mouse embryonic stem cells. *Nat. Struct. Mol. Biol.* **16,** 1021–1025.

Han, Y. C., Park, C. Y., Bhagat, G., Zhang, J., Wang, Y., Fan, J. B., Liu, M., Zou, Y., Weissman, I. L., and Gu, H. (2010). microRNA-29a induces aberrant self-renewal capacity in hematopoietic progenitors, biased myeloid development, and acute myeloid leukemia. *J. Exp. Med.* **207,** 475–489.

He, L. (2010). Posttranscriptional regulation of PTEN dosage by noncoding RNAs. *Sci. Signal.* **3,** pe39.

Hornstein, E., and Shomron, N. (2006). Canalization of development by microRNAs. *Nat. Genet.* **38**(Suppl.), S20–S24.

Hu, Y. L., Fong, S., Largman, C., and Shen, W. F. (2010). HOXA9 regulates miR-155 in hematopoietic cells. *Nucleic Acids Res.* **38,** 5472–5478.

Hunter, M. P., Ismail, N., Zhang, X., Aguda, B. D., Lee, E. J., Yu, L., Xiao, T., Schafer, J., Lee, M. L., Schmittgen, T. D., Nana-Sinkam, S. P., Jarjoura, D., *et al.* (2008). Detection of microRNA expression in human peripheral blood microvesicles. *PLoS One* **3,** e3694.

Johnnidis, J. B., Harris, M. H., Wheeler, R. T., Stehling-Sun, S., Lam, M. H., Kirak, O., Brummelkamp, T. R., Fleming, M. D., and Camargo, F. D. (2008). Regulation of progenitor cell proliferation and granulocyte function by microRNA-223. *Nature* **451,** 1125–1129.

Joslin, J. M., Fernald, A. A., Tennant, T. R., Davis, E. M., Kogan, S. C., Anastasi, J., Crispino, J. D., and Le Beau, M. M. (2007). Haploinsufficiency of EGR1, a candidate gene in the del(5q), leads to the development of myeloid disorders. *Blood* **110,** 719–726.

Judson, R. L., Babiarz, J. E., Venere, M., and Blelloch, R. (2009). Embryonic stem cell-specific microRNAs promote induced pluripotency. *Nat. Biotechnol.* **27,** 459–461.

Kamimura, K., Ohi, H., Kubota, T., Okazuka, K., Yoshikai, Y., Wakabayashi, Y., Aoyagi, Y., Mishima, Y., and Kominami, R. (2007). Haploinsufficiency of Bcl11b for suppression of lymphomagenesis and thymocyte development. *Biochem. Biophys. Res. Commun.* **355,** 538–542.

Kedde, M., Strasser, M. J., Boldajipour, B., Oude Vrielink, J. A., Slanchev, K., le Sage, C., Nagel, R., Voorhoeve, P. M., van Duijse, J., Orom, U. A., Lund, A. H., Perrakis, A., *et al.* (2007). RNA-binding protein Dnd1 inhibits microRNA access to target mRNA. *Cell* **131,** 1273–1286.

Khanna, A., and Stamm, S. (2010). Regulation of alternative splicing by short non-coding nuclear RNAs. *RNA Biol.* **7,** 480–485.

Khurana, J. S., and Theurkauf, W. (2010). piRNAs, transposon silencing, and Drosophila germline development. *J. Cell Biol.* **191,** 905–913.

Kim, H. H., Kuwano, Y., Srikantan, S., Lee, E. K., Martindale, J. L., and Gorospe, M. (2009). HuR recruits let-7/RISC to repress c-Myc expression. *Genes Dev.* **23,** 1743–1748.

Kinjyo, I., Hanada, T., Inagaki-Ohara, K., Mori, H., Aki, D., Ohishi, M., Yoshida, H., Kubo, M., and Yoshimura, A. (2002). SOCS1/JAB is a negative regulator of LPS-induced macrophage activation. *Immunity* **17,** 583–591.

Klein, U., Lia, M., Crespo, M., Siegel, R., Shen, Q., Mo, T., Ambesi-Impiombato, A., Califano, A., Migliazza, A., Bhagat, G., and Dalla-Favera, R. (2010). The DLEU2/miR-15a/16-1 cluster controls B cell proliferation and its deletion leads to chronic lymphocytic leukemia. *Cancer Cell* **17,** 28–40.

Koralov, S. B., Muljo, S. A., Galler, G. R., Krek, A., Chakraborty, T., Kanellopoulou, C., Jensen, K., Cobb, B. S., Merkenschlager, M., Rajewsky, N., and Rajewsky, K. (2008). Dicer ablation affects antibody diversity and cell survival in the B lymphocyte lineage. *Cell* **132,** 860–874.

Krek, A., Grun, D., Poy, M. N., Wolf, R., Rosenberg, L., Epstein, E. J., MacMenamin, P., da Piedade, I., Gunsalus, K. C., Stoffel, M., and Rajewsky, N. (2005). Combinatorial microRNA target predictions. *Nat. Genet.* **37,** 495–500.

Kuchen, S., Resch, W., Yamane, A., Kuo, N., Li, Z., Chakraborty, T., Wei, L., Laurence, A., Yasuda, T., Peng, S., Hu-Li, J., Lu, K., *et al.* (2010). Regulation of microRNA expression and abundance during lymphopoiesis. *Immunity* **32,** 828–839.

Kuipers, H., Schnorfeil, F. M., Fehling, H. J., Bartels, H., and Brocker, T. (2010). Dicer-dependent microRNAs control maturation, function, and maintenance of Langerhans cells in vivo. *J. Immunol.* **185,** 400–409.

Kumar, M. S., Pester, R. E., Chen, C. Y., Lane, K., Chin, C., Lu, J., Kirsch, D. G., Golub, T. R., and Jacks, T. (2009). Dicer1 functions as a haploinsufficient tumor suppressor. *Genes Dev.* **23,** 2700–2704.

Landthaler, M., Gaidatzis, D., Rothballer, A., Chen, P. Y., Soll, S. J., Dinic, L., Ojo, T., Hafner, M., Zavolan, M., and Tuschl, T. (2008). Molecular characterization of human Argonaute-containing ribonucleoprotein complexes and their bound target mRNAs. *RNA* **14,** 2580–2596.

Le Toriellec, E., Despouy, G., Pierron, G., Gaye, N., Joiner, M., Bellanger, D., Vincent-Salomon, A., and Stern, M. H. (2008). Haploinsufficiency of CDKN1B contributes to leukemogenesis in T-cell prolymphocytic leukemia. *Blood* **111,** 2321–2328.

Leung, A. K., Calabrese, J. M., and Sharp, P. A. (2006). Quantitative analysis of Argonaute protein reveals microRNA-dependent localization to stress granules. *Proc. Natl. Acad. Sci. USA* **103,** 18125–18130.

Li, Q. J., Chau, J., Ebert, P. J., Sylvester, G., Min, H., Liu, G., Braich, R., Manoharan, M., Soutschek, J., Skare, P., Klein, L. O., Davis, M. M., *et al.* (2007). miR-181a is an intrinsic modulator of T cell sensitivity and selection. *Cell* **129,** 147–161.

Li, Z., Lu, J., Sun, M., Mi, S., Zhang, H., Luo, R. T., Chen, P., Wang, Y., Yan, M., Qian, Z., Neilly, M. B., Jin, J., *et al.* (2008). Distinct microRNA expression profiles in acute myeloid leukemia with common translocations. *Proc. Natl. Acad. Sci. USA* **105,** 15535–15540.

Linnstaedt, S. D., Gottwein, E., Skalsky, R. L., Luftig, M. A., and Cullen, B. R. (2011). Virally induced cellular microRNA miR-155 plays a key role in B-cell immortalization by Epstein-Barr virus. *J. Virol.* **84,** 11670–11678.

Lodish, H. F., Zhou, B., Liu, G., and Chen, C. Z. (2008). Micromanagement of the immune system by microRNAs. *Nat. Rev. Immunol.* **8,** 120–130.

Lu, J., Guo, S., Ebert, B. L., Zhang, H., Peng, X., Bosco, J., Pretz, J., Schlanger, R., Wang, J. Y., Mak, R. H., Dombkowski, D. M., Preffer, F. I., *et al.* (2008a). MicroRNA-mediated control of cell fate in megakaryocyte-erythrocyte progenitors. *Dev. Cell* **14,** 843–853.

Lu, J., Shen, Y., Wu, Q., Kumar, S., He, B., Shi, S., Carthew, R. W., Wang, S. M., and Wu, C. I. (2008b). The birth and death of microRNA genes in Drosophila. *Nat. Genet.* **40,** 351–355.

Lu, L. F., Thai, T. H., Calado, D. P., Chaudhry, A., Kubo, M., Tanaka, K., Loeb, G. B., Lee, H., Yoshimura, A., Rajewsky, K., and Rudensky, A. Y. (2009). Foxp3-dependent microRNA155 confers competitive fitness to regulatory T cells by targeting SOCS1 protein. *Immunity* **30,** 80–91.

Lu, L. F., Boldin, M. P., Chaudhry, A., Lin, L. L., Taganov, K. D., Hanada, T., Yoshimura, A., Baltimore, D., and Rudensky, A. Y. (2010). Function of miR-146a in controlling Treg cell-mediated regulation of Th1 responses. *Cell* **142,** 914–929.

Malecova, B., and Morris, K. V. (2010). Transcriptional gene silencing through epigenetic changes mediated by non-coding RNAs. *Curr. Opin. Mol. Ther.* **12,** 214–222.

Mattick, J. S. (2007). A new paradigm for developmental biology. *J. Exp. Biol.* **210,** 1526–1547.

Medina, P. P., Nolde, M., and Slack, F. J. (2010). OncomiR addiction in an in vivo model of microRNA-21-induced pre-B-cell lymphoma. *Nature* **467,** 86–90.

Melton, C., Judson, R. L., and Blelloch, R. (2010). Opposing microRNA families regulate self-renewal in mouse embryonic stem cells. *Nature* **464,** 126.

Muljo, S. A., Ansel, K. M., Kanellopoulou, C., Livingston, D. M., Rao, A., and Rajewsky, K. (2005). Aberrant T cell differentiation in the absence of Dicer. *J. Exp. Med.* **202,** 261–269.

Nagai, Y., Garrett, K. P., Ohta, S., Bahrun, U., Kouro, T., Akira, S., Takatsu, K., and Kincade, P. W. (2006). Toll-like receptors on hematopoietic progenitor cells stimulate innate immune system replenishment. *Immunity* **24,** 801–812.

O'Connell, R. M., Taganov, K. D., Boldin, M. P., Cheng, G., and Baltimore, D. (2007). MicroRNA-155 is induced during the macrophage inflammatory response. *Proc. Natl. Acad. Sci. USA* **104,** 1604–1609.

O'Connell, R. M., Rao, D. S., Chaudhuri, A. A., Boldin, M. P., Taganov, K. D., Nicoll, J., Paquette, R. L., and Baltimore, D. (2008). Sustained expression of microRNA-155 in hematopoietic stem cells causes a myeloproliferative disorder. *J. Exp. Med.* **205,** 585–594.

O'Connell, R. M., Chaudhuri, A. A., Rao, D. S., and Baltimore, D. (2009). Inositol phosphatase SHIP1 is a primary target of miR-155. *Proc. Natl. Acad. Sci. USA* **106,** 7113–7118.

O'Connell, R. M., Chaudhuri, A. A., Rao, D. S., Gibson, W. S., Balazs, A. B., and Baltimore, D. (2010a). MicroRNAs enriched in hematopoietic stem cells differentially regulate long-term hematopoietic output. *Proc. Natl. Acad. Sci. USA* **107,** 14235–14240.

O'Connell, R. M., Kahn, D., Gibson, W. S., Round, J. L., Scholz, R. L., Chaudhuri, A. A., Kahn, M. E., Rao, D. S., and Baltimore, D. (2010b). MicroRNA-155 promotes autoimmune inflammation by enhancing inflammatory T cell development. *Immunity* **33,** 607–619.

O'Connell, R. M., Rao, D. S., Chaudhuri, A. A., and Baltimore, D. (2010c). Physiological and pathological roles for microRNAs in the immune system. *Nat. Rev. Immunol.* **10,** 111–122.

O'Neill, L. A., Sheedy, F. J., and McCoy, C. E. (2011). MicroRNAs: The fine-tuners of Toll-like receptor signalling. *Nat. Rev. Immunol.* **11,** 163–175.

Ooi, A. G., Sahoo, D., Adorno, M., Wang, Y., Weissman, I. L., and Park, C. Y. (2010). MicroRNA-125b expands hematopoietic stem cells and enriches for the lymphoid-balanced and lymphoid-biased subsets. *Proc. Natl. Acad. Sci. USA* **107,** 21505–21510.

Orkin, S. H., and Zon, L. I. (2008). Hematopoiesis: An evolving paradigm for stem cell biology. *Cell* **132,** 631–644.

Patrick, D. M., Zhang, C. C., Tao, Y., Yao, H., Qi, X., Schwartz, R. J., Jun-Shen Huang, L., and Olson, E. N. (2010). Defective erythroid differentiation in miR-451 mutant mice mediated by 14-3-3zeta. *Genes Dev.* **24,** 1614–1619.

Pauley, K. M., Cha, S., and Chan, E. K. (2009). MicroRNA in autoimmunity and autoimmune diseases. *J. Autoimmun.* **32,** 189–194.

Peterson, K. J., Dietrich, M. R., and McPeek, M. A. (2009). MicroRNAs and metazoan macroevolution: Insights into canalization, complexity, and the Cambrian explosion. *Bioessays* **31,** 736–747.

Pontes, O., and Pikaard, C. S. (2008). siRNA and miRNA processing: New functions for Cajal bodies. *Curr. Opin. Genet. Dev.* **18,** 197–203.

Popovic, R., Riesbeck, L. E., Velu, C. S., Chaubey, A., Zhang, J., Achille, N. J., Erfurth, F. E., Eaton, K., Lu, J., Grimes, H. L., Chen, J., Rowley, J. D., *et al.* (2009). Regulation of mir-196b by MLL and its overexpression by MLL fusions contributes to immortalization. *Blood* **113,** 3314–3322.

Puebla-Osorio, N., Miyahara, Y., Coimbatore, S., Limon-Flores, A. Y., Kazimi, N., Ullrich, S. E., and Zhu, C. (2011). Induction of B-cell lymphoma by UVB radiation in p53 haploinsufficient mice. *BMC Cancer* **11,** 36.

Raaijmakers, M. H., Mukherjee, S., Guo, S., Zhang, S., Kobayashi, T., Schoonmaker, J. A., Ebert, B. L., Al-Shahrour, F., Hasserjian, R. P., Scadden, E. O., Aung, Z., Matza, M., *et al.* (2010). Bone progenitor dysfunction induces myelodysplasia and secondary leukaemia. *Nature* **464,** 852–857.

Ramachandran, V., and Chen, X. (2008). Degradation of microRNAs by a family of exoribonucleases in Arabidopsis. *Science* **321,** 1490–1492.

Rao, D. S., O'Connell, R. M., Chaudhuri, A. A., Garcia-Flores, Y., Geiger, T. L., and Baltimore, D. (2010). MicroRNA-34a perturbs B lymphocyte development by repressing the forkhead box transcription factor Foxp1. *Immunity* **33,** 48–59.

Rasmussen, K. D., Simmini, S., Abreu-Goodger, C., Bartonicek, N., Di Giacomo, M., Bilbao-Cortes, D., Horos, R., Von Lindern, M., Enright, A. J., and O'Carroll, D. (2010). The miR-144/451 locus is required for erythroid homeostasis. *J. Exp. Med.* **207,** 1351–1358.

Robb, L. (2007). Cytokine receptors and hematopoietic differentiation. *Oncogene* **26,** 6715–6723.

Rodriguez, A., Vigorito, E., Clare, S., Warren, M. V., Couttet, P., Soond, D. R., van Dongen, S., Grocock, R. J., Das, P. P., Miska, E. A., Vetrie, D., Okkenhaug, K., *et al.* (2007). Requirement of bic/microRNA-155 for normal immune function. *Science* **316,** 608–611.

Rossi, D. J., Jamieson, C. H., and Weissman, I. L. (2008). Stems cells and the pathways to aging and cancer. *Cell* **132,** 681–696.

Ruggiero, T., Trabucchi, M., De Santa, F., Zupo, S., Harfe, B. D., McManus, M. T., Rosenfeld, M. G., Briata, P., and Gherzi, R. (2009). LPS induces KH-type splicing regulatory protein-dependent processing of microRNA-155 precursors in macrophages. *FASEB J.* **23,** 2898–2908.

Salmena, L., Carracedo, A., and Pandolfi, P. P. (2008). Tenets of PTEN tumor suppression. *Cell* **133,** 403–414.

Sandberg, R., Neilson, J. R., Sarma, A., Sharp, P. A., and Burge, C. B. (2008). Proliferating cells express mRNAs with shortened 3′ untranslated regions and fewer microRNA target sites. *Science* **320,** 1643–1647.

Schraml, E., Voglauer, R., Fortschegger, K., Sibilia, M., Stelzer, I., Grillari, J., and Schauenstein, K. (2008). Haploinsufficiency of senescence evasion factor causes defects of hematopoietic stem cells functions. *Stem Cells Dev.* **17,** 355–366.

Sernandez, I. V., de Yebenes, V. G., Dorsett, Y., and Ramiro, A. R. (2008). Haploinsufficiency of activation-induced deaminase for antibody diversification and chromosome translocations both in vitro and in vivo. *PLoS One* **3,** e3927.

Sheedy, F. J., Palsson-McDermott, E., Hennessy, E. J., Martin, C., O'Leary, J., Ruan, Q., Johnson, D. P., Chen, Y., and O'Neill, L. A. (2009). Negative regulation of TLR4 via targeting of the proinflammatory tumor suppressor PDCD4 by the microRNA miR-21. *Nat. Immunol.* **11,** 141–147.

Shomron, N., Golan, D., and Hornstein, E. (2009). An evolutionary perspective of animal microRNAs and their targets. *J. Biomed. Biotechnol.* **2009,** 594738.

Sportoletti, P., Grisendi, S., Majid, S. M., Cheng, K., Clohessy, J. G., Viale, A., Teruya-Feldstein, J., and Pandolfi, P. P. (2008). Npm1 is a haploinsufficient suppressor of myeloid and lymphoid malignancies in the mouse. *Blood* **111,** 3859–3862.

Sugatani, T., and Hruska, K. A. (2009). Impaired micro-RNA pathways diminish osteoclast differentiation and function. *J. Biol. Chem.* **284,** 4667–4678.

Sun, W., and Downing, J. R. (2004). Haploinsufficiency of AML1 results in a decrease in the number of LTR-HSCs while simultaneously inducing an increase in more mature progenitors. *Blood* **104,** 3565–3572.

Taganov, K. D., Boldin, M. P., Chang, K. J., and Baltimore, D. (2006). NF-kappaB-dependent induction of microRNA miR-146, an inhibitor targeted to signaling proteins of innate immune responses. *Proc. Natl. Acad. Sci. USA* **103,** 12481–12486.

Takizawa, H., Regoes, R. R., Boddupalli, C. S., Bonhoeffer, S., and Manz, M. G. (2011). Dynamic variation in cycling of hematopoietic stem cells in steady state and inflammation. *J. Exp. Med.* **208,** 273–284.

Thai, T. H., Calado, D. P., Casola, S., Ansel, K. M., Xiao, C., Xue, Y., Murphy, A., Frendewey, D., Valenzuela, D., Kutok, J. L., Schmidt-Supprian, M., Rajewsky, N., *et al.* (2007). Regulation of the germinal center response by microRNA-155. *Science* **316,** 604–608.

Trabucchi, M., Briata, P., Garcia-Mayoral, M., Haase, A. D., Filipowicz, W., Ramos, A., Gherzi, R., and Rosenfeld, M. G. (2009). The RNA-binding protein KSRP promotes the biogenesis of a subset of microRNAs. *Nature* **459,** 1010–1014.

Ueda, Y., Kondo, M., and Kelsoe, G. (2005). Inflammation and the reciprocal production of granulocytes and lymphocytes in bone marrow. *J. Exp. Med.* **201,** 1771–1780.

Ventura, A., Young, A. G., Winslow, M. M., Lintault, L., Meissner, A., Erkeland, S. J., Newman, J., Bronson, R. T., Crowley, D., Stone, J. R., Jaenisch, R., Sharp, P. A., *et al.* (2008). Targeted deletion reveals essential and overlapping functions of the miR-17 through 92 family of miRNA clusters. *Cell* **132,** 875–886.

Vigorito, E., Perks, K. L., Abreu-Goodger, C., Bunting, S., Xiang, Z., Kohlhaas, S., Das, P. P., Miska, E. A., Rodriguez, A., Bradley, A., Smith, K. G., Rada, C., *et al.* (2007). microRNA-155 regulates the generation of immunoglobulin class-switched plasma cells. *Immunity* **27,** 847–859.

Wang, J., Fernald, A. A., Anastasi, J., Le Beau, M. M., and Qian, Z. (2010). Haploinsufficiency of Apc leads to ineffective hematopoiesis. *Blood* **115,** 3481–3488.

Wu, H., Neilson, J. R., Kumar, P., Manocha, M., Shankar, P., Sharp, P. A., and Manjunath, N. (2007). miRNA profiling of naive, effector and memory CD8 T cells. *PLoS One* **2,** e1020.

Wu, S., Huang, S., Ding, J., Zhao, Y., Liang, L., Liu, T., Zhan, R., and He, X. (2010). Multiple microRNAs modulate p21Cip1/Waf1 expression by directly targeting its 3′ untranslated region. *Oncogene* **29,** 2302–2308.

Xiang, Z., Luo, H., Payton, J. E., Cain, J., Ley, T. J., Opferman, J. T., and Tomasson, M. H. (2010). Mcl1 haploinsufficiency protects mice from Myc-induced acute myeloid leukemia. *J. Clin. Invest.* **120,** 2109–2118.

Xiao, C., and Rajewsky, K. (2009). MicroRNA control in the immune system: Basic principles. *Cell* **136,** 26–36.

Xiao, C., Calado, D. P., Galler, G., Thai, T. H., Patterson, H. C., Wang, J., Rajewsky, N., Bender, T. P., and Rajewsky, K. (2007). MiR-150 controls B cell differentiation by targeting the transcription factor c-Myb. *Cell* **131,** 146–159.

Xiao, C., Srinivasan, L., Calado, D. P., Patterson, H. C., Zhang, B., Wang, J., Henderson, J. M., Kutok, J. L., and Rajewsky, K. (2008). Lymphoproliferative disease and autoimmunity in mice with increased miR-17-92 expression in lymphocytes. *Nat. Immunol.* **9,** 405–414.

Xu, G., Fewell, C., Taylor, C., Deng, N., Hedges, D., Wang, X., Zhang, K., Lacey, M., Zhang, H., Yin, Q., Cameron, J., Lin, Z., *et al.* (2010). Transcriptome and targetome analysis in MIR155 expressing cells using RNA-seq. *RNA* **16,** 1610–1622.

Yang, W., Chendrimada, T. P., Wang, Q., Higuchi, M., Seeburg, P. H., Shiekhattar, R., and Nishikura, K. (2006). Modulation of microRNA processing and expression through RNA editing by ADAR deaminases. *Nat. Struct. Mol. Biol.* **13,** 13–21.

Yekta, S., Shih, I. H., and Bartel, D. P. (2004). MicroRNA-directed cleavage of HOXB8 mRNA. *Science* **304,** 594–596.

Zandi, S., Mansson, R., Tsapogas, P., Zetterblad, J., Bryder, D., and Sigvardsson, M. (2008). EBF1 is essential for B-lineage priming and establishment of a transcription factor network in common lymphoid progenitors. *J. Immunol.* **181,** 3364–3372.

Zhang, Y., Liu, D., Chen, X., Li, J., Li, L., Bian, Z., Sun, F., Lu, J., Yin, Y., Cai, X., Sun, Q., Wang, K., *et al.* (2010). Secreted monocytic miR-150 enhances targeted endothelial cell migration. *Mol. Cell* **39,** 133–144.

Zhao, J. L., Rao, D. S., Boldin, M. P., Taganov, K. D., O'Connell, R. M., and Baltimore, D. (2011). NF-kappaB dysregulation in microRNA-146a-deficient mice drives the development of myeloid malignancies. *Proc. Natl. Acad. Sci.* **108,** 9184–9189.

CHAPTER SEVEN

A miR Image of Stem Cells and Their Lineages

Rui Yi* *and* Elaine Fuchs†

Contents

Abstract

MicroRNAs (miRs) comprise a class of tiny (~19–24 nucleotide), noncoding RNAs that regulate gene expression posttranscriptionally. Since the discovery of the founding members *lin-4* and *let-7* as key regulators in the developing nematode, miRs have been found throughout the eukaryotic kingdom. Functions for miRs are wide-ranging and encompass embryogenesis, stem cell

* Department of Molecular, Cellular, and Developmental Biology, University of Colorado at Boulder, Boulder, Colorado, USA
† Laboratory of Mammalian Cell Biology and Development, Howard Hughes Medical Institute, The Rockefeller University, New York, USA

Current Topics in Developmental Biology, Volume 99
ISSN 0070-2153, DOI: 10.1016/B978-0-12-387038-4.00007-0

biology, tissue differentiation, and human diseases including cancers. In this chapter, we begin by acquainting our readers with miRs and introducing them to their biogenesis. Then, we focus on the roles of miRs in stem cells during tissue development and homeostasis. We use mammalian skin as our main paradigm, but we also consider miR functions in several different types of adult stem cells. We conclude by discussing future challenges that will lead to a comprehensive understanding of miR functions in stem cells and their lineages.

1. Introduction

MicroRNAs (miRs) comprise a class of noncoding, regulatory RNAs that are widely expressed in both plants and animals (reviewed by Ambros, 2004; Bartel, 2009). In animals, the miR pathway constitutes small RNAs and their protein partners, including *Dicer* and *Argonaute* proteins. Their origins can be traced back to sponges, making them one of the most ancient pathways to regulate output of the transcriptome (Grimson *et al.*, 2008).

MiRs regulate mRNA stability and protein production by recruiting the RNA-induced silencing complex (RISC) to their cognate target sites (reviewed by Bartel, 2009). It is estimated that more than one-third of protein-coding mRNAs are regulated by miRs (Friedman *et al.*, 2009; Lewis *et al.*, 2005). In turn, miR-mediated regulation is believed to have a widespread impact both on protein output of the transcriptome (Baek *et al.*, 2008; Selbach *et al.*, 2008) and on evolution of gene regulatory networks (Farh *et al.*, 2005; Stark *et al.*, 2005). miR's recognition of its mRNA targets is primarily mediated by base-pairing between nucleotides no. 2–8 of the miR, commonly referred to as its "seed sequences," and complementary mRNA sequences that are often located within the 3′-untranslated region (3′UTR) (reviewed by Bartel, 2009). Although the targets of a miR cannot be identified simply by whether an mRNA contains a sequence of perfect complementarity to the miR's seed sequence, there is no doubt that a miR's seed sequence plays a critical role in target recognition. Such evidence includes not only preferential conservation of the seed sequences of miRs and their cognate sequences in the 3′UTR of mRNA but also the crystal structures of these sequences complexed with the Argonaute protein (Wang *et al.*, 2009; Xie *et al.*, 2005). However, because a single miR can recognize >100 targets, a single mRNA can be simultaneously regulated by multiple miRs, and the homology match is not always a perfect one, miR-mediated gene regulation networks are extraordinarily complex.

The complexity of miR-mediated regulation and its potential impact on the expression of a large number of proteins has drawn increasing attention to these tiny riboregulators. Since the discovery a decade ago of *let-7*, much progress has been made in deciphering miR expression patterns and functions in embryonic development, differentiation, maintenance of stem cells, and human diseases (reviewed by Ambros, 2004; Bartel, 2009). The existence of a hitherto unappreciated dimension of gene regulation mediated by miRs led to an explosion of new studies, which have not only provided new insights into our understanding of developmental and stem cell biology but also pointed to new directions for therapeutic applications of stem cells.

In this review, we consider the most recent progress toward understanding the biological role of miRs in mammalian skin development and in various adult stem cells. Because of the significant interest in this topic, many excellent reviews have been published in recent years and we also refer our readers to these chapters for a comprehensive view of the field (Gangaraju and Lin, 2009; Ivey and Srivastava, 2010; Martinez and Gregory, 2010).

2. The Discovery of miRs and Their Roles in *Caenorhabditis elegans*

The founding members of the miR family, namely, *lin-4* and *let-7*, were originally identified as heterochronic genes that govern developmental transitions in the nematode *C. elegans* (Lee *et al.*, 1993; Pasquinelli *et al.*, 2000; Reinhart *et al.*, 2000; Wightman *et al.*, 1993). In those early studies, *lin-4* and *let-7* were both found to negatively regulate the translation of master regulators of differentiation such as lin-14, lin-28, and lin-41. Intriguingly, these proteins were found not only to maintain an early developmental lineage but also then be downregulated as the animal transitions to a later lineage. Thus for example, lin-4 functions in downregulating lin-14 and lin-28 through the first larval stage (L1), while let-7 downregulates hbl-1 and lin-41 when the animal develops from the fourth larval stage (L4) to the adult stage (Lee *et al.*, 1993; Reinhart *et al.*, 2000).

The negative regulatory networks involving lin-4, let-7, and their mRNA targets provided the basis for the paradigm, whereby miRs help deplete protein production by mRNAs that were inherited from an earlier stage in the lineage, but which impair progression to the next step. In doing so, miRs facilitate a precise, robust transition through the developmental program. Indeed, in many of the recent investigations which we discuss in the chapter, and which are aimed at exploring miR regulation of mammalian stem cell lineages and development, miRs appear to act in accordance to such model.

Although lin-4 was identified in 1993, it took the subsequent discovery of let-7 in 2000 and the rapid realization that it is highly conserved before miRs began to attract broader interest from the scientific community. A decade later, as the sequencing of genomes and their miR components soared to an unprecedented pace, it has become apparent that lin-4 is also deeply conserved (Christodoulou *et al.*, 2010). However, unlike let-7 whose sequence is almost identical between *C. elegans* and human, the conservation between lin-4 and its human orthologues miR-125a/b is limited to the 5′end of the miR. Despite these variations, the miR pathway has emerged as one of the most ancient pathways that are deeply involved in animal evolution (Christodoulou *et al.*, 2010; Grimson *et al.*, 2008). Notably, miR-100 is likely to be the most deeply conserved miR as its orthologues are identified from *Nematostella* to human (Christodoulou *et al.*, 2010; Grimson *et al.*, 2008).

Because of the ease for genetic manipulation, *C. elegans* has long served as a fertile ground for functional characterization of novel genes by the loss-of-function studies. Recently, the roles of most individual miRs and miR families have been investigated by examining developmental defects and viability in a large collection of miR knockout (KO) models in *C. elegans* (Alvarez-Saavedra and Horvitz, 2010; Miska *et al.*, 2007). Surprisingly, very few miRs either individually (7 out of 95) or collectively as a family (3 out of 15 families) were found to be essential for development and viability in these studies (Alvarez-Saavedra and Horvitz, 2010; Miska *et al.*, 2007). However, when challenged with sensitized genetic background, many miR mutants (25 out of 31 miRs) manifest strong defects (Brenner *et al.*, 2010). These interesting findings are also echoed by studies with miR KO mouse models where miR KO animals often only show discernible defects under stress conditions (Liu and Olson, 2010). Together, these results support the view that the miR-mediated regulatory network is complex and suggest that rather than having absolute functions, miRs may instead operate to more rapidly reinforce or enhance a pathway, a feature that may only surface under circumstances like stress, where the sense of urgency in responding is more acutely felt by a tissue.

3. The Biogenesis Pathways of miR

In the past decade, the miR biogenesis pathway has been extensively investigated. For the clarity of this chapter, we reiterate some of the key steps that are important to our discussion but refer our readers to several recent reviews for detailed analysis (Kim *et al.*, 2009; Krol *et al.*, 2010).

In mammals, most miRs are transcribed by RNA polymerase II but some of them are generated by RNA polymerase III (Borchert *et al.*, 2006; Cai *et al.*, 2004). During the transcription of the primary transcript, the flanking sequences of the miR fold into a hairpin structure, called pre-miR, characteristic of miR coding genes (Fig. 7.1). For stereotypical miRs that are represented by most abundantly expressed miRs, the hairpin is first processed by the

Drosha–DiGeorge syndrome critical region gene 8 (Dgcr8) microprocessor complex in the nucleus (Denli *et al.*, 2004; Gregory *et al.*, 2004; Han *et al.*, 2004; Landthaler *et al.*, 2004). However, in a few exceptions where the hairpin is generated directly from transcription of short-hairpin RNA or derived from splicing of a certain type of intron that is capable of folding into the hairpin structure (also known as the mirtron pathway), the biogenesis of these

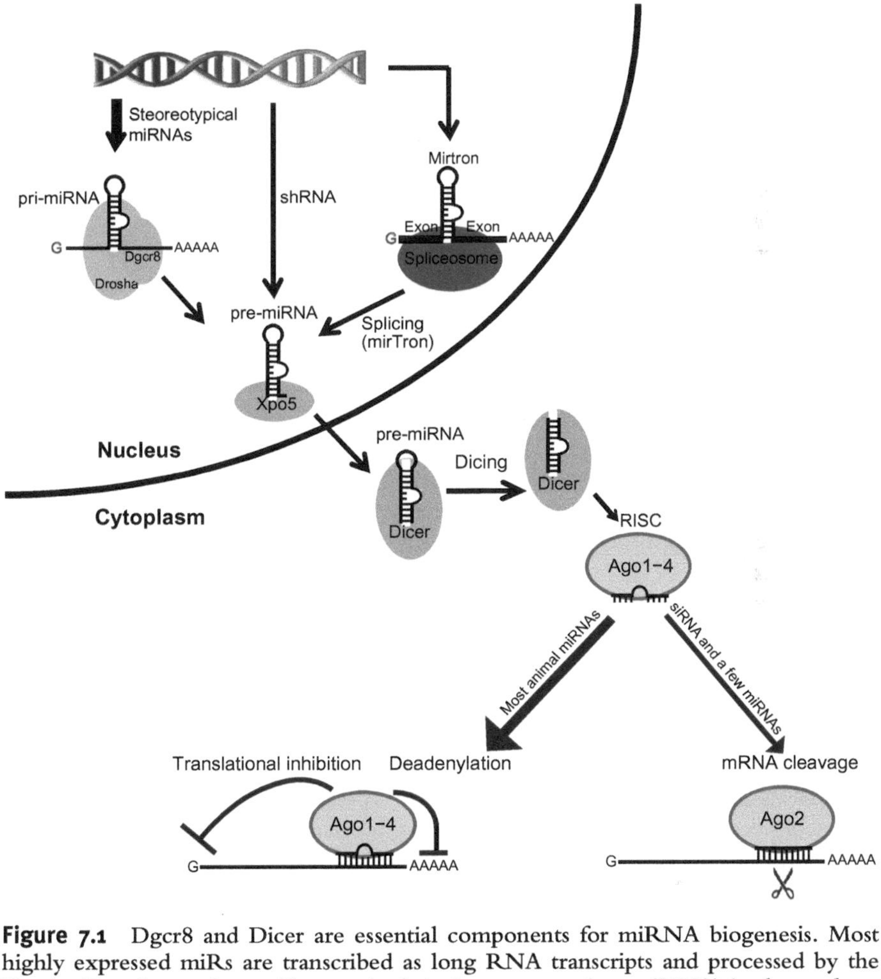

Figure 7.1 Dgcr8 and Dicer are essential components for miRNA biogenesis. Most highly expressed miRs are transcribed as long RNA transcripts and processed by the Drosha/Dgcr8 complex to liberate the hairpin precursor (pre-miRNA) in the nucleus. However, a few miRs are either directly transcribed as short-hairpin (sh) RNA or processed via splicing (the mirtron pathway). Thus these miRs are independent of the Drosha/Dgcr8 processing for the generation of the hairpin. After transported to the cytoplasm by Xpo5, pre-miRNAs are further processed by Dicer. One strand of the double-stranded RNA duplex is complexed with Ago proteins to form the RISC. (For color version of this figure, the reader is referred to the Web version of this chapter.)

noncanonical miRs can bypass the requirement of Drosha/Dgcr8 (Babiarz *et al.*, 2008). In all cases, the liberated hairpin is then exported by Xpo5 (Lund *et al.*, 2004; Yi *et al.*, 2003) to the cytoplasm where it undergoes a second cleavage by Dicer (Hutvagner *et al.*, 2001). After the processing by Dicer, one strand of the double-stranded duplex is selectively incorporated into an RISC composed of one of the four Argonaute proteins and their associated cofactors (Schwarz *et al.*, 2003). Given the nearly universal mechanism for miR biogenesis, it is possible to explore the global function of the miR pathway by genetically deleting either the Drosha-Dgcr8 complex or the Dicer.

4. Mammalian Skin as a Model to Uncover How miRs Regulate Stem Cell Biology and Tissue Morphogenesis

With its easy accessibility and established genetic tools for conditional targeting, the mouse skin epithelium provides an excellent system to explore functions of miR regulatory pathways in mammalian stem cell biology and tissue morphogenesis. The architecture of the skin is spatially and temporally well defined. The skin epithelium is separated from the underlying dermis by a basement membrane rich in extracellular matrix. In the epidermis, an inner (basal) layer of progenitors periodically withdraws from the cell cycle and commits to upward program of terminal differentiation, involving three distinct stages: spinous, granular, and stratum corneum (Fig. 7.2A). Transcriptional activity ceases at the transition after the granular stage, as terminally differentiated dead squames are sloughed from the skin surface, continually replaced by inner cells moving outward.

By contrast, hair follicles (HFs) and their associated sebaceous glands (SGs) are appendages of the epidermis (Fig. 7.2B). Hair growth is cyclical, fueled by epithelial stem cells that reside within a niche called the bulge, located just below the SG. At the base of the bulge is an extension of SCs referred to as the secondary hair germ (HG). The HG maintains the closest proximity to an essential cluster of mesenchymal cells, the dermal papilla (DP), which undergoes stimulatory crosstalk with the stem cells. Once activated (the telogen to anagen transition), a new HF emerges from the HG as the DP is pushed downward. At the base of the mature HF and in contact with the DP, matrix cells rapidly but transiently divide and then differentiate in concentric upward cylinders of cells to form the hair shaft at the center and the channel or inner root sheath (IRS) surrounded by the outer root sheath (ORS). Following the anagen growth phase, most of the lower two-thirds of the HF undergo apoptosis and the remaining epithelial strand retracts upward with the DP to begin the cycle anew (Fig. 7.2C).

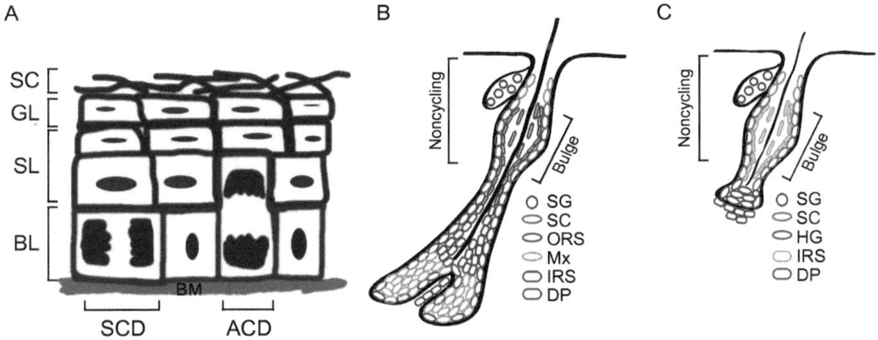

Figure 7.2 Illustration of skin lineages. (A) The epidermis comprises of four layers: basal layer (red, BL), spinous layer (black, SL), granular layer (black, GL), and stratum corneum (black, SC). Stem cells are located in the basal layer, residing on a basement membrane (BM). They undergo symmetric cell division (SCD) to double their population or asymmetric cell division (ACD) to self-renew and generate another daughter for terminal differentiation. (B) Hair follicle and their associated sebaceous glands are appendages of the epidermis. Stem cells (orange, SC) reside in the bulge stem cell niche. They move upward to become progenitor cells for the sebaceous gland (light blue, SG-P) and subsequently terminally differentiated to sebaceous gland cells (dark blue, SG-M). In the hair follicle lineage, stem cells give rise to outer root sheath (dark green, ORS), matrix (light green, Mx), inner root sheath (purple, IRS), and eventually hair shaft. Derma papilla (DP) is a cluster of mesenchymal cells surrounded by the hair bulb. (C) At the telogen, the secondary hair germ (HG) is formed at the base of the bulge and directly interacts with the DP. Inner root sheath (IRS) is generated directly from the stem cells. (See Color Insert.)

With over two decades of molecular genetic analyzes of the epidermis and HFs, there is extensive knowledge of the signaling circuitry and physiologically relevant changes in gene expression that occur, not only during normal adult homeostasis and embryonic development but also in wound repair and tumorigenesis. This provides an exceptional foundation for dissecting how miR-mediated regulation fine-tunes these molecular events in mammalian stem cell and tissue biology.

5. Insights into the Global Role of miRs in the Skin Epithelium: Conditional Ablation of *Dicer*

Not surprisingly, the first insights of the importance of miRs in the skin epithelium came from profiling the ones that are expressed in this tissue and examining the consequences of conditionally ablating *Dicer* in this tissue

(Andl *et al.*, 2006; Yi *et al.*, 2006). Temporal profiling during skin development unveiled a myriad of miRs which are differentially expressed in the epidermis and its notable appendage, the HF (Yi *et al.*, 2006). Interestingly, this included not only miRs such as miR-100, miR-125, and let-7, which are conserved throughout the animal kingdom, but also several vertebrate-specific miRs such as miR-203 and miR-205. As more and more species were analyzed for tissue-specific expression of miRs, it became evident that these miRs are characteristic of the vertebrate skin epithelium.

Complete ablation of *Dicer* in all mouse tissues results in the arrest of development by embryonic day 7.5 (E7.5) (Bernstein *et al.*, 2003). Epidermal stratification does not take place until E13.5→E16.5, and HFs form in waves from E14.5→P0 (Schmidt-Ullrich and Paus, 2005). Thus to study Dicer function in skin, several groups have generated *Dicer* conditional KO (cKO) mice by using a floxed Dicer allele and a *Cre* recombinase transgene driven by a skin epithelial-specific *keratin-14* (K14) promoter (Andl *et al.*, 2006; Yi *et al.*, 2006). Like its coexpressed partner *K5*, *K14* becomes strongly active in embryonic skin progenitor cells by ~E15 (Byrne *et al.*, 1994; Vassar *et al.*, 1989). As such, the K14-Cre-mediated Dicer ablation depletes Dicer and miRs in all skin lineages.

Because the loss of mature miRs is secondary to the ablation of *Dicer* and the relatively long half-life of miRs, mature miRs were not completely depleted until E17.5 (Yi *et al.*, 2006) or birth (Andl *et al.*, 2006) depending most likely upon strain-specific differences. However, more than 100 miRs were lost by then, and yet the cKO pups initially appeared normal in size and appearance, and the epidermis, HFs, and SGs were correctly specified. Thus, unlike well-established master regulatory circuits, for example, BMP, Wnt, and Notch signaling pathways, the miR pathway did not seem to be required for lineage specification during skin development.

Closer inspection revealed that the epidermis generated the expected architectural and morphological signs of differentiation (Yi *et al.*, 2006). By contrast, developing *Dicer*-null HFs evaginated upward and arrested within the epidermis, rather than invaginating inward into the dermis (Andl *et al.*, 2006; Yi *et al.*, 2006) (Fig. 7.3). Strikingly, the evaginating HFs attracted the DP into the epidermis, thereby maintaining the mesenchymal–epithelial association typically requisite to execute the program of terminal differentiation. However, only concentric rings of matrix and IRS cells surrounded the basement membrane around the DP, with no traces of hair shaft formation.

Exactly, how miR-deficiency-induced dysregulation in epithelial gene expression might cause these gross abnormalities remains unclear. However, even when *Dicer* was ablated after the HFs have formed, they subsequently degenerated, leaving behind only cyst-like structures, and interestingly no bulge compartment (Andl *et al.*, 2006). Thus, the arrest in HFs cannot simply be due to the architectural constraints associated with evagination, but rather must reflect perturbations in the differentiation program itself.

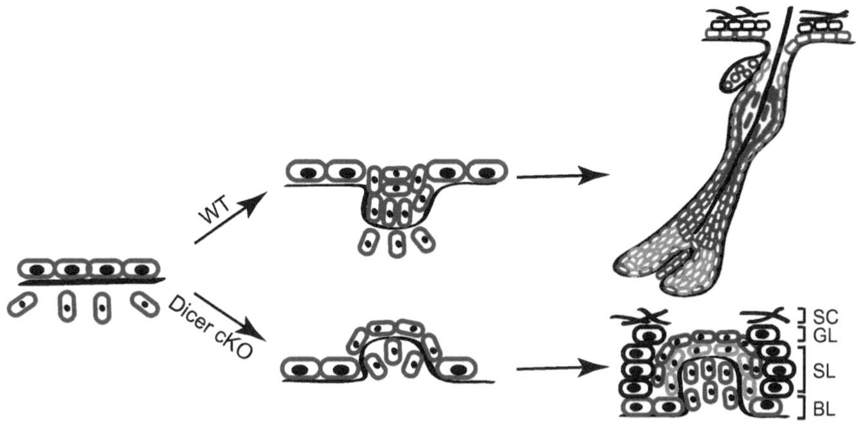

Figure 7.3 Hair follicle evagination in the Dicer cKO skin. In the Dicer-null skin, developing hair follicles evaginate upward and arrest within the epidermis. Notably, the epidermal differentiation is intact except in the region where the epidermal integrity is disrupted by evaginating hair follicles. The same phenotype was also observed in the Dgcr8-null skin, suggesting a causative role of the loss of stereotypical miRs for the defects in hair morphogenesis. (See Color Insert.)

These collective studies on *Dicer* cKO skin present the view that miRs in the skin might be required to maintain the appropriate output of signaling pathway(s) and, in turn, the finely tuned pathway(s) that are essential for the maintenance of HF stem cells and the mesenchymal–epithelial cross talk that orchestrates their proper downgrowth and lineage progression.

In striking contrast to hypoproliferation within the HF and the depletion of its stem cells, hyperproliferation was observed in the mature *Dicer* cKO epidermis (Andl *et al.*, 2006). These defects suggest that the specific miRs differentially expressed by these tissues may have functionally distinct roles. If so, HF miRs would appear to control stem cell survival and maintenance, while epidermal miRs seem more likely to govern cell cycle exit and/or the balance between proliferation and differentiation.

Another differential feature of ablating *Dicer* in the skin was the preferential increase in apoptosis within the HFs, and particularly so within the highly proliferative matrix compartment (Andl *et al.*, 2006; Yi *et al.*, 2006). The specific localization of apoptotic cells was suggestive of either a general requirement of Dicer (miRs) in rapidly dividing cells, consistent with apoptotic phenotypes observed in Dicer cKO limb and T cells (Harfe *et al.*, 2005; Muljo *et al.*, 2005), or a specific requirement for Dicer (miRs) in the hair bulb. Recently, an exciting study has identified the *C. elegans* Dicer homolog as a caspase substrate, and upon cleavage, the liberated Dicer RNase III domain translocates to the nucleus and becomes a DNase that is critical for the DNA fragmentation during apoptosis (Nakagawa *et al.*, 2010). It is of great interest to examine mammalian Dicer homologues share this function that is independent of the miR pathway, and if so, how it might be differentially regulated in the epidermis and HF.

6. Conditional Ablation of *Dgcr8*

While studies with *Dicer* cKO skin provided a glimpse of how miRs may be functioning in the skin, two key questions arise: (1) Are the phenotypes observed in the *Dicer* cKO truly caused by depletion of mature miRs? and (2) Are miRs the only Dicer products in the skin? To address these questions, researchers engineered the equivalent *K14-Cre*-driven cKO of *Dgcr8*, thereby targeting an essential nuclear cofactor for miR processing. Moreover, while Dicer has been reported to control the processing of both miRs- and mRNA-derived small RNAs (Babiarz *et al.*, 2008; Tam *et al.*, 2008; Watanabe *et al.*, 2008), Dgcr8 is specific for stereotypical miRs (Babiarz *et al.*, 2008; Yi *et al.*, 2009).

When comparing the number of deep sequencing reads that can be mapped to different classes of small RNAs, stereotypical miRs emerged as the most abundant species in the skin (Yi *et al.*, 2009). In comparing the reads between the *K14-Cre X Dicer* and *K14-Cre* X *Dgcr8* cKO skins, the production of the overwhelming majority of miRs was dependent upon both Dicer and Dgcr8, while only a few hairpin miR- and mRNA-derived small RNAs showed dependency only upon Dicer and not Dgcr8 (Yi *et al.*, 2009).

Most importantly, both *Dicer* and *Dgcr8* skin cKO animals displayed indistinguishable phenotypes including evaginating HFs, enriched apoptosis in hair bulbs, rough and dehydrated skin, and neonatal lethality. Thus, these results confirmed that the previously reported *Dicer* cKO skin phenotypes (Andl *et al.*, 2006; Yi *et al.*, 2006) were indeed *bona fide* consequences of the loss of miRs in the skin and firmly established that stereotypical miRs are the key Dicer products in skin.

7. Dissecting the Complexities of the Differential Expression of miRs in Skin

By investigating *Dicer* and *Dgcr8* skin cKOs, the global importance of miRs in skin development was established and the differential effects on the epidermis versus HF pointed to a physiological relevance to their differential expression. With this information at hand, the next step was to begin to unearth the functions of individual miRs in the skin.

miR-203 is highly expressed in a spatiotemporal-specific manner in vertebrate epidermis (Aberdam *et al.*, 2008; Lena *et al.*, 2008; Sonkoly *et al.*, 2007; Yi *et al.*, 2008), rendering it an obvious first choice for exploring the functional significance of an individual miR in the skin. *In situ* hybridizations revealed miR-203 in the differentiating suprabasal and not the basal progenitors of the epidermis (Yi *et al.*, 2009), making it tempting to speculate that the hyperproliferative epidermal phenotype observed in the *Dicer* cKO skin (Andl *et al.*, 2006; Yi *et al.*, 2006, 2008) might be attributable specifically to the loss of miR-203.

Functional studies revealed miR-203's ability to unleash its potent inhibitory powers on proliferative potential. When miR-203 was precociously expressed in the epidermal stem/progenitor cells either *in vivo* or *in vitro*, keratinocytes exited the cell cycle and displayed significantly reduced capacity to form colonies (Lena *et al.*, 2008; Yi *et al.*, 2008). Conversely, when epidermal cells or neonatal mice were treated with the chemically modified antisense oligonucleotide (antagomir) against miR-203, an increase was observed in actively dividing cells within the suprabasal epidermal layers (Yi *et al.*, 2008). Consistent with its function in inhibiting cell proliferation in mammalian skin, miR-203 potently inhibited the growth of the repairing fin when overexpressed in Zebrafish skin (Thatcher *et al.*, 2008).

To understand how miR-203 exerts these effects, it is critical to know its physiological targets. Although systematic and unbiased target identification approaches are yet to be established, a few miR-203 targets have now been documented (Lena *et al.*, 2008; Yi *et al.*, 2008). The best characterized is the transcription factor ΔNp63α (also known as p73-like in human). Over the past decade, ΔNp63α has been extensively investigated for its essential function in the maintenance of "stemness" in the skin and in other stratified epithelia (Mills *et al.*, 1999; Senoo *et al.*, 2007; Yang *et al.*, 1999). Intriguingly and in a fashion expected of a miR's action on its targets, ΔNp63α and miR-203 have mutually exclusive expression patterns, opposite functions, and evolutionarily conserved regulatory relationships (Yi *et al.*, 2008).

Importantly in the miR-203 gain- and loss-of-function studies, ΔNp63α was diminished when miR-203 was prematurely expressed basally, and conversely, ΔNp63α expanded aberrantly into the suprabasal layers when animals were treated with miR-203 antagomir (Yi *et al.*, 2008). Although it is not yet clear whether ΔNp63α is the only target that mediates the function of miR-203, this study provided an important example of miR's ability to sharpen a developmental transition by targeting a master stem cell regulator at the juncture at which they become induced to differentiate (Yi *et al.*, 2008) (Fig. 7.4A).

An additional note of intrigue is that whereas miR-203 potently repressed epidermal stem cell proliferation, many structural markers of terminal differentiation were not ectopically activated in the stem cells when miR-203 was precociously expressed (Yi *et al.*, 2008). Thus in this

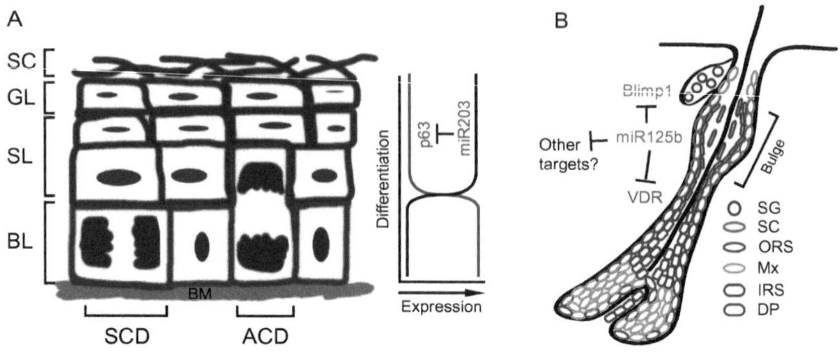

Figure 7.4 Two distinct models for miR's functions in skin stem cells. (A) Reciprocal spatial expression of miR-203 and its target, p63, in the epidermis. In the basal stem cells (BL), p63 is highly expressed to maintain the stemness. Upon differentiation, miR-203 is rapidly upregulated to repress p63 expression in the suprabasal layers including spinous (SL) and granular (GL) layers. (B) miR-125b fine-tunes the input of hair follicle stem cells into sebaceous gland and hair follicle lineages by targeting Blimp1 and VDR, respectively. (See Color Insert.)

case, the exit of somatic stem cell cycling and the induction of terminal differentiation appeared to be regulated by distinct mechanisms. This model is in a close agreement with the paradigm established by the lin-4 and let-7 studies in *C. elegans* where miRs function to promote developmental transitions by inhibiting key molecules that are required for early developmental stages, but which must later be downregulated to progress through the differentiation program.

8. miR-203 as a Tumor Suppressor?

Because of the potent inhibition of miR-203 to cell proliferation, miR-203 might be expected to function as a tumor suppressor. Indeed, miR-203 was induced in a head and neck squamous cell carcinoma cell line exposed to UVC irradiation, a treatment that results in cell death and cell cycle exit (Lena *et al.*, 2008). Conversely, an elegant study in human hematopoietic tumors revealed that frequent silencing of miR-203 either genetically or epigenetically is correlated with T cell lymphomas (Bueno *et al.*, 2008). Moreover, forced expression of miR-203 directly resulted in the downregulation of the oncogene, BCR-ABL1, and blocked cancer cell proliferation (Bueno *et al.*, 2008). Oddly, however, and in contrast to the epidermis, miR-203 is not detected in normal T cell lineages (Landgraf *et al.*, 2007; Neilson *et al.*, 2007; Wu *et al.*, 2007). Even more paradoxical are the findings that miR-203 is upregulated in psoriasis, a common human hyperproliferative skin disease involving a proinflammatory response, and in a few epithelial tumors (Bandres *et al.*, 2006; Gottardo *et al.*, 2007; Iorio *et al.*, 2007; Sonkoly *et al.*, 2007; Szafranska *et al.*, 2007). While further studies will be necessary to reconcile these differences, one possibility is that the expression of miR-203 can be independently induced under stress conditions, for example, cancer, psoriasis, or other diseases, perhaps as a negative and in some cases futile feedback mechanism to suppress the proliferative state.

A final interesting twist to the possible tumor suppressive roles of miR-203 comes from recent studies on the human papillomaviruses (HPVs), which do not encode their own miRs (Cai *et al.*, 2006) but do modulate expression of cellular miRNAs to regulate the activities of cellular proteins. Recently, it was discovered that one of HPV's two oncoproteins, E7, downregulates miR-203 expression upon epidermal differentiation (Melar-New and Laimins, 2010), while the other, E6, downregulates miR-203 by compromising p53 function (McKenna *et al.*, 2010). Moreover, HPV-positive cells maintain significantly higher levels of ΔNp63α than normal keratinocytes do, and when miR-203 was introduced into keratinocytes that stably maintain HPV episomes, the HPV was rapidly lost upon subsequent passage. Together, these findings suggest that miR-203 is inhibitory to HPV amplification and that HPV oncoproteins act in part by suppressing miR-203 in differentiating cells to disrupt the balance between proliferation and differentiation and allow productive HPV replication and propagation.

9. Characterizing and Defining the Functions of Specific miRs that are Differentially Expressed by HF Stem Cells

Most recently, a study for miR in HF stem cells provides another dimension to miR function. MiR-125b, a *lin-4* homologue, was identified as a markedly upregulated miR in HF stem cells relative to any of the three other proliferating compartments within the skin epithelium (Zhang *et al.*, 2011). Moreover, as judged by *in situ* hybridizations, miR-125b rapidly waned once stem cells exited their niche and became ORS progenitors. Thereafter, miR-125b remained off as ORS cells progressed further to become the matrix cells (Zhang *et al.*, 2011).

To understand how this switch in miR-125b expression relates to the ability of SCs to embark upon the HF lineage, Zhang and Fuchs took a doxycycline (tetracycline) inducible, transgenic strategy and sustained miR-125b expression in the proliferative progeny of HF-SCs. The outcome was a shift in the balance between the stem cells and their committed progeny.

Thus, when the HF-SCs were unable to downregulate miR-125b, the upper ORS became hyperthickened with cells that appeared to be uncommitted HF-SCs. Similarly, the entire process of lineage progression appeared to be delayed when miR-125b could not be properly downregulated (Zhang *et al.*, 2011). The overall outcome was eventual baldness. Similarly, the SG progenitors were expanded as a consequence of miR-125b overexpression, leading to enlarged SGs.

Intriguingly, the maintenance of HF-SCs and SG progenitors appeared to be intact, as upon withdrawal of doxycycline, the block to differentiation was lifted and HFs and SGs were restored even after up to 4 months of continuous miR-125b induction (Zhang *et al.*, 2011). These observations suggest a new model for the role of miRs in stem cells, namely, as a rheostat to precisely govern their input into the differentiation program (Fig. 7.4B).

In *C. elegans*, one of lin-4's major targets is lin-28, which binds endogenous primary let-7 transcripts and blocks its biogenesis into a miR. Although let-7 is expressed by skin progenitors, lin-28 is not expressed anywhere in the normal skin epithelium, suggesting that miR-125b/lin4 in the HF-SCs must have other targets besides lin-28, and that let-7 must be regulated by mechanisms that go beyond lin-28 (Zhang *et al.*, 2011). Microarray profiling of miR-125b-induced skin epithelium and subsequent target analyzes revealed several downstream targets that could explain the phenotype: the *VdR* gene encoding the vitamin D receptor, whose ablation results in baldness (Li *et al.*, 1997; Palmer *et al.*, 2008) and the *Blimp1* gene encoding a transcriptional repressor of *c-Myc*, whose ablation results in SG enlargement (Horsley *et al.*, 2006). Interestingly, both *VdR* and *Blimp1* have been identified as miR-125b targets in other cell types and tissues (Gururajan *et al.*, 2010; Malumbres *et al.*, 2009; Mohri *et al.*, 2009). That said, the skin phenotype of the miR-125b-induced mouse is complex, and additional targets are likely to be involved. Future studies will be needed to address these issues.

10. Delving Further into miR Functions in Stem Cells: The Hematopoietic System

Hematopoietic stem cells (HSCs), with their well-defined cell lineages, offer an ideal system to map the lineage-specific expression patterns of miRs and decipher their roles in differentiation. In one of the earliest studies of mammalian miRs, miR-181, miR-223, and miR-142 were all found to be upregulated when bone marrow progenitor cells differentiate toward B lymphocytes (Chen *et al.*, 2004). Moreover, forced expression of miR-181 in those progenitors enhanced B-lymphoid development at the expense of T-lymphoid differentiation. In another important study, it was discovered that miR-155 is expressed in mature B and T lymphocytes, and that it is also significantly upregulated in a variety of lymphomas (Eis *et al.*, 2005). Loss-of-function studies in mice revealed that miR-155 is a critical *in vivo* regulator of specific differentiation processes in the immune response (Rodriguez *et al.*, 2007; Thai *et al.*, 2007).

In an effort to identify miR-155 targets, upregulated mRNAs in *miR-155* null T lymphocytes were analyzed for the presence of seed matches to miR-155 (Rodriguez *et al.*, 2007). Even though ~65% of the upregulated genes contained miR-155 seed matches in their 3′UTRs, the key miR-155 targets appeared to be mRNAs encoding cytokines, providing an explanation for the remarkable impact and specificity of this miR on immune cell lineages.

Another elegant study of hematopoietic-specific miRs focused on miR-150 and provided an example of a single mRNA:miR regulatory pair that functions critically in lymphocyte development (Xiao *et al.*, 2007). Like miR-155, miR-150 is primarily expressed in mature lymphocytes rather than the progenitor cells. By loss- and gain-of-function studies, miR-150 was shown to regulate lymphocyte terminal differentiation (Xiao *et al.*, 2007). Importantly, the expression of c-Myb, a key transcription factor and the top predicted target of miR-150, inversely correlates with that of miR-150, suggesting that miR-150 might be both necessary and sufficient to control c-Myb expression. Interestingly, the ensuing *c-Myb* heterozygous KO mouse showed phenotypes similar to those observed in a *miR-150* transgenic mouse, lending support to the argument that miR-150 carries out its function through the accurate control of c-Myb expression (Xiao *et al.*, 2007).

Similar to miR-150 and miR-155, miR-223 is also expressed at a low level in the HSCs but gradually upregulated during the differentiation toward mature peripheral blood granulocytes (Johnnidis *et al.*, 2008). When genetically ablated, miR-223 KO mice have an increased population of granulocyte progenitors. This phenotype is likely due to the derepression of one of miR-223 direct targets, *Mef2c*, a transcription factor that promotes proliferation of the myeloid progenitors. Remarkably, when the miR-223 KO mice were across to *Mef2c* KO mice, the loss of Mef2c was able to repress the enhanced progenitor proliferation. This result strongly suggests that the downregulation of Mef2c by miR-223 during granulocyte differentiation is important to suppress the proliferative capacity of the differentiating progenitors. It is also interesting that although Mef2c is dispensable for the hematopoietic homeostasis, its regulation by miR-223 appears to be critical to modulate the pool of progenitor cells during the granulocytic differentiation.

Studies of the erythroid lineage have also revealed an inverse relationship between miR expression and progenitor. When erythroid progenitors begin to differentiate into red blood cells, the miR-144/miR-451 locus is directly activated by a key erythroid transcription factor, Gata-1 (Dore *et al.*, 2008). Loss-of-function studies in both Zebrafish and mouse revealed that this conserved miR locus is required for erythroblast maturation (Dore *et al.*, 2008; Rasmussen *et al.*, 2010). Interestingly, the mRNA of 14-3-3ζ, a key regulator of cytokine signaling, was identified as a direct target of miR-451. More importantly, knocking down 14-3-3ζ in *miR-451* null HSCs rescued these defects during erythroid differentiation *in vitro,* providing another example where a single target may be primarily responsible for mediating the effects of a miR in a defined cellular context (Patrick *et al.*, 2010; Yu *et al.*, 2010).

11. miR Function in Neural Stem Cells

In the adult brain, Dicer ablation results in massive hypotrophy of the cortex due to neuronal apoptosis, accompanied by a dramatic impairment of neuronal differentiation (De Pietri Tonelli *et al.*, 2008). Remarkably, the neuroepithelial cells and the neurogenic progenitors derived from them were not grossly affected by depletion of Dicer, suggesting that like the epidermis, the progenitors are less dependent on miRNAs than their differentiated progeny.

miR-9 is interesting in that it is expressed specifically in the subventricular zone (SVZ) of the brain known to harbor the neural stem cells (NSCs). However, miR-9 appears to negatively regulate NSC proliferation and accelerate neural differentiation (Zhao *et al.*, 2009). Intriguingly, TLX, requisite for NSC self-renewal, is a target for miR-9, but, in turn, it also antagonizes miR-9's expression by directly reducing the transcription of *pri-miR-9* (Zhao *et al.*, 2009). Thus, TLX and miR-9 form a negative feedback regulatory network to balance both proliferation and differentiation of NSCs. In this regard, miR-9's effects on NSCs appear to be distinct from the effects of either miR-125b or miR-203 on skin SCs.

In contrast, miR-124, one of the most specific and abundant miRs in the brain, more closely mirrors the behavior of miR-203 in skin. Like miR-203, miR-124 is expressed at low levels in the SVZ stem cell compartment but is sharply upregulated in mature granule and periglomerular neurons (Cheng *et al.*, 2009). Similarly, gain-of-function of miR-124 induces cell cycle exit, while inhibition of miR-124 by antagomir *in vivo* results in an increase in the population of precursor cells in the SVZ. Moreover, Sox9, a key transcription factor whose downregulation is required for neural differentiation, has been identified as a direct target of miR-124 (Cheng *et al.*, 2009). In this regard, the effect of miR-124 on NSCs seems to be a mirror image of those of miR-125b on HF-SCs.

Finally, it is worth mentioning that miRs have also been implicated in the regulation of stem cell aging. Hmga2, a key transcription factor for the self-renewal of several types of stem cells, is highly expressed in fetal NSCs, but its levels decline by more than 99% during the mouse's lifespan (Nishino *et al.*, 2008). This decline in Hmga2 is accompanied by a reduction in self-renewal capacity of NSCs and appears to be in part caused by a ~30-fold age-induced increase in the miR let-7b, known to target and inhibit Hmga2 expression (Mayr *et al.*, 2007; Nishino *et al.*, 2008). The specific disruption of Hmga2 regulation with a truncated 3′UTR refractory to let-7 targeting resulted in a significant rescue of the self-renewal capacity in an *in vitro* culture assay, supporting the causative role of let-7b in the downregulation of Hmga2 expression (Nishino *et al.*, 2008).

12. miR Function in Muscle Stem Cells

A recent study of skeletal muscle satellite cells provides another interesting parallel to that of the skin stem cell lineages. Pax7 is a critical transcription factor that is highly expressed in quiescent muscle stem (satellite) cells and is required to maintain the stem cell population. Upon injury, satellite stem cells are activated to repair the wound, and Pax7 is concomitantly downregulated to allow these cells to progress to differentiate and contribute to muscle regeneration. Both miR-1 and miR-206 target Pax7 mRNA and are markedly upregulated concomitant with downregulation of Pax7 and satellite cell differentiation (Chen *et al.*, 2010). Conversely, specific inhibition of miR-1 and miR-206 by antagomirs delays Pax7 downregulation and interferes with the differentiation program (Chen *et al.*, 2010). Overall, such findings are remarkably similar to the role of miR-203 in regulating ΔNp63α in epidermal stem cells (Yi *et al.*, 2008) and together, point to the view that a number of somatic miRs expressed at the transition between the stem cells and their differentiation lineage function in fine-tuning the switch.

13. miR Function in Cancer Stem Cells

The role of miRs extends beyond normal development and is particularly intriguing in human cancer. The miR pathway is often dampened in tumor cells (Lu *et al.*, 2005). Indeed, reduced miR expression in Dicer heterozygous animals has been shown a causative role in driving tumorigenesis and cellular transformation in several mouse models (Kumar *et al.*, 2009), and in the skin, the hyperproliferative epidermal phenotype is suggestive of an increased sensitivity (Andl *et al.*, 2006). In addition, the shortening of 3′UTRs of proto-oncogenes in cancer cell lines has been found to enable the mRNA to escape the repression by miRs and result in oncogene activation (Mayr and Bartel, 2009). Together, these studies raise the hypothesis that the miR pathway often functions in tumor suppression, as suggested earlier in this review.

It is also interesting that let-7 miRs appear to be powerful tumor suppressors in their ability to target multiple critical oncogenes including RAS, c-Myc, Hmga2, and Lin28 (Johnson *et al.*, 2005; Mayr *et al.*, 2007; Roush and Slack, 2008; Viswanathan *et al.*, 2009). Several miRs that are highly expressed in normal skin, including miR-200 and miR-205, have also been implicated in epithelial cancers. Both miR-200 miRs and miR-205 are highly expressed in normal skin, where they specifically target the expression of the mRNAs encoding transcriptional repressors of E-cadherin, ZEB1, and ZEB2 (Christoffersen *et al.*, 2007; Gregory *et al.*, 2008; Korpal *et al.*, 2008; Park *et al.*, 2008). By doing so, miR-200 and miR-205 promote the upregulated expression of the intercellular adhesion protein E-cadherin. Conversely, downregulation of the miR-200 family or miR-205 leads to the inhibition of E-cadherin, thereby promoting an epithelial–mesenchymal transition (Gregory *et al.*, 2008; Korpal *et al.*, 2008; Park *et al.*, 2008). Consistent with these studies, miR-205 is often significantly downregulated in human epithelial tumors when comparing to the normal tissues (Childs *et al.*, 2009; Feber *et al.*, 2008). That said, there are profiling studies of human epithelial cancers, where miR-200 family and/or miR-205 were found to be upregulated (Iorio *et al.*, 2007; Tran *et al.*, 2007; Yu *et al.*, 2008). These apparently contrasting results suggest that miRs may have different targets and roles in different cells.

The identification of cancer stem cells in numerous tumors including those derived from brain, breast, prostate, colon, head and neck, pancreas, and skin necessitates an investigation into the underlying molecular mechanisms involved. Given the diversity of miR expression patterns and functions, it is perhaps not surprising that miRs are critically involved in cancer stem cells. For instance, in the tumor initiating cells of breast cancer, let-7 is significantly downregulated to allow the expression of H-Ras and Hmga2 (Yu *et al.*, 2007). Forced expression of let-7 in the tumor initiating cells inhibits self-renewal through the downregulation of H-Ras and promotes differentiation by repressing Hmga2 (Yu *et al.*, 2007). Similarly, miR-34a, a miR transcriptionally activated by p53 (He *et al.*, 2007), functions to inhibit clonogenic expansion of prostate cancer stem cells and metastasis by directly repressing CD44 (Liu *et al.*, 2011). In breast cancer stem cells, the miR-200 family members including miR-200b and miR-200c have multiple functions in restricting tumor development, including repressing self-renewal by repressing a key polycomb protein complex 1 (PRC1) member, BMI1 (Shimono *et al.*, 2009; Wellner *et al.*, 2009). These miRs also play important roles in EMTs (Wellner *et al.*, 2009), as well as repress a key epigenetic component of the PRC2 complex, Suz12, which together with PRC1 functions in epigenetic silencing of genes through H3K27me3 trimethylation (Iliopoulos *et al.*, 2010).

Together, these findings provide critical mechanisms that link miRs to cancer stem cells. Recently, a chemical screen specifically targeting breast cancer stem cells led to the identification of compounds that are particular effective to reduce the cancer stem cell population (Gupta *et al.*, 2009). With the emerging roles of miRs in cancer stem cells, it is tempting to speculate that studies specifically focusing on miRs in these often rare cell populations could yield novel insights toward diagnosis and treatment in the future.

14. Closing Remarks and Future Challenges

Functional characterization of miRs in mammalian development and adult stem cells is still in its infancy. Much of our knowledge of the just how critical miRs are to mammalian development can still be traced to the global miR depletion studies in which *Dicer* and *Dgcr8* were ablated specifically in either embryonic stem cells or adult tissues. The strong defects in differentiation and minor defects in proliferation appear to be a recurring theme, whether for embryonic or adult SCs (Kanellopoulou *et al.*, 2005; Murchison *et al.*, 2005; Wang *et al.*, 2007). However, ESCs and adult tissues do show some notable differences, for instance, those showing that the differences between *Dicer-* and *Dgcr8*-deficient ESCs and mouse oocytes are significantly greater than for skin lineages (Suh *et al.*, 2010; Wang *et al.*, 2007). That said, it is notable that endo-siRNAs, accounting for much of the difference between *Dicer* and *Dgcr8* null phenotypes, have only been detected in ESCs and oocytes and not in adult tissues in mammals (Babiarz *et al.*, 2011; Suh *et al.*, 2010; Tam *et al.*, 2008; Watanabe *et al.*, 2008). In this regard, it is also interesting to note that lin-28, a key downstream target of lin-4 in *C. elegans*, is also primarily restricted to mammalian ESCs and is generally not seen in adult SCs. Whether these differences are reflective of the unique pluripotent state of the ESCs remains to be addressed in the future.

Among the myriad of key issues that remain unresolved is the question regarding the role of a single miR or miR family in the regulation of mammalian development and stem cells. When *C. elegans* miRs are knocked out individually or by the whole family that share the same seed sequences combinatorially, only a few exhibit developmental defects (Alvarez-Saavedra and Horvitz, 2010; Miska *et al.*, 2007). Similarly, with more than 30 miRs having been individually knocked out to date without affecting viability in mice, it is evident that the loss of a single miR or even a whole family might not cause the severe developmental defects in cell fate specification that would be expected from ablation of a gene encoding a master regulator. However, when these mutant mice are subject to physiological stresses like injury or DNA damage, significant phenotypes surface (reviewed by Leung and Sharp, 2010; Liu and Olson, 2010). These observations are yet further indicators that miR-mediated regulation functions by fine-tuning responses and reinforcing the robustness of biological systems that may not be readily manifested in a well-maintained laboratory condition (reviewed by Herranz and Cohen, 2010). It is worth noting that because of the central role of stem cells during the development and homeostasis of adult tissues, dysregulation in stem cells could be magnified and manifested as defects in their differentiated daughters. This illustrates a critical need for the future to investigate how stress signals regulate miR expression and how miR-mediated regulation, in turn, balances the output of gene expression and protects stem cells from the various and diverse stresses that they encounter throughout the life.

A second key issue to address will be how a single miR can execute its physiological function by regulating its targets. Directly related to this is the question whether a single miR can perform different roles when placed in distinct cellular contexts. As miR functions in mammalian stem cells begin to unfold, some of the most challenging questions to be answered will be which mRNAs are targeted by a specific miR and which of its many targets are key in a given cellular context.

A third key area to be addressed in the future is the question of how miR-mediated regulation integrates with other regulatory mechanisms, for example, transcriptional regulation, to modulate stem cell fate. Although biologists now appreciate the importance of the regulatory layer provided by miRs, the interactions between the miR pathway and other regulatory mechanisms must be elucidated to fully understand how miRs work. Further, it will be important to unveil the characteristics of the miR pathway that are distinct from other mechanisms such as cellular context-dependent function.

Finally, miR's functions in human diseases especially cancer are particularly intriguing. Because miRs' functions are highly dependent upon local mRNA content, it is conceivable that their functions in cancer cells could dramatically differ from their functions during normal development. Thus, characterization of individual miRs' roles at different stage of tumorigenesis will be important to understand the underlying biology of tumor development. In particular, because of the central roles of cancer stem cells in tumor development and relapse, it is critical to focus on miR-mediated regulation in these cells. Answers to these fascinating questions are certainly not only going to provide significant new insights into miR functions but also point to new directions to utilize and target miRs in the manipulation of stem cells for regenerative medicine as well as in the development of cancer therapies.

ACKNOWLEDGMENTS

This publication was made possible by grants R00AR054704 and R01AR059697 (to R. Y.), and R01AR031737 (to E. F.) from NIAMS/NIH. E. F. is a Howard Hughes Medical Institute Investigator.

REFERENCES

Aberdam, D., Candi, E., Knight, R. A., and Melino, G. (2008). *Trends Biochem. Sci.* **33,** 583–591.

Alvarez-Saavedra, E., and Horvitz, H. R. (2010). *Curr. Biol.* **20,** 367–373.

Ambros, V. (2004). *Nature* **431,** 350–355.

Andl, T., Murchison, E. P., Liu, F., Zhang, Y., Yunta-Gonzalez, M., Tobias, J. W., Andl, C. D., Seykora, J. T., Hannon, G. J., and Millar, S. E. (2006). *Curr. Biol.* **16,** 1041–1049.

Babiarz, J. E., Ruby, J. G., Wang, Y., Bartel, D. P., and Blelloch, R. (2008). *Genes Dev.* **22,** 2773–2785.

Babiarz, J. E., Hsu, R., Melton, C., Thomas, M., Ullian, E. M., and Blelloch, R. (2011). *RNA* **17,** 1489–1501.

Baek, D., Villen, J., Shin, C., Camargo, F. D., Gygi, S. P., and Bartel, D. P. (2008). *Nature* **455,** 64–71.

Bandres, E., Cubedo, E., Agirre, X., Malumbres, R., Zarate, R., Ramirez, N., Abajo, A., Navarro, A., Moreno, I., Monzo, M., and Garcia-Foncillas, J. (2006). *Mol. Cancer* **5,** 29.

Bartel, D. P. (2009). *Cell* **136,** 215–233.

Bernstein, E., Kim, S. Y., Carmell, M. A., Murchison, E. P., Alcorn, H., Li, M. Z., Mills, A. A., Elledge, S. J., Anderson, K. V., and Hannon, G. J. (2003). *Nat. Genet.* **35,** 215–217.

Borchert, G. M., Lanier, W., and Davidson, B. L. (2006). *Nat. Struct. Mol. Biol.* **13,** 1097–1101.

Brenner, J. L., Jasiewicz, K. L., Fahley, A. F., Kemp, B. J., and Abbott, A. L. (2010). *Curr. Biol.* **20,** 1321–1325.

Bueno, M. J., Perez de Castro, I., Gomez de Cedron, M., Santos, J., Calin, G. A., Cigudosa, J. C., Croce, C. M., Fernandez-Piqueras, J., and Malumbres, M. (2008). *Cancer Cell* **13,** 496–506.

Byrne, C., Tainsky, M., and Fuchs, E. (1994). *Development* **120,** 2369–2383.

Cai, X., Hagedorn, C. H., and Cullen, B. R. (2004). *RNA* **10,** 1957–1966.

Cai, X., Li, G., Laimins, L. A., and Cullen, B. R. (2006). *J. Virol.* **80,** 10890–10893.

Chen, C. Z., Li, L., Lodish, H. F., and Bartel, D. P. (2004). *Science* **303,** 83–86.

Chen, J. F., Tao, Y., Li, J., Deng, Z., Yan, Z., Xiao, X., and Wang, D. Z. (2010). *J. Cell Biol.* **190,** 867–879.

Cheng, L. C., Pastrana, E., Tavazoie, M., and Doetsch, F. (2009). *Nat. Neurosci.* **12,** 399–408.

Childs, G., Fazzari, M., Kung, G., Kawachi, N., Brandwein-Gensler, M., McLemore, M., Chen, Q., Burk, R. D., Smith, R. V., Prystowsky, M. B., Belbin, T. J., and Schlecht, N. F. (2009). *Am. J. Pathol.* **174,** 736–745.

Christodoulou, F., Raible, F., Tomer, R., Simakov, O., Trachana, K., Klaus, S., Snyman, H., Hannon, G. J., Bork, P., and Arendt, D. (2010). *Nature* **463,** 1084–1088.

Christoffersen, N. R., Silahtaroglu, A., Orom, U. A., Kauppinen, S., and Lund, A. H. (2007). *RNA* **13,** 1172–1178.

De Pietri Tonelli, D., Pulvers, J. N., Haffner, C., Murchison, E. P., Hannon, G. J., and Huttner, W. B. (2008). *Development* **135,** 3911–3921.
Denli, A. M., Tops, B. B., Plasterk, R. H., Ketting, R. F., and Hannon, G. J. (2004). *Nature* **432,** 231–235.
Dore, L. C., Amigo, J. D., Dos Santos, C. O., Zhang, Z., Gai, X., Tobias, J. W., Yu, D., Klein, A. M., Dorman, C., Wu, W., Hardison, R. C., Paw, B. H., *et al.* (2008). *Proc. Natl. Acad. Sci. USA* **105,** 3333–3338.
Eis, P. S., Tam, W., Sun, L., Chadburn, A., Li, Z., Gomez, M. F., Lund, E., and Dahlberg, J. E. (2005). *Proc. Natl. Acad. Sci. USA* **102,** 3627–3632.
Farh, K. K., Grimson, A., Jan, C., Lewis, B. P., Johnston, W. K., Lim, L. P., Burge, C. B., and Bartel, D. P. (2005). *Science* **310,** 1817–1821.
Feber, A., Xi, L., Luketich, J. D., Pennathur, A., Landreneau, R. J., Wu, M., Swanson, S. J., Godfrey, T. E., and Litle, V. R. (2008). *J. Thorac. Cardiovasc. Surg.* **135,** 255–260. (discussion 260).
Friedman, R. C., Farh, K. K., Burge, C. B., and Bartel, D. P. (2009). *Genome Res.* **19,** 92–105.
Gangaraju, V. K., and Lin, H. (2009). *Nat. Rev. Mol. Cell Biol.* **10,** 116–125.
Gottardo, F., Liu, C. G., Ferracin, M., Calin, G. A., Fassan, M., Bassi, P., Sevignani, C., Byrne, D., Negrini, M., Pagano, F., Gomella, L. G., Croce, C. M., *et al.* (2007). *Urol. Oncol.* **25,** 387–392.
Gregory, R. I., Yan, K. P., Amuthan, G., Chendrimada, T., Doratotaj, B., Cooch, N., and Shiekhattar, R. (2004). *Nature* **432,** 235–240.
Gregory, P. A., Bert, A. G., Paterson, E. L., Barry, S. C., Tsykin, A., Farshid, G., Vadas, M. A., Khew-Goodall, Y., and Goodall, G. J. (2008). *Nat. Cell Biol.* **10,** 593–601.
Grimson, A., Srivastava, M., Fahey, B., Woodcroft, B. J., Chiang, H. R., King, N., Degnan, B. M., Rokhsar, D. S., and Bartel, D. P. (2008). *Nature* **455,** 1193–1197.
Gupta, P. B., Onder, T. T., Jiang, G., Tao, K., Kuperwasser, C., Weinberg, R. A., and Lander, E. S. (2009). *Cell* **138,** 645–659.
Gururajan, M., Haga, C. L., Das, S., Leu, C. M., Hodson, D., Josson, S., Turner, M., and Cooper, M. D. (2010). *Int. Immunol.* **22,** 583–592.
Han, J., Lee, Y., Yeom, K. H., Kim, Y. K., Jin, H., and Kim, V. N. (2004). *Genes Dev.* **18,** 3016–3027.
Harfe, B. D., McManus, M. T., Mansfield, J. H., Hornstein, E., and Tabin, C. J. (2005). *Proc. Natl. Acad. Sci. USA* **102,** 10898–10903.
He, L., He, X., Lim, L. P., de Stanchina, E., Xuan, Z., Liang, Y., Xue, W., Zender, L., Magnus, J., Ridzon, D., Jackson, A. L., Linsley, P. S., *et al.* (2007). *Nature* **447,** 1130–1134.
Herranz, H., and Cohen, S. M. (2010). *Genes Dev.* **24,** 1339–1344.
Horsley, V., O'Carroll, D., Tooze, R., Ohinata, Y., Saitou, M., Obukhanych, T., Nussenzweig, M., Tarakhovsky, A., and Fuchs, E. (2006). *Cell* **126,** 597–609.
Hutvagner, G., McLachlan, J., Pasquinelli, A. E., Balint, E., Tuschl, T., and Zamore, P. D. (2001). *Science* **293,** 834–838.
Iliopoulos, D., Lindahl-Allen, M., Polytarchou, C., Hirsch, H. A., Tsichlis, P. N., and Struhl, K. (2010). *Mol. Cell* **39,** 761–772.
Iorio, M. V., Visone, R., Di Leva, G., Donati, V., Petrocca, F., Casalini, P., Taccioli, C., Volinia, S., Liu, C. G., Alder, H., Calin, G. A., Menard, S., *et al.* (2007). *Cancer Res.* **67,** 8699–8707.
Ivey, K. N., and Srivastava, D. (2010). *Cell Stem Cell* **7,** 36–41.
Johnnidis, J. B., Harris, M. H., Wheeler, R. T., Stehling-Sun, S., Lam, M. H., Kirak, O., Brummelkamp, T. R., Fleming, M. D., and Camargo, F. D. (2008). *Nature* **451,** 1125–1129.
Johnson, S. M., Grosshans, H., Shingara, J., Byrom, M., Jarvis, R., Cheng, A., Labourier, E., Reinert, K. L., Brown, D., and Slack, F. J. (2005). *Cell* **120,** 635–647.

Kanellopoulou, C., Muljo, S. A., Kung, A. L., Ganesan, S., Drapkin, R., Jenuwein, T., Livingston, D. M., and Rajewsky, K. (2005). *Genes Dev.* **19,** 489–501.

Kim, V. N., Han, J., and Siomi, M. C. (2009). *Nat. Rev. Mol. Cell Biol.* **10,** 126–139.

Korpal, M., Lee, E. S., Hu, G., and Kang, Y. (2008). *J. Biol. Chem.* **283,** 14910–14914.

Krol, J., Loedige, I., and Filipowicz, W. (2010). *Nat. Rev. Genet.* **11,** 597–610.

Kumar, M. S., Pester, R. E., Chen, C. Y., Lane, K., Chin, C., Lu, J., Kirsch, D. G., Golub, T. R., and Jacks, T. (2009). *Genes Dev.* **23,** 2700–2704.

Landgraf, P., Rusu, M., Sheridan, R., Sewer, A., Iovino, N., Aravin, A., Pfeffer, S., Rice, A., Kamphorst, A. O., Landthaler, M., Lin, C., Socci, N. D., *et al.* (2007). *Cell* **129,** 1401–1414.

Landthaler, M., Yalcin, A., and Tuschl, T. (2004). *Curr. Biol.* **14,** 2162–2167.

Lee, R. C., Feinbaum, R. L., and Ambros, V. (1993). *Cell* **75,** 843–854.

Lena, A. M., Shalom-Feuerstein, R., di Val Cervo, P. R., Aberdam, D., Knight, R. A., Melino, G., and Candi, E. (2008). *Cell Death Differ.* **15,** 1187–1195.

Leung, A. K., and Sharp, P. A. (2010). *Mol. Cell* **40,** 205–215.

Lewis, B. P., Burge, C. B., and Bartel, D. P. (2005). *Cell* **120,** 15–20.

Li, Y. C., Pirro, A. E., Amling, M., Delling, G., Baron, R., Bronson, R., and Demay, M. B. (1997). *Proc. Natl. Acad. Sci. USA* **94,** 9831–9835.

Liu, N., and Olson, E. N. (2010). *Dev. Cell* **18,** 510–525.

Liu, C., Kelnar, K., Liu, B., Chen, X., Calhoun-Davis, T., Li, H., Patrawala, L., Yan, H., Jeter, C., Honorio, S., Wiggins, J. F., Bader, A. G., *et al.* (2011). *Nat. Med.* **17,** 211–215.

Lu, J., Getz, G., Miska, E. A., Alvarez-Saavedra, E., Lamb, J., Peck, D., Sweet-Cordero, A., Ebert, B. L., Mak, R. H., Ferrando, A. A., Downing, J. R., Jacks, T., *et al.* (2005). *Nature* **435,** 834–838.

Lund, E., Guttinger, S., Calado, A., Dahlberg, J. E., and Kutay, U. (2004). *Science* **303,** 95–98.

Malumbres, R., Sarosiek, K. A., Cubedo, E., Ruiz, J. W., Jiang, X., Gascoyne, R. D., Tibshirani, R., and Lossos, I. S. (2009). *Blood* **113,** 3754–3764.

Martinez, N. J., and Gregory, R. I. (2010). *Cell Stem Cell* **7,** 31–35.

Mayr, C., and Bartel, D. P. (2009). *Cell* **138,** 673–684.

Mayr, C., Hemann, M. T., and Bartel, D. P. (2007). *Science* **315,** 1576–1579.

McKenna, D. J., McDade, S. S., Patel, D., and McCance, D. J. (2010). *J. Virol.* **84,** 10644–10652.

Melar-New, M., and Laimins, L. A. (2010). *J. Virol.* **84,** 5212–5221.

Mills, A. A., Zheng, B., Wang, X. J., Vogel, H., Roop, D. R., and Bradley, A. (1999). *Nature* **398,** 708–713.

Miska, E. A., Alvarez-Saavedra, E., Abbott, A. L., Lau, N. C., Hellman, A. B., McGonagle, S. M., Bartel, D. P., Ambros, V. R., and Horvitz, H. R. (2007). *PLoS Genet.* **3,** e215.

Mohri, T., Nakajima, M., Takagi, S., Komagata, S., and Yokoi, T. (2009). *Int. J. Cancer* **125,** 1328–1333.

Muljo, S. A., Ansel, K. M., Kanellopoulou, C., Livingston, D. M., Rao, A., and Rajewsky, K. (2005). *J. Exp. Med.* **202,** 261–269.

Murchison, E. P., Partridge, J. F., Tam, O. H., Cheloufi, S., and Hannon, G. J. (2005). *Proc. Natl. Acad. Sci. USA* **102,** 12135–12140.

Nakagawa, A., Shi, Y., Kage-Nakadai, E., Mitani, S., and Xue, D. (2010). *Science* **328,** 327–334.

Neilson, J. R., Zheng, G. X., Burge, C. B., and Sharp, P. A. (2007). *Genes Dev.* **21,** 578–589.

Nishino, J., Kim, I., Chada, K., and Morrison, S. J. (2008). *Cell* **135,** 227–239.

Palmer, H. G., Anjos-Afonso, F., Carmeliet, G., Takeda, H., and Watt, F. M. (2008). *PLoS One* **3,** e1483.

Park, S. M., Gaur, A. B., Lengyel, E., and Peter, M. E. (2008). *Genes Dev.* **22,** 894–907.
Pasquinelli, A. E., Reinhart, B. J., Slack, F., Martindale, M. Q., Kuroda, M. I., Maller, B., Hayward, D. C., Ball, E. E., Degnan, B., Muller, P., Spring, J., Srinivasan, A., *et al.* (2000). *Nature* **408,** 86–89.
Patrick, D. M., Zhang, C. C., Tao, Y., Yao, H., Qi, X., Schwartz, R. J., Jun-Shen Huang, L., and Olson, E. N. (2010). *Genes Dev.* **24,** 1614–1619.
Rasmussen, K. D., Simmini, S., Abreu-Goodger, C., Bartonicek, N., Di Giacomo, M., Bilbao-Cortes, D., Horos, R., Von Lindern, M., Enright, A. J., and O'Carroll, D. (2010). *J. Exp. Med.* **207,** 1351–1358.
Reinhart, B. J., Slack, F. J., Basson, M., Pasquinelli, A. E., Bettinger, J. C., Rougvie, A. E., Horvitz, H. R., and Ruvkun, G. (2000). *Nature* **403,** 901–906.
Rodriguez, A., Vigorito, E., Clare, S., Warren, M. V., Couttet, P., Soond, D. R., van Dongen, S., Grocock, R. J., Das, P. P., Miska, E. A., Vetrie, D., Okkenhaug, K., *et al.* (2007). *Science* **316,** 608–611.
Roush, S., and Slack, F. J. (2008). *Trends Cell Biol.* **18,** 505–516.
Schmidt-Ullrich, R., and Paus, R. (2005). *Bioessays* **27,** 247–261.
Schwarz, D. S., Hutvagner, G., Du, T., Xu, Z., Aronin, N., and Zamore, P. D. (2003). *Cell* **115,** 199–208.
Selbach, M., Schwanhausser, B., Thierfelder, N., Fang, Z., Khanin, R., and Rajewsky, N. (2008). *Nature* **455,** 58–63.
Senoo, M., Pinto, F., Crum, C. P., and McKeon, F. (2007). *Cell* **129,** 523–536.
Shimono, Y., Zabala, M., Cho, R. W., Lobo, N., Dalerba, P., Qian, D., Diehn, M., Liu, H., Panula, S. P., Chiao, E., Dirbas, F. M., Somlo, G., *et al.* (2009). *Cell* **138,** 592–603.
Sonkoly, E., Wei, T., Janson, P. C., Saaf, A., Lundeberg, L., Tengvall-Linder, M., Norstedt, G., Alenius, H., Homey, B., Scheynius, A., Stahle, M., and Pivarcsi, A. (2007). *PLoS One* **2,** e610.
Stark, A., Brennecke, J., Bushati, N., Russell, R. B., and Cohen, S. M. (2005). *Cell* **123,** 1133–1146.
Suh, N., Baehner, L., Moltzahn, F., Melton, C., Shenoy, A., Chen, J., and Blelloch, R. (2010). *Curr. Biol.* **20,** 271–277.
Szafranska, A. E., Davison, T. S., John, J., Cannon, T., Sipos, B., Maghnouj, A., Labourier, E., and Hahn, S. A. (2007). *Oncogene* **26,** 4442–4452.
Tam, O. H., Aravin, A. A., Stein, P., Girard, A., Murchison, E. P., Cheloufi, S., Hodges, E., Anger, M., Sachidanandam, R., Schultz, R. M., and Hannon, G. J. (2008). *Nature* **453,** 534–538.
Thai, T. H., Calado, D. P., Casola, S., Ansel, K. M., Xiao, C., Xue, Y., Murphy, A., Frendewey, D., Valenzuela, D., Kutok, J. L., Schmidt-Supprian, M., Rajewsky, N., *et al.* (2007). *Science* **316,** 604–608.
Thatcher, E. J., Paydar, I., Anderson, K. K., and Patton, J. G. (2008). *Proc. Natl. Acad. Sci. USA* **105,** 18384–18389.
Tran, N., McLean, T., Zhang, X., Zhao, C. J., Thomson, J. M., O'Brien, C., and Rose, B. (2007). *Biochem. Biophys. Res. Commun.* **358,** 12–17.
Vassar, R., Rosenberg, M., Ross, S., Tyner, A., and Fuchs, E. (1989). *Proc. Natl. Acad. Sci. USA* **86,** 1563–1567.
Viswanathan, S. R., Powers, J. T., Einhorn, W., Hoshida, Y., Ng, T. L., Toffanin, S., O'Sullivan, M., Lu, J., Phillips, L. A., Lockhart, V. L., Shah, S. P., Tanwar, P. S., *et al.* (2009). *Nat. Genet.* **41,** 843–848.
Wang, Y., Medvid, R., Melton, C., Jaenisch, R., and Blelloch, R. (2007). *Nat. Genet.* **39,** 380–385.
Wang, Y., Juranek, S., Li, H., Sheng, G., Wardle, G. S., Tuschl, T., and Patel, D. J. (2009). *Nature* **461,** 754–761.

Watanabe, T., Totoki, Y., Toyoda, A., Kaneda, M., Kuramochi-Miyagawa, S., Obata, Y., Chiba, H., Kohara, Y., Kono, T., Nakano, T., Surani, M. A., Sakaki, Y., *et al.* (2008). *Nature* **453,** 539–543.

Wellner, U., Schubert, J., Burk, U. C., Schmalhofer, O., Zhu, F., Sonntag, A., Waldvogel, B., Vannier, C., Darling, D., zur Hausen, A., Brunton, V. G., Morton, J., *et al.* (2009). *Nat. Cell Biol.* **11,** 1487–1495.

Wightman, B., Ha, I., and Ruvkun, G. (1993). *Cell* **75,** 855–862.

Wu, H., Neilson, J. R., Kumar, P., Manocha, M., Shankar, P., Sharp, P. A., and Manjunath, N. (2007). *PLoS One* **2,** e1020.

Xiao, C., Calado, D. P., Galler, G., Thai, T. H., Patterson, H. C., Wang, J., Rajewsky, N., Bender, T. P., and Rajewsky, K. (2007). *Cell* **131,** 146–159.

Xie, X., Lu, J., Kulbokas, E. J., Golub, T. R., Mootha, V., Lindblad-Toh, K., Lander, E. S., and Kellis, M. (2005). *Nature* **434,** 338–345.

Yang, A., Schweitzer, R., Sun, D., Kaghad, M., Walker, N., Bronson, R. T., Tabin, C., Sharpe, A., Caput, D., Crum, C., and McKeon, F. (1999). *Nature* **398,** 714–718.

Yi, R., Qin, Y., Macara, I. G., and Cullen, B. R. (2003). *Genes Dev.* **17,** 3011–3016.

Yi, R., O'Carroll, D., Pasolli, H. A., Zhang, Z., Dietrich, F. S., Tarakhovsky, A., and Fuchs, E. (2006). *Nat. Genet.* **38,** 356–362.

Yi, R., Poy, M. N., Stoffel, M., and Fuchs, E. (2008). *Nature* **452,** 225–229.

Yi, R., Pasolli, H. A., Landthaler, M., Hafner, M., Ojo, T., Sheridan, R., Sander, C., O'Carroll, D., Stoffel, M., Tuschl, T., and Fuchs, E. (2009). *Proc. Natl. Acad. Sci. USA* **106,** 498–502.

Yu, F., Yao, H., Zhu, P., Zhang, X., Pan, Q., Gong, C., Huang, Y., Hu, X., Su, F., Lieberman, J., and Song, E. (2007). *Cell* **131,** 1109–1123.

Yu, J., Ryan, D. G., Getsios, S., Oliveira-Fernandes, M., Fatima, A., and Lavker, R. M. (2008). *Proc. Natl. Acad. Sci. USA* **105,** 19300–19305.

Yu, D., dos Santos, C. O., Zhao, G., Jiang, J., Amigo, J. D., Khandros, E., Dore, L. C., Yao, Y., D'Souza, J., Zhang, Z., Ghaffari, S., Choi, J., *et al.* (2010). *Genes Dev.* **24,** 1620–1633.

Zhang, L., Stokes, N., Polak, L., and Fuchs, E. (2011). *Cell Stem Cell* **8,** 294–308.

Zhao, C., Sun, G., Li, S., and Shi, Y. (2009). *Nat. Struct. Mol. Biol.* **16,** 365–371.

CHAPTER EIGHT

Exploiting *Drosophila* Genetics to Understand MicroRNA Function and Regulation

Qi Dai,[1] Peter Smibert,[1] *and* Eric C. Lai

Contents

Department of Developmental Biology, Sloan-Kettering Institute, New York, USA

[1] These authors contributed equally.

Current Topics in Developmental Biology, Volume 99
ISSN 0070-2153, DOI: 10.1016/B978-0-12-387038-4.00008-2

Abstract

Although a great deal is known about the identity, biogenesis, and targeting capacity of microRNAs (miRNAs) in animal cells, far less is known about their functional requirements at the organismal level. Much remains to be understood about the necessity of miRNAs for overt phenotypes, the identity of critical miRNA targets, and the control of miRNA transcription. In this review, we provide an overview of genetic strategies to study miRNAs in the *Drosophila* system, including loss- and gain-of-function techniques, genetic interaction strategies, and transgenic reporters of miRNA expression and activity. As we illustrate the usage of these techniques in intact *Drosophila*, we see certain recurrent themes for miRNA functions, including energy homeostasis, apoptosis suppression, growth control, and regulation of core cell signaling pathways. Overall, we hope that this exposition of *Drosophila* genetic techniques, well known to the legions of fly geneticists and used to study all genes, can inform the general miRNA community that focuses on other biochemical, molecular, computational, and structural avenues. Clearly, it is the combination of these myriad techniques that has accelerated miRNA research to its extraordinary pace.

1. Introduction

MicroRNAs (miRNAs) are short regulatory RNAs that mediate broad networks of posttranscriptional repression, with consequences for diverse aspects of development and physiology (Flynt and Lai, 2008). Correspondingly, there is growing appreciation of how human disease and cancer is driven by aberrant or dysfunctional miRNA activity. The majority of miRNAs are generated by a characteristic biogenesis pathway (Kim *et al.*, 2009). In animal cells, this consists of stepwise processing of hairpin precursor transcripts by the Drosha and Dicer RNase III enzymes to yield a ~22nucleotide (nt) small RNA duplex, of which one strand is preferentially loaded into an Argonaute protein and guides it to regulatory targets. In addition, several alternate pathways generate miRNAs via Drosha-independent or Dicer-independent pathways (Yang and Lai, 2011).

As with other classes of Argonaute-mediated small RNAs, that is, siRNAs and piRNAs, the miRNA serves as sequence-specific guide that recruits the Argonaute complex to target transcripts (Czech and Hannon, 2010). In animal cells, the key information for miRNA target identification lies within the 5′ end of the small RNA (the "seed"), since ~7nt complements to positions 2–8 of the miRNA are capable of mediating substantial repression (Brennecke *et al.*, 2005; Doench and Sharp, 2004; Lai, 2002). Consequently, miRNA targets can be predicted genome-wide via conserved miRNA seed matches (Bartel, 2009); location within the transcript, local sequence bias and secondary structure, and other RNA-binding proteins can also influence target efficacy. The abundance of sequenced and aligned genome alignments provides evidence that a large fraction of well-studied metazoan transcripts bear conserved miRNA target sites (Friedman *et al.*, 2009; Jan *et al.*, 2011; Mangone *et al.*, 2010; Ruby *et al.*, 2007b). The endogenous impact of miRNA-mediated repression should be larger still, in light of the facts that many functional sites are not conserved and that at least certain types of nonseed sites can confer repression (Brodersen and Voinnet, 2009).

The wealth of information from comparative genomics, as well as transcriptome- and proteome-based analyses (Baek *et al.*, 2008; Guo *et al.*, 2010; Lim *et al.*, 2005), provides ever-increasing information on the scope of miRNA-mediated repression. Nevertheless, such studies have not provided a straightforward route toward predicting the phenotypic consequences of altering miRNA activity in the context of the whole organism (Smibert and Lai, 2008). Ironically, some of the best-understood biological usages of miRNAs derived from studies conducted prior to the formal recognition of miRNAs. In particular, *Caenorhabditis elegans* genetics permitted the first (lin-4) and second (let-7) identified miRNAs to be placed within regulatory hierarchies that control developmental timing and identified their key direct target genes (Lee *et al.*, 1993; Moss *et al.*, 1997; Reinhart *et al.*, 2000; Wightman *et al.*, 1993). In addition, genetic studies of the *Drosophila* Notch pathway identified key miRNA target genes prior to the cloning of miRNAs (Lai and Posakony, 1997, 1998; Lai *et al.*, 1998) and led to the concept of 7 nt complements to miRNA 5′ ends as animal miRNA-binding sites (Lai, 2002). Therefore, genetic analysis was central to revealing the existence and mechanism of miRNAs.

The genome and transcriptome of *Drosophila melanogaster* have been extensively scoured for miRNA genes, and its current state of annotation is perhaps the deepest among any animal species (Berezikov *et al.*, 2011; Chung *et al.*, 2011) and includes loci derived from several noncanonical pathways (Flynt *et al.*, 2010; Okamura *et al.*, 2007; Ruby *et al.*, 2007a). Deletion mutations of over 30 well-conserved *Drosophila* miRNA genes, comprising 18 genomic loci/clusters, have been described and collectively reveal critical biological requirements for miRNAs. Many other miRNA loci have been associated with compelling gain-of-function phenotypes, and still others are "interesting" from the point of view of cell- or tissue-specific expression patterns (Aboobaker *et al.*, 2005; Ruby *et al.*, 2007b), conserved targeting of well-studied protein-coding genes (Ruby *et al.*, 2007b; Stark *et al.*, 2007), or principles that relate cohorts of miRNA target genes (Lai, 2002; Stark *et al.*, 2005). Altogether, studies in the *Drosophila* model have richly illuminated our understanding of miRNA-mediated regulation (Smibert and Lai, 2010).

In this review, we concentrate on what has been learned about the biological usage of miRNAs from fly genetics. The *Drosophila* system is privileged to have had many researchers devoted to technology development over the decades, resulting in a battery of powerful methods for genetic analysis. We do not intend to present an exhaustive compilation of fly techniques; rather, we summarize the genetic toolbox that has been most utilized for miRNA research. We hope that these detailed illustrations of how specific techniques have illuminated *Drosophila* miRNA biology may provide a useful reference and comparison for those studying other animal models (e.g., *C. elegans*, zebrafish, mouse, and cultured mammalian cells). As well, in the course of discussing landmark and recent *Drosophila* miRNA literature, we specifically point out general implications from fly genetics that have particular bearing on the collective endeavor of studying miRNA functions.

2. Generation of miRNA Mutants

The gold standard for studying gene function in model organisms is by mutant/deletion alleles within the intact organism. In *Drosophila* and *C. elegans*, large-scale genetic screens using chemical mutagens or transposons were instrumental in elucidating genes and pathways that control development. The small size of miRNAs makes them poor mutagenesis targets. Only through extremely deep genetic screens, and heroic and stubborn positional cloning efforts, have rare point mutations in miRNA genes been recognized in nematodes (Fig. 8.1A; Chalfie *et al.*, 1981; Johnston and Hobert, 2003; Lee *et al.*, 1993; Reinhart *et al.*, 2000). Similarly, a handful of *Drosophila* miRNA loci were associated with notable loss-of-function phenotypes prior to realization of their encoded small RNAs (Brennecke *et al.*, 2003; Cayirlioglu *et al.*, 2008; Hardiman *et al.*, 2002; Hipfner *et al.*, 2002; Raisin *et al.*, 2003; Xu *et al.*, 2003). However, in flies as in all other species, by far the vast majority of miRNA mutants have been studied by reverse genetics. We begin by reviewing the methods by which *Drosophila* miRNA mutant alleles have been generated.

2.1. Transposon-induced miRNA mutants from forward genetic screening

The "taming" of transposable elements for experimental use has been invaluable for the creation of mutant alleles, provide entry points to clone genes, probe expression patterns, and of course facilitate a majority of *Drosophila* genetic experiments. The general utility of transposon insertion collections for forward phenotypic screening, as well as serving as an allele

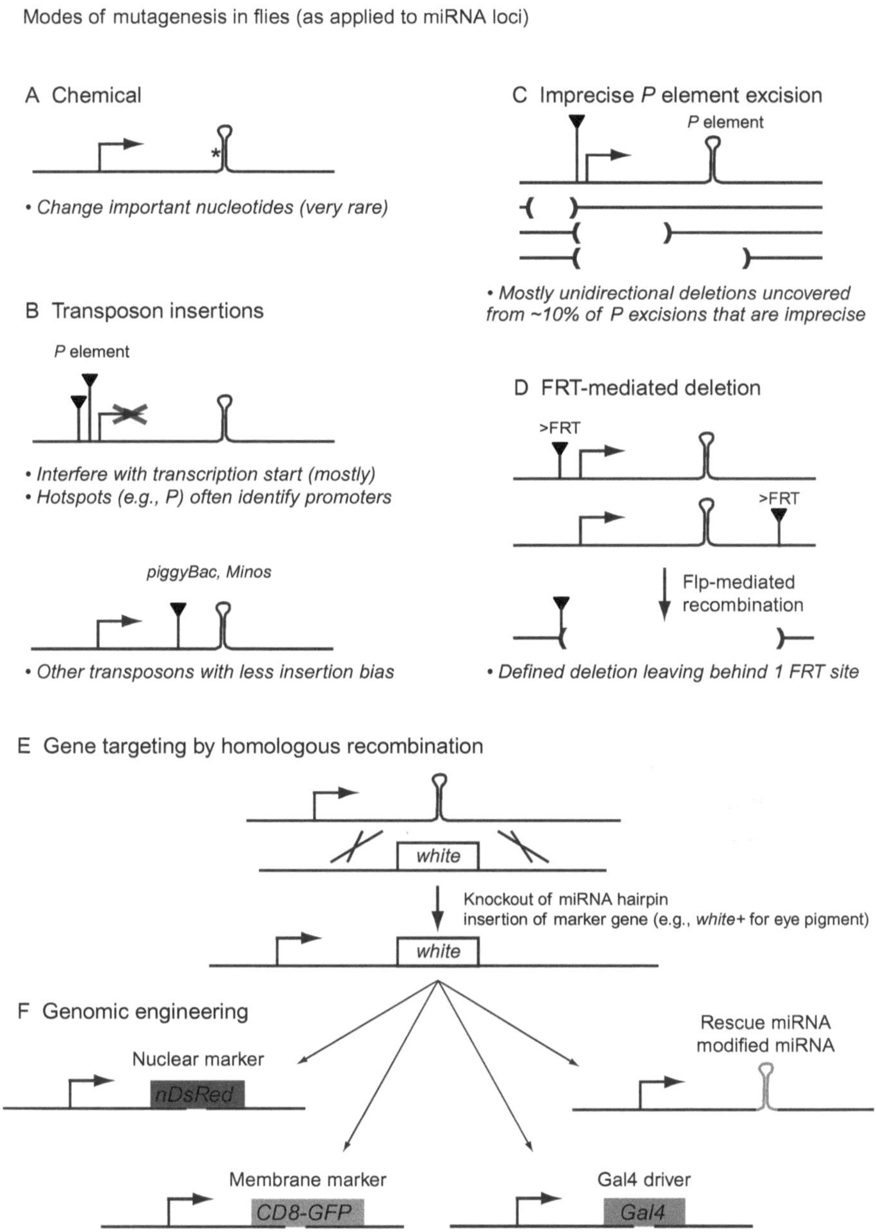

Figure 8.1 Making fly miRNA mutants. The major methods for generating mutations in fly miRNA loci are described. (A) Chemical mutagens, though very useful for generating mutant alleles of protein-coding genes, have been of limited use in investigating *Drosophila* miRNA genes. The only published example is a point mutant in miR-278 that was recovered as a revertant of a gain-of-function phenotype (see Section 3.1). (B) Transposon insertions can interfere with transcription and therefore

repository, has resulted in mapping and curation of several hundred thousand transposon mobilization events over the years. The 20-year-old *Drosophila* Gene Disruption Project (GDP) is a prominent community resource project that seeks to isolate insertions in all *Drosophila* genes. In addition to an immense work from the core project members (Bellen *et al.*, 2011), the GDP has also consolidated transposon collections from many other laboratories and even companies (e.g., the Exelixis and GenExel collections; Bellen *et al.*, 2004; Deak *et al.*, 1997; Lee *et al.*, 2005; Spradling *et al.*, 1999; Thibault *et al.*, 2004).

Some transposons, most notably the *P* element, prefer to insert in active promoters (Fig. 8.1B). When viewed alongside genome annotation tracks (e.g., http://flybase.org/), this preference becomes quite obvious, with many insertions piling up at the 5′ ends of annotated genes. Of course, not every insertion represents a promoter, and many insertions are not associated with annotated genes. However, in the case of hotspots associated with multiple insertions, one can infer that some type of promoter has been tagged. A number of these "orphan" insertion hotspots were later recognized to identify alternative 5′ exons of protein-coding genes (Manak *et al.*, 2006), but a number of them proved to identify miRNA genes (Brennecke *et al.*, 2003; Cayirlioglu *et al.*, 2008; Hipfner *et al.*, 2002; Raisin *et al.*, 2003).

Transposons frequently affect the expression of the inserted loci, thereby acting as mutant alleles (Fig. 8.1B). One of the very first *Drosophila* miRNAs reported was recovered on the basis of a *P* insertion in the *mir-14* locus, which is indeed a *P* hotspot (Xu *et al.*, 2003). In this study, flies were sensitized by expression of the proapoptotic gene reaper in the developing eye (*GMR>reaper*), which induces small, rough eyes. This stock was crossed to a collection of lethal *P* insertions to find ones that could modify the rough eye phenotype, thus representing loci that putatively affect cell death. An insertion in *mir-14* enhanced *GMR>reaper*, yielding smaller eyes and increased pupal lethality. This proved to be due to loss of *mir-14*, since the

have potential to be mutagenic for miRNA loci. The *P* element strongly prefers to insert in promoters and can disrupt transcription. The *Minos* and *piggyBac* elements, whose insertion preferences are less biased, can disrupt miRNA primary transcripts. (C) *P* elements (and to a lesser extent *Minos* elements, but not *piggyBacs*) can be used to generate local deletions by imprecise excision. (D) FRT-mediated deletion. Flp recombinase can induce deletions between different FRT-containing transposons located *in trans*. This can result in the precise deletion of the intervening region. (E) Homologous recombination (HR) allows the precisely engineered generation of a mutant allele. The miRNA hairpin can be replaced with a sequence of choice. (F) Recently, the advent of genomic engineering has made it possible to generate a founder knockout line by HR that contains a phage attachment site (not shown) which enables relatively easily modification of the founder line as desired, for example, adding various markers, Gal4 transgenes or modified hairpins to the endogenous locus. (For color version of this figure, the reader is referred to the Web version of this chapter.)

enhancement of *GMR>reaper* was reverted by precise excision of the *P* insertion, recapitulated in a specific deletion of the *mir-14* locus, and rescued by reintroduction of *mir-14* genomic DNA. In fact, overexpression of miR-14 potently suppressed *GMR>reaper*, rescuing eyes almost back to wild type. This was not specific to reaper, since ectopic miR-14 also strongly rescued the small, rough eyes induced by other proapoptotic transgenes such as *GMR-hid* and *GMR-grim* (Xu *et al.*, 2003). These genetic tests established an intimate connection between miR-14 and suppression of proapoptotic factors.

Another *Drosophila* locus identified on the basis of forward loss-of-function screening of *P* insertions was *mir-279* (Cayirlioglu *et al.*, 2008). This mutant was identified by clonal screening for mutations that affected CO_2-sensing neurons in the olfactory system. These neurons normally arise only in the antenna, but a *P* element insertion near *mir-279* caused ectopic CO_2 sensing to be specified in the maxillary palp. As with *mir-14*, the *mir-279* insertion acted as a loss-of-function mutant, since these phenotypes were recapitulated by independent deletions of the *mir-279* locus. Therefore, miR-279 suppresses target genes in the maxillary palp to prevent the specification of CO_2-sensitive neurons. One of its more critical direct targets is the zinc finger transcription factor encoded by *nerfin-1*, since heterozygosity for this gene substantially decreased the ectopic CO_2-sensing neurons in *mir-279* mutant palps (Cayirlioglu *et al.*, 2008). This implied that derepression of nerfin-1 was critical for generating these neurons; nevertheless, there must be other relevant miR-279 targets since misexpression of nerfin-1 was not sufficient to generate CO_2-sensing neurons in palps.

These examples highlight how the context of forward genetics provides insight regarding the *in vivo* function of miRNAs. Even to this day, despite great improvements in our understanding of parameters underlying miRNA target regulation and comprehensive set of sequenced Drosophilid genomes (Bartel, 2009; Stark *et al.*, 2007), and substantial additional study of miR-14 genetics (Varghese and Cohen, 2007; Varghese *et al.*, 2010), the direct connection of miR-14 to the apoptosis pathway remains unclear. Nevertheless, the dose sensitivity of *mir-14* for reaper-mediated cell death, as well as the potent capacity of miR-14 to suppress the activity of multiple other proapoptotic triggers, supports an intimate connection of miR-14 and cell death. This remains to be deciphered in the future. Likewise, *mir-279* was known for some time to be expressed specifically in the nervous system from embryo studies (Aboobaker *et al.*, 2005), but a specific and essential functional connection to a relatively obscure population of CO_2-sensing neurons in the brain certainly could not have been anticipated from target predictions. Therefore, as was so elegantly illustrated by studies of *C. elegans* miRNAs to emerge from forward genetics—lin-4 and let-7 that control heterochronic timing (Lee *et al.*, 1993; Reinhart *et al.*, 2000) and lsy-6 that controls left–right asymmetry of ASE neurons (Johnston and Hobert,

2003)—phenotype-based genetic studies can lay an invaluable foundation for the biological study of miRNAs.

2.2. Deletion of miRNA loci by imprecise excision of *P* elements

Unfortunately, rather few loss-of-function mutants of miRNA loci have had the privilege of emerging from forward genetic studies. Consequently, studies of miRNA biology have been dominated by reverse genetic approaches. Although transposon movement is generally mediated by specific recombination events, a useful feature of *P* elements is their tendency to excise imprecisely at low frequency, thereby inducing local deletions (Fig. 8.1C). Therefore, even if the starting insertion does not disrupt expression, as with the original insertions in *mir-14* and *mir-279*, one can utilize imprecise *P* excision to delete neighboring DNA yielding null alleles. This is commonly used to delete protein-coding genes and has similarly been applied to miRNA genes where possible. Since many miRNA genes have been tagged by *P* elements, this has permitted them to be deleted by imprecise excisions, including *bantam* (Hipfner *et al.*, 2002), *mir-7* (Li and Carthew, 2005), *mir-8* (Karres *et al.*, 2007), *mir-14* (Xu *et al.*, 2003), *mir-279* (Cayirlioglu *et al.*, 2008), *mir-310/311/312/313* cluster (Tsurudome *et al.*, 2010), and *mir-263a* (Hardiman *et al.*, 2002); a deletion of *mir-iab-4/8* was generated by *P* element-induced gene conversion (Bender, 2008). Moreover, perusal of *P* insertions curated by FlyBase (http://flybase.org/) reveals a number of other miRNA insertion hotspots (e.g., *mir-275/305*, *mir-276a*, *mir-282*) for which deletions could presumably be generated in a straightforward manner.

Studies of *mir-263a* illustrate a particularly interesting case of how reverse genetics uncovered a key developmental role of a miRNA (Hilgers *et al.*, 2010). Over 20 years ago, the *P* element *lacZ* enhancer trap, *E8-2-46*, was isolated on the basis of its sensory organ-specific expression (Bier *et al.*, 1989). Molecular characterization of the region identified several long transcripts lacking substantial open reading frames (Hardiman *et al.*, 2002). Nevertheless, generation of deletion alleles by *P* excision revealed that this locus, *bereft*, was important for at least one class of sensory organs, the interommatidial bristles (IOBs) of the eye. Subsequent computational discovery of miRNAs elucidated the related, but unlinked, loci *mir-263a/mir-263b* (Lai *et al.*, 2003). The former was located just downstream of *bereft*, raising the possibility of a relation between these loci.

This indeed proved to be the case. Although the original *bereft* deletions did not remove the *mir-263a* hairpin itself, Northern analysis showed they lacked mature miR-263a just the same (Hilgers *et al.*, 2010). Moreover, specific mutants of *mir-263a* generated by homologous recombination (HR) (a technique discussed in Section 2.5) failed to complement the original *bereft* alleles and recapitulated *bereft* phenotypes in IOB organs. Further analysis of pupal development revealed that *bereft* mutants specify the sensory organ precursors for IOBs and also execute this cell lineage normally. However, the shaft cells of IOB organs succumb to apoptosis, indicating that miR-263a is antiapoptotic. This interpretation is supported by the fact that loss of *mir-263a* can be compensated by misexpression of the antiapoptotic viral protein p35.

The proapoptotic gene *hid* was found to be a key target gene of miR-263a/b, since heterozygosity for *hid* could substantially rescue *mir-263* mutants, and miR-263 could suppress luciferase-*hid* 3′UTR sensors (Hilgers *et al.*, 2010). Curiously, miR-263a and miR-263b are rather unusual family members in that their seed regions have diverged (at position 2). According to current knowledge, continuous Watson–Crick seed pairing is the major determinant for animal miRNA target recognition so that seed changes are expected to redirect targeting capacity (Bartel, 2009). The relevant miR-236a/b target sites in the *hid* 3′UTR are atypical in that none of the sites are canonical 2–8 matches. This might be compensated by the existence of multiple target sites; for example, miR-263a pairs to 4 sites in *hid* exhibiting 3–9 pairing (and 1G:U), 1–7 pairing, or 2–7 pairing (2 sites).

Two aspects of the miR-263 story are particularly instructive with respect to general miRNA studies. First, it shows that a miRNA locus was actually deleted and shown to have a compelling developmental defect in the "pre-miRNA era" but was not initially deciphered due to lack of knowledge of the associated small RNA. One may wonder whether other miRNA loci were studied genetically over the years, but the associated small RNA was not appreciated. Second, it highlights how genetics led to elucidation of a miRNA:target relationship that is critical for normal animal development, but that could not have been found by standard miRNA target predictions, owing to its noncanonical sites. Thus, one should be circumspect in utilizing genome-wide target predictions, which comprise powerful information but may not include all critical targeting relationships.

2.3. Other transposons

Owing to the insertion site preference of *P* elements, the yield of new genomic insertions has steadily decreased over the years. The use of other transposons with distinct insertion mechanisms can broaden the distribution and coverage of these collections. One widely used element is piggyBac, which exhibits broader insertion range across the genome relative to *P*, including the capacity to insert within genes and introns (Fig. 8.1B; Bellen *et al.*, 2004). Because of this, piggyBac has broader possibilities to disrupt gene expression or function. However, a disadvantage of piggyBac is that it essentially only excises precisely. This disadvantage can be ameliorated by adding other functionalities to the vector backbone. For example, one can place recombinogenic sequences such as FRT onto the transposon. When placed *in trans* with another FRT-bearing insertion, and in the presence of FLP recombinase, one can induce deletions of genomic DNA between the FRT sites (Fig. 8.1D). The possibility to generate molecularly defined deletions has motivated the creation of extensive FRT-piggyBac collections for custom deletions (Parks *et al.*, 2004; Ryder *et al.*, 2007; Thibault *et al.*, 2004). Several *Drosophila* miRNAs, including *mir-1* (Kwon *et al.*, 2005), *mir-184* (Iovino *et al.*, 2009), and *mir-284* (Karr *et al.*, 2009), were deleted in this fashion. In the future, "finishing" of the *Drosophila* GDP to 95% of all genes will be facilitated by yet other transposons. In particular, Minos combines a broad capacity for random insertion throughout the genome with the possibility for imprecise excisions (Fig. 8.1B; Bellen *et al.*, 2011).

2.4. miRNA alleles induced by homologous recombination

Since the vast majority of miRNA loci lack useful transposon insertions, or reside in introns where deletions would affect both miRNA- and protein-coding genes, targeted methods must be used to generate alleles (Fig. 8.1E). Of course, HR has been widely used in unicellular organisms and mice for quite some time, but a strategy to perform this in *Drosophila* was only developed a decade ago (Rong and Golic, 2000, 2001; Rong *et al.*, 2002). The main limitation in fly, compared to mouse, is the inability to regenerate an intact organism from an ES-like cell that can be propagated and manipulated in culture. HR requires the introduction of linear molecules, which unlike circular or supercoiled forms are recombinogenic. This was finally solved by a two-step procedure for generating the HR substrate *in vivo*, in the germline, by using an integrated vector with flanking FRT sites and an internal rare restriction endonuclease site (I-SceI). Expression of FLP recombinase in the germline excises the targeting vector, converting it into a circular form that is then linearized by I-SceI cleavage. HR is still a relatively rare event, requiring extensive screening to identify targeted alleles. Nevertheless, it has proven to be a reasonably reliable technique that has been adapted to generate many designer alleles, including a number of miRNA deletions.

A growing number of *Drosophila* miRNA mutants have been made by HR, including *mir-1* (Sokol and Ambros, 2005), *mir-309/3/286/4/5/6-1,2,3* cluster (Bushati *et al.*, 2008), *mir-278* (Teleman *et al.*, 2006), *let-7/mir-125/mir-100* cluster (Caygill and Johnston, 2008; Sokol *et al.*, 2008), *mir-9a* (Li *et al.*, 2006), *mir-12/283/304* cluster (Friggi-Grelin *et al.*, 2008), *mir-31a* (Weng *et al.*, 2009), *mir-263a*, and *mir-263b* (Hilgers *et al.*, 2010). Their study has uncovered diverse aspects of development and physiology regulated by miRNAs, many of which will be discussed in subsequent sections. We highlight here one example from targeted knockout of *mir-9a*.

HR-induced deletions of *mir-9a* exhibit several mutant phenotypes that manifest in adult morphology (Li *et al.*, 2006). First, these mutants develop some ectopic mechanosensory bristle organs on their back (notum). Although this effect is somewhat variable and mild, the specification of extra organs indicates that the full program of peripheral neurogenesis has been activated ectopically. More notably, *mir-9a* mutant animals exhibit fully penetrant loss of wing tissue, preferably along the posterior edge; this is a very easily noticeable phenotype. Two main target genes have been assigned to these phenotypes, the zinc finger transcription factor *senseless* and the LIM domain transcription cofactor *dLMO* (Bejarano *et al.*, 2010; Biryukova *et al.*, 2009; Li *et al.*, 2006).

Interestingly, deregulation of both targets contributes to both phenotypes, although not equally. *senseless* and *dLMO* are both proneural genes, so it makes sense that their derepression can promote neurogenesis; the extent of ectopic bristles is mitigated in *mir-9a* mutant animals that are also heterozygous for either *senseless* or *dLMO* (Bejarano *et al.*, 2010; Li *et al.*, 2006). Both of these targets induce loss of wing tissue when misexpressed, although the wing is especially sensitive to dLMO. High level misexpression of Senseless is required to induce loss of wing margin (Nolo *et al.*, 2001), whereas mere duplication of the *dLMO* locus can induce loss of wing margin (Lifschytz and Green, 1979). Consequently, heterozygosity for *senseless* provides mild rescue to the *mir-9a* wing defect, but heterozygosity for *dLMO* provides a complete rescue (Bejarano *et al.*, 2010; Biryukova *et al.*, 2009; Li *et al.*, 2006).

There are many substantial conclusions relevant to general miRNA biology to be drawn from studies of miR-9a. First, it is important to recognize that bristles and wings are two of the most intensely studied aspects of *Drosophila* development. Nevertheless, reverse genetics of *mir-9a* revealed its requirement in these two well-studied systems, indicating that it was somehow "missed" by 100 years of genetic screening. Second, it has been posited that many miRNA mutants may have subtle defects due to overlapping functions of related family members. This has proven the case with some miRNA families, although not for many others (Abbott *et al.*, 2005; Alvarez-Saavedra and Horvitz, 2010). miR-9a has two nearly identical paralogs in *Drosophila*, miR-9b and miR-9c, yet the single *mir-9a* deletion exhibits compelling phenotypes. We do not know whether multiple *mir-9* mutants may exhibit stronger effects, yet it is clear that mutants in individual family members can have strong effects. In fact, the very first miRNAs recognized, lin-4 and let-7, are both members of miRNA families (Ambros *et al.*, 2003), yet they clearly have potent nonredundant functions. Third, it is worth commenting that deciphering reverse genetics can depend strongly on the phenotypic richness of the system. The wing defect of *mir-9a* mutants is unmistakable; however, the few extra dorsocentral bristle organs are evident only because of the precision with which they are specified. One may have easily missed this phenotype, and one wonders whether an analogous vertebrate phenotype, say, having a few extra whiskers on the face of a mouse, might be noticeable. Finally, it is salient to mention that derepression of different targets of a miRNA may have distinct consequences in different locations of miRNA expression. We will return to this point later in Section 4.3.

2.5. On the importance of rescuing mutant phenotypes

It cannot be overemphasized that the careful practice of genetics involves a certain amount of legwork to prove that a given mutation is causal to an observed phenotype. Beyond the notes written on a stock label, there is always the possibility of additional unanticipated mutations that could underlie a phenotype of interest. First, many *Drosophila* miRNA alleles were generated by transposon-mediated aberrations. *P* elements are known to incur a reasonably high frequency of hit-and-run events, such that the gene tagged by a transposon is not necessarily the same locus responsible for an associated phenotype. There may be other genes that were mutated before the transposon landed in its final spot, or transposon mobilization may have left damaged copies in the genome that are not easily detected by routine PCR checks. Second, while it is more effort to induce alleles by HR, and these aberrations are generated more precisely than by random transposon-induced deletions, it is well documented that the process of HR has a high frequency of inducing second-site mutations (O'Keefe *et al.*, 2007). If one is unlucky, these might occur in some biased fashion leading to their presence on independent HR events. Worse yet, the homology arms themselves might affect neighboring genes in a systematic fashion. Third, the practice of keeping *Drosophila* stocks in a balanced state against chromosomes that suppress recombination permits unlinked mutations to accumulate over time.

Therefore, while one may be fortunate to be able to order miRNA loci tagged with *P* elements from public stock centers, or to have generated one's own miRNA alleles using HR, one should be cautious in relating observed phenotypes to the mutated locus. A few tests can be performed. In the case of transposon alleles, one can ask if a precise excision reverts the phenotype; if the phenotype persists, it may be a sign that it is due to some unlinked mutation. For example, in the original *mir-14 P* lethal stock identified as a suppressor of *GMR-reaper*, lethality was due to a background mutation and not by loss of *mir-14 per se* (Xu *et al.*, 2003); of course, the apoptosis-modifying activity proved *mir-14* dependent.

One may also seek to obtain independent alleles and show that the phenotype is recapitulated in all of them, as was done with *mir-279* (Cayirlioglu *et al.*, 2008). This is not a foolproof test, and in unlucky cases, independent *Drosophila* stocks have been found to contain the same, unlinked mutation that was responsible for the observed phenotype (Roegiers *et al.*, 2009). Generally speaking, though, this test provides a good measure of confidence. A powerful resource for "second alleles" is the *Drosophila* deficiency collection (Lindsley *et al.*, 1972). This comprises a collection of balanced fly stocks that remove defined portions across >95% of the euchromatic genome. Therefore, it is a simple matter to place one's mutation of interest *in trans* to a deficiency of the region, which effectively acts as a null allele.

However, the most convincing test that an identified mutation is causal to phenotype is to rescue the mutant by reintroducing the gene product. One method is to reexpress the miRNA in the mutant background; often this might be done using the Gal4–UAS system (see Section 3). However, as misexpression of miRNAs often induces dominant phenotypes, it can be difficult to achieve the "right" amount of expression *in vivo*. Another option is to reintroduce the genomic locus as a transgene so that miRNA expression is subject to endogenous transcriptional control. Historically, this was challenged by the paucity of miRNA promoters and *cis*-regulatory elements known (see also Section 5) and the size limitation of inserts for efficient *P* element transgenesis (~15 kb). However, these issues have been obviated by the development of phiC31 transgenesis systems capable of integrating large inserts with high efficiency. Recently, collections of BAC (Venken *et al.*, 2009) and fosmid (Ejsmont *et al.*, 2009) inserts that tile across the *Drosophila* genome were made publicly available. This makes it convenient to simply order a genomic rescue for nearly every miRNA (and protein-coding) gene; the limitation being rare loci that cover in excess of >120 kb.

As many miRNA loci have relatively subtle phenotypes, and are implicated in maintaining organismal robustness (see Chapter 9), it becomes increasingly critical to determine that any observed difference of interest between a mutant and a wild-type reference is definitively due to loss of the miRNA, as opposed to strain background mutations. Therefore, one may wish to obtain such a rescuing genomic transgene early in the process of generating or analyzing miRNA mutants.

3. The Gal4–UAS System for Transgene Activation

The binary Gal4–UAS system comprises the backbone of a great deal of *Drosophila* research (Brand and Perrimon, 1993). This flexible, and simple, system permits researchers to conduct sophisticated genetics with little more base knowledge than the ability to sort females from males to set up crosses. The system comprises the transgene encoding the yeast transcription factor Gal4 as a "driver" and a "responder" transgene linked to

Gal4-binding sites (upstream activating sequence, or "UAS" sites). These components are kept in separate stocks, and Gal4 by itself does not usually induce obvious defects; only by crossing to bring the driver and responder transgenes into the same animal is the responder transgene expressed for phenotypic evaluation. There are many hundreds of fly stocks expressing Gal4 in various spatial and temporal patterns, facilitating diverse tissue-specific misexpression experiments. The Gal4 collections are complemented by many thousands of fly stocks containing UAS-responsive genes, either generated by directed cloning of individual genes or from mobilization of UAS-bearing transposons around the genome (often termed "EP" elements). These resources permit a mind-boggling number of *in vivo* misexpression experiments to be conducted with ease (Rørth *et al.*, 1998).

3.1. Misexpression of miRNAs from genomic EP insertions

While the Gal4–UAS system was originally developed for the purpose of conducting gain-of-function analysis of protein-coding genes (Brand and Perrimon, 1993), it has found wide adoption for the misexpression of RNAi transgenes for knockdowns (Dietzl *et al.*, 2007; Kennerdell and Carthew, 2000). Moreover, as mentioned above, a number of miRNA loci serve as *P* hotspots and are thus represented in EP or other UAS-responsive insertion collections (Fig. 8.2A). A number of these miRNA insertions routinely score as hits in diverse gain-of-function screens (including *mir-7*, *mir-8*, *mir-14*, *mir-282*, *mir-278*, *mir-310-313* cluster, and *bantam*), implying that this system is effective for misexpression of miRNAs. This is probably due to the endogenous transcription of most miRNAs by RNA Polymerase II (Lee *et al.*, 2004), as with protein-coding genes.

Indeed, many of these miRNA hotspots emerged from gain-of-function screens conducted prior to the general recognition of miRNAs in *Drosophila* in 2001 (Lagos-Quintana *et al.*, 2001) or prior to the initial large-scale annotations of *Drosophila* miRNAs in 2003 (Aravin *et al.*, 2003; Lai *et al.*, 2003) and thus were not likely recognized as miRNA loci at the time of their genetic isolation. Examples of miRNA hits in pre-miRNA era screens include that a gain-of-function screen for loci that could affect adult bristle sensory organs recovered *mir-7*, *mir-278*, and *bantam* (Abdelilah-Seyfried *et al.*, 2000); that a misexpression screen for genes affecting motor axons and synaptogenesis identified *bantam* (Kraut *et al.*, 2001); and that an ectopic expression study for modifiers of dorsal thorax formation hit *mir-278* (Pena-Rangel *et al.*, 2002). Now that miRNAs are better annotated in the *Drosophila* genome, nearly all genome-wide misexpression studies end up recovering one or more of the set of EP-miRNA hotspots.

Most of these miRNA misexpression hits have yet to be followed up, but they provide possible entry points to understanding miRNA biology. In addition, EP-induced dominant phenotypes provide a genetic entry point

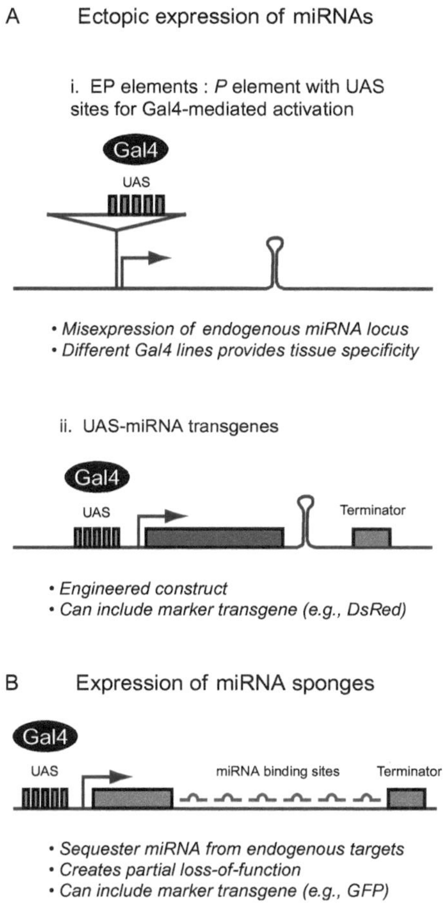

Figure 8.2 Modulating miRNA activity using the Gal4/UAS system. Gal4/UAS is a modular system for overexpressing transcripts under the control of the yeast Gal4 transcription factor, via "upstream activating sequences" (UASs) linked to a target gene. Many spatially/temporally restricted Gal4 transgenes are available. (A) Ectopic expression of miRNAs. As with protein-coding genes, ectopic expression of miRNAs can lead to dominant phenotypes. Many miRNA loci were discovered by genetic screens of collections of EP elements (i) transposons bearing UAS sites, which permit activation of neighboring genes in a Gal4-dependent manner. Engineered UAS-miRNA transgenes (ii) often contain the hairpin flanked by ~200 nt of endogenous sequence, as well as a marker transgene that reports on miRNA-expressing cells. (B) Inducible miRNA sponge transgenes, bearing multiple copies of a given miRNA target site, enable tissue-specific knockdown of miRNA activity.

to generate specific miRNA mutants. Nearly all miRNA alleles are straight deletions, but having a range of point mutations could provide interesting insights, as with protein-coding genes. In the case of *mir-278*, its misexpression generates tissue overgrowths (Nairz *et al.*, 2006), a property that was used in a reversion screen. *EP-mir-278* flies were mutagenized and screened for loss of the ability to induce large eyes. A single-point mutant in the seed region of miR-278 was isolated as a revertant from ~10,000 animals (Nairz *et al.*, 2006), a substantial but by no means daunting number for *Drosophila* genetics (Fig. 8.1A). One could imagine that additional reversion screening of this and other EP-miRNA loci could produce insights regarding the necessity of 3′ regions of miRNAs, for example.

3.2. Directed analysis of UAS-miRNA transgenes

The realization that miRNAs can be ectopically expressed effectively using the Gal4–UAS system has led to the production of many UAS-miRNA transgenes and extensive illustrations of the detrimental consequences of ectopic miRNAs (Fig. 8.2A). One notable set of phenotypes from miRNA gain-of-function emerged from studies of the *Drosophila* Bithorax-Complex (BX-C) locus *mir-iab-4/mir-iab-8*, which generates miRNAs from bidirectional transcription and processing of the same genomic hairpin locus (Bender, 2008; Stark *et al.*, 2008; Tyler *et al.*, 2008). Among the targets of these miRNAs are several homeobox genes in the BX-C, which specify the identities of various abdominal segments (Lewis, 1978). One such gene is *Ultrabithorax* (*Ubx*), which normally represses the wing development program in the segment that generates the haltere; thus *Ubx* mutants bear homeotic transformation of halteres into an extra pair of wings. Misexpression of *mir-iab-4* can directly repress *Ubx* in developing halteres, causing them to transform partially into wings (Ronshaugen *et al.*, 2005); while many miRNAs can target Hox genes, this was the first demonstration that a miRNA could induce a homeotic segment transformation in the animal. More strikingly, *mir-iab-8* has even stronger capacity to repress *Ubx*, and correspondingly, its ectopic expression generates a fuller haltere-to-wing transformation (Stark *et al.*, 2008; Tyler *et al.*, 2008). Notably, ectopic expression of other miRNAs with conserved seed matches in the *Ubx* 3′UTR does not generate extra wings, indicating that there is a practical difference between the existence of conserved miRNA-binding sites and their ability to mediate sufficient downregulation to yield mutant phenotypes.

Another particularly compelling set of miRNA misexpression phenotypes comes from those that inhibit apoptosis or that promote tissue growth. Both of these activities are central to the process of oncogenesis and are mediated by distinct effectors, since the inhibition of apoptosis by itself does not promote tissue overgrowth. Reciprocally, promotion of cell proliferation is usually accompanied by excess cell death.

Misexpression of the bantam miRNA provided early evidence for the notion of a miRNA oncogene, since this miRNA locus could both induce cell proliferation and inhibit apoptosis (Brennecke *et al.*, 2003). miR-278 similarly exhibits joint ability to promote tissue growth and inhibit apoptosis (Nairz *et al.*, 2006; Teleman *et al.*, 2006), and as mentioned, miR-14 (Xu *et al.*, 2003) and miR-263 (Hilgers *et al.*, 2010) are also antiapoptotic. Finally, members of the extensive K box family of miRNAs (Lai, 2002; Lai *et al.*, 1998), the largest in *Drosophila* (*mir-2* family, *mir-6* family, *mir-11*, *mir-13* family, *mir-308*), share capacity to directly repress multiple members of the proapoptotic reaper/grim/hid/sickle family (Brennecke *et al.*, 2005). Therefore, it seems rather common among *Drosophila* miRNAs to be able to inhibit apoptosis and/or to promote tissue overgrowth.

These and other examples from the literature indicate that despite propensity of animal miRNAs to subtly repress large numbers of targets, many *Drosophila* miRNAs elicit specific and interpretable phenotypes when misexpressed *in vivo*. Moreover, many of these effects are not predictable from target predictions. Therefore, directed phenotypic screening for miRNA gain-of-function phenotypes may be a profitable strategy to gain insight into the *in vivo* activities of miRNAs.

3.3. miRNA sponges

The improvements to HR notwithstanding, it is not a trivial effort to generate mutant alleles in *Drosophila*. Thus, an easier route toward preliminary evidence of *in vivo* miRNA function is desirable. In mammalian systems, the vast majority of loss-of-function studies rely upon modified antisense oligonucleotides, termed "antagomirs" (Krutzfeldt *et al.*, 2005). While this method is powerful and in wide use, it is also worth considering that the large set of antagomir-induced miRNA phenotypes in *Drosophila* (Leaman *et al.*, 2005) were not phenocopied by subsequent null alleles. The nature of these discrepancies remains to be understood, but the possibility of off-target effects cannot be discounted. Importantly, one can use neither the lack of phenotypes induced by scrambled antagomirs nor the apparent rescue by sense small RNAs, as compelling evidence for on-target inhibition. Although such criteria are popularly used to control mammalian studies, the former would alleviate specific off-target effects while the latter could represent a titration effect of the antagomir away from off-target substrates. Even if antagomirs are truly specific, they have limited options for tissue-specific delivery. Therefore, additional methods for miRNA sequestration are desirable.

One promising strategy is to use a decoy target transcript bearing multiple imperfect binding sites for a given miRNA, often termed as miRNA sponge (Fig. 8.2B; Ebert *et al.*, 2007). These are proposed to act as competitive inhibitors that distract endogenous miRNAs from regulating bona fide targets. miRNA sponges have shown efficacy in lentiviral infections of mammalian cells (Gentner *et al.*, 2009) and were recently shown to induce phenocopies of certain *Drosophila* miRNA knockouts (Loya *et al.*, 2009). For example, transgenic sponges to miR-9a could induce posterior wing notching, as shown for the *mir-9a* deletion (Bejarano *et al.*, 2010; Biryukova *et al.*, 2009; Li *et al.*, 2006), and the miR-8 sponge induces leg deformities characteristic of *mir-8* deletions (Karres *et al.*, 2007). A caveat of these studies was that it was necessary to reduce the endogenous dosage of the "sponged" miRNA, by making the animal heterozygous for the miRNA locus in order to obtain more consistent phenotypes (Loya *et al.*, 2009). Potential improvement to the system may come from increasing the number of sponge sites or the dosage of sponge transcripts.

A central feature of the Gal4–UAS system is the ability to assess autonomous and nonautonomous effects with ease. This was nicely illustrated using the miRNA sponge technique to study miR-8 function at the neuromuscular junction (NMJ). Analysis of *mir-8* mutants indicated a defect in NMJ morphogenesis (Loya *et al.*, 2009). This might reflect an autonomous requirement of miR-8 in CNS motor neurons, or alternatively, a nonautonomous function of miR-8 in target muscles that are innervated by these neurons. Tissue-specific expression of the miR-8 sponge showed that it was tolerated in neurons, whereas inhibition of miR-8 in muscles induced the NMJ defect, indicating that miR-8 has a nonautonomous function in controlling NMJ morphology (Loya *et al.*, 2009).

4. Genetic Interactions and Epistasis Tests

4.1. Double mutant analysis to uncover redundant functions

A trademark of genetic analysis is to study the consequences of combining mutations in the same organism. A few principles deserve emphasis. First is the notion of genetic interactions, that is, situations in which a given mutation exhibits a phenotype only in the background of another mutation. For example, one can imagine that null conditions for two or more genes whose loss does not result in a phenotype individually but does in concert. Here, *C. elegans* provides the best illustrations of this principle with respect to miRNAs, owing to the large collection of extant miRNA mutants (Miska *et al.*, 2007). For example, analysis of a triple mutant of let-7-related miRNAs (*mir-48*, *mir-84*, and *mir-241*) reveals their redundant control over the transition from the second to third larval stages, via cotargeting of *hbl-1* (Abbott *et al.*, 2005). Similarly, systematic deletion of miRNA families revealed additional examples of synthetic phenotypes (Alvarez-Saavedra and Horvitz, 2010). In *Drosophila*, a deletion of *mir-263b* did not cause obvious phenotypes; nevertheless, it could enhance the loss of eye bristles seen in the *mir-263a* knockout, implying that the function of these two miRNAs is partially redundant during eye development (Hilgers *et al.*, 2010).

4.2. Epistasis analysis

Epistasis means "standing upon" and, in a genetic sense, refers to a situation where the presence of one mutation masks the phenotype usually associated with another mutation. An epistatic relationship associated with two oppositely directed mutants, as opposed to an intermediate phenotype of the double mutant, can provide strong evidence that two genes act in a common pathway. Moreover, the direction of epistasis can inform which gene acts upstream or downstream of the other. If mutant phenotypes are strictly epistatic, then the double mutant will resemble the single mutant of the downstream factor.

One of the most striking examples of miRNA epistasis involves bantam. As mentioned, bantam is overtly essential for proliferation of imaginal disc tissues as well as to suppress apoptosis of certain cell populations, and ectopic bantam reciprocally drives strong disc overgrowths and prevents apoptosis. Both activities are opposite to the function of the Hippo pathway, a highly conserved signaling system that restricts tissue and organ size in flies and vertebrates (Pan, 2010). In brief, a major function of Hippo signaling is to repress the activity of the Yorkie transcription cofactor. The first two transcriptional targets of Yorkie elucidated were *diap1* and *cyclinE*, which makes sense given that the former prevents cell death and the latter promotes cell cycle.

These opposite activities of Hippo signaling and bantam set the stage for a relatively clean epistatic test: what is the phenotype of imaginal disc clones that are doubly mutant for a Hippo pathway member (that gives disc overgrowth) and bantam (which normally fail to grow)? The answer is that bantam is epistatic, indicating that the disc overgrowths in Hippo pathway mutant clones are driven by bantam function (Nolo *et al.*, 2006; Thompson and Cohen, 2006). Reciprocally, constitutive activation of Hippo signaling results in apoptosis and reduced proliferation. Strikingly, this can be substantially rescued by forced activation of *bantam*, but not by *diap1* or *cyclinE* (Nolo *et al.*, 2006; Thompson and Cohen, 2006).

Evidently, the combined activity of bantam as a progrowth, antiapoptotic factor defines it as a key downstream target for repression by hippo signaling. Therefore, even though it is surely the case that Yorkie has many targets genome-wide, the bantam miRNA must be one of its more important effector molecules. A piece of the puzzle remains, however, since we do not yet know of any relevant targets of bantam that can explain its pro-proliferative capacity. Recall that suppression of apoptosis is not by itself sufficient to explain tissue growth (Brennecke *et al.*, 2003). Presumably, epistatic analysis with the appropriate growth-suppressing molecules may shed light on this issue.

4.3. Rescue of miRNA mutants by target heterozygosity

Another general way to test for genetic interactions is to ask whether the mutant phenotype associated with one locus can be suppressed by modulation of another locus. In particular, cases of strong modification of one mutant by heterozygosity of another mutation can serve as compelling evidence to link the respective gene functions. This is especially the case as relatively few genes exhibit obvious phenotypes when only one allele is lost. This test, either by heterozygosity of classical mutant alleles of potential targets or by RNAi-mediated knockdown, has proven to be one of the cornerstones of linking specific target genes to miRNAs in *Drosophila* (Fig. 8.3A). As mentioned, finding "the" target gene of a miRNA is a challenging if not potentially futile task, given that most conserved miRNAs have tens if not hundreds of conserved targets. If the biological role of a miRNA is to slightly repress 100 equivalent targets to tune the transcriptome of a given cell, then one would not expect to observe genetic interactions with any individual target. However, if there are specific targets of particular genetic importance, this situation may be fulfilled. Especially compelling would be the rescue of a miRNA mutant by heterozygosity of a given target, which would suggest that derepression of that target plays an important role in the etiology of the miRNA mutant phenotype.

Although it might be imagined that this is an exceptional genetic situation, in fact it has proven to be rather common in *Drosophila* (Smibert and Lai, 2010). We have already mentioned how rescue of *mir-9a* and *mir-279* mutant phenotypes was achieved by target heterozygosity. Beyond these, metabolic defects in *mir-278* mutants, including insulin resistance concomitant with elevated circulating sugar, could be suppressed by heterozygosity for the FERM domain protein encoded by *expanded* (Teleman *et al.*, 2006). NMJ morphology defects observed in a *let-7*, *mir-125* double mutants could be rescued by heterozygosity for the direct let-7 target *abrupt*, encoding a transcription factor that regulates NMJ development (Caygill and Johnston, 2008). Additionally, the enhanced neurotransmitter release seen in the NMJ of deletion mutants of the *mir-310/311/312/313* cluster, all of which encoded seed-related miRNAs, was rescued by heterozygosity for *khc-73*, a neural-specific kinesin family member (Tsurudome *et al.*, 2010).

These findings suggest that it may be fairly common for the derepression of specific miRNA target genes to be of disproportionate phenotypic impact. We certainly do not wish to imply that these particular targets are the "only" targets of these miRNAs, and phenotypic rescue by target reduction is still compatible with the notion that miRNAs might commonly have hundreds of target genes. It merely highlights that these are not necessarily of equal functional regulatory consequence. This is actually reminiscent of the situation of transcription factor target genes. With modern molecular profiling methods, it is clear that typical transcription

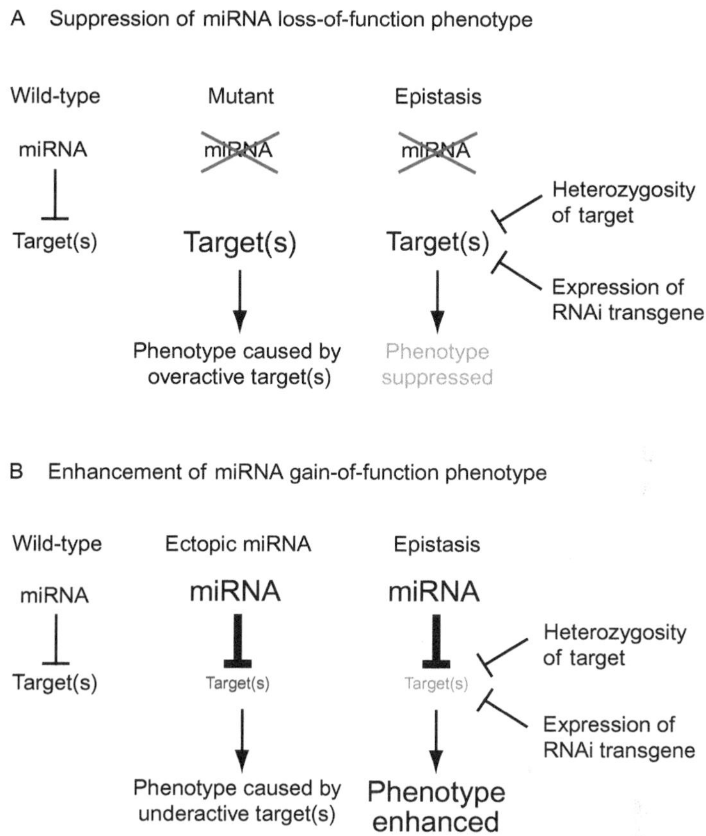

Figure 8.3 Usage of genetic interactions to elucidate miRNA function. Since miRNAs act by repressing target genes, the phenotypic outcome of modulating miRNA activity can be sensitive to the dose of the key miRNA target(s). (A) In a miRNA loss-of-function condition, crucial target(s) are derepressed, leading to a phenotype. Lowering the dose of these target(s) by heterozygosity or RNAi can reduce the target levels to below a phenotypic threshold and suppress the phenotype. This situation can only be fulfilled when there are individual target genes that contribute substantially to organismal phenotype; nevertheless, this seems to be common among *Drosophila* miRNA mutants. (B) In a miRNA gain-of-function condition, a phenotype can arise due to crucial target(s) being lowered to levels below a phenotypic threshold. Heterozygosity or RNAi knockdown of these target (s) can enhance this phenotype. Note that for both loss- and gain-of-function miRNA conditions, the observation of genetic interactions with target heterozygosity is more powerful than the usage of RNAi transgenes. Target knockdown often generates phenotypes on its own, whereas heterozygosity is usually phenotypically benign. (For color version of this figure, the reader is referred to the Web version of this chapter.)

factors typically bind thousands of genomic regions (Gerstein *et al.*, 2010; Macquarrie *et al.*, 2011; Roy *et al.*, 2010), and that these are often conserved (Birney *et al.*, 2007; He *et al.*, 2011; Stark *et al.*, 2007). Nevertheless, it is obvious that some of these regulatory connections play more critical roles in development and physiology than many others do. It seems likely that the same consideration applies to miRNAs.

It is common for regulatory proteins to be recycled during developmental and physiological processes, and it is similarly observed that individual miRNAs function in multiple settings (Smibert and Lai, 2010). A curious emerging theme is that many miRNAs exhibit different key targets in different locations. For example, excess brain apoptosis and associated behavioral defects in *mir-8* mutants are suppressed by heterozygosity for the direct target *atrophin*, encoding a transcriptional coactivator (Karres *et al.*, 2007). However, miR-8 also functions in the fat body to activate insulin signaling by repressing the zinc finger protein encoded by *u-shaped* (Hyun *et al.*, 2009). Knockdown of *u-shaped*, but not *atrophin*, in the fat body rescued *mir-8* mutant phenotypes. In another example, miR-14 directly restricts the activity of the nuclear receptor encoded by the *Ecdysone receptor* during metamorphosis (Varghese and Cohen, 2007), whereas it also functions in insulin neurosecretory cells of the adult brain to directly repress *sugarbabe*, a zinc finger protein that regulates insulin gene expression (Varghese *et al.*, 2010). It is plausible to consider that miRNAs may have so many targets, in part due to their frequent acquisition of new compelling functions in different settings via novel target genes.

4.4. Dominant modifier screens for miRNA interactors

In most cases, directed tests for genetic interactions in a homozygous mutant background require the generation of complicated stocks or recombinant animals, which limits the throughput of analysis. However, one can exploit this test in forward genetic screening by starting with a dominant phenotype. A simple way to do this is by sensitizing the animal by misexpressing a gene to generate a mutant phenotype and then to ask whether the removal of one copy of any other gene can modify this phenotype (Fig. 8.3B). This approach, termed dominant enhancer/suppressor screening, was codified in studies of the Ras signaling pathway 20 years ago (Simon *et al.*, 1991). Here, misexpression of oncogenic Ras in the developing *Drosophila* eye yielded a rough eye, presumably due to hyperactivation of its downstream pathway. Loci for which removal of one gene copy reduced the activity of oncogenic Ras identified several downstream members of the pathway (Simon *et al.*, 1991) and helped to elucidate the signaling mechanism of this important cancer pathway. Since effects of removing only one gene copy are assessed, dominant enhancer/suppressor screening can be executed on a massive scale, as illustrated by the subsequent examination of 850,000 mutagenized animals in the *Ras1* eye screen (Karim *et al.*, 1996), which collectively hit most of the key components of Ras signal transduction. Thus, dominant modifier screening is genuinely a "genome-wide" technique. Such an approach using chemical mutagenesis or transposon screening has yet to be applied to *Drosophila* miRNAs, but there is no particular reason to believe that it would not be profitable.

A quick-and-dirty version of the dominant screen utilizes the *Drosophila* deficiency collection, which allows one to assess genome-wide, and within only a few hundred crosses, the effects of removing one gene copy on a genetically sensitized phenotype. This strategy was recently used to great effect with the *Drosophila* ortholog of *mir-1* (King *et al.*, 2011), a miRNA that is deeply conserved across the animals both in sequence and in specific expression in heart and muscle. Previously, misexpression of *mir-1* in the center of the developing wing, using *dpp-Gal4*, was found to generate patterning and morphogenesis defects including cell-autonomous loss of wing tissue (Kwon *et al.*, 2005). As this presumably resulted from the ectopic repression of miR-1 target genes, such as the Notch ligand *Delta*, it was hypothesized that further reducing the dosage of relevant target genes should enhance the severity of wing phenotypes. Therefore, a *dpp-Gal4>UAS-mir-1* stock was crossed to the deficiency kit and scored for dominant enhancers that exhibited greater loss of wing tissue.

Out of 284 deficiency lines covering about 70% of the euchromatic genome, 32 enhancers were recovered (King *et al.*, 2011). These included two conserved targets of miR-1, *Delta* and *mirror*, and novel interacting partners such as *kayak*, encoding a leucine-zipper transcription factor and a component of the planar cell polarity (PCP) pathway. Although the wing is clearly not a physiological location of miR-1 function, these interactors could be examined in the relevant tissues that endogenously express miR-1. Indeed, alteration of miR-1 dosage by null mutants or by overexpression disrupted cardioblast cell polarity in the fly heart, similar to *kayak* mutants. Importantly, heterozygosity of *kayak* could mitigate the loss of miR-1, suggesting that dysregulation of Kayak partly accounts for these defects in *mir-1* null mutants. These findings illustrate the power of *Drosophila* genetics to uncover unexpected functional connections, in an unbiased and systematic manner.

5. Detecting miRNA Expression and Activity *In Vivo*

Temporal-, tissue-, or cell-specificity of miRNA expression can be assessed by several strategies. These include Northern analysis or small RNA library cloning from different sources (Berezikov *et al.*, 2011; Leaman *et al.*, 2005; Ruby *et al.*, 2007b), *in situ* hybridization to primary miRNA transcripts (Aboobaker *et al.*, 2005; Kosman *et al.*, 2004) or mature miRNAs (Li and Carthew, 2005; Sokol and Ambros, 2005). We focus this section on a few transgenic strategies that have been used to great effect in *Drosophila*, including enhancer–reporter constructs and genetically encoded miRNA sensors.

5.1. miRNA promoters "identified" by *P* elements

It is first worth commenting on the fact that *P* elements tend to insert in active promoters. Although information on miRNA promoters and transcript models from microarray analysis following knockdown of nuclear miRNA biogenesis factors (Kadener *et al.*, 2009b), or from RNA-seq and CAGE analyses (Enderle *et al.*, 2011; Graveley *et al.*, 2011), is growing, our knowledge is far from complete. It is therefore humbling to realize that *P* elements were far ahead of humans in identifying not only miRNA genes but also their promoters. In some cases, the miRNA promoter is located quite distantly from the miRNA gene itself. For example, the *mir-278* hairpin is associated with a cluster of *P* elements located 45kb upstream, which identifies its promoter (Nairz *et al.*, 2006; Teleman *et al.*, 2006). Similarly, the *bantam* miRNA hairpin is associated with multiple *P* element insertion clusters, which likely signify complexity of its transcriptional control via multiple promoters; the most distal promoter is some 40kb away from the *bantam* hairpin (Graveley *et al.*, 2011; Peng *et al.*, 2009). Importantly, *P* elements that contain marker genes can report on the transcriptional output of the promoter within which they are inserted (Fig. 8.4A) through expression of the marker gene (e.g., *bereft/mir-263a:lacZ*; Section 2.2). This is particularly useful for miRNA genes, as it provides a relatively stable readout of primary transcript expression of the tagged miRNA locus.

5.2. miRNA enhancer:reporter transgenes

"Enhancer bashing" is a time-honored tradition in *Drosophila*, an exercise in which one attempts to identify fragments of genomic DNA that can drive a reporter gene (e.g., *lacZ, GFP, DsRed*, etc.) in a pattern that recapitulates known aspects of the endogenous transcript (Fig. 8.4A). An *in situ* hybridization survey of primary miRNA transcripts revealed a diversity of spatially restricted patterns in *Drosophila* embryos (Aboobaker *et al.*, 2005), and relevant *cis*-regulatory modules (CRMs) for several embryonically expressed miRNAs have been isolated, including for the *mir-309/3/286/3/4/5/6* cluster in the early blastoderm embryo (Biemar *et al.*, 2005), for *mir-1* in the mesoderm and muscles (Biemar *et al.*, 2005; Kwon *et al.*, 2005; Sokol and Ambros, 2005), and for *mir-124* in the central nervous system (Xu *et al.*, 2008). Similarly, a *mir-7* CRM located within the intron of its host transcript *bancal* (Li and Carthew, 2005; Li *et al.*, 2009) and a CRM upstream of *mir-279* (Cayirlioglu *et al.*, 2008) have proven useful to probe

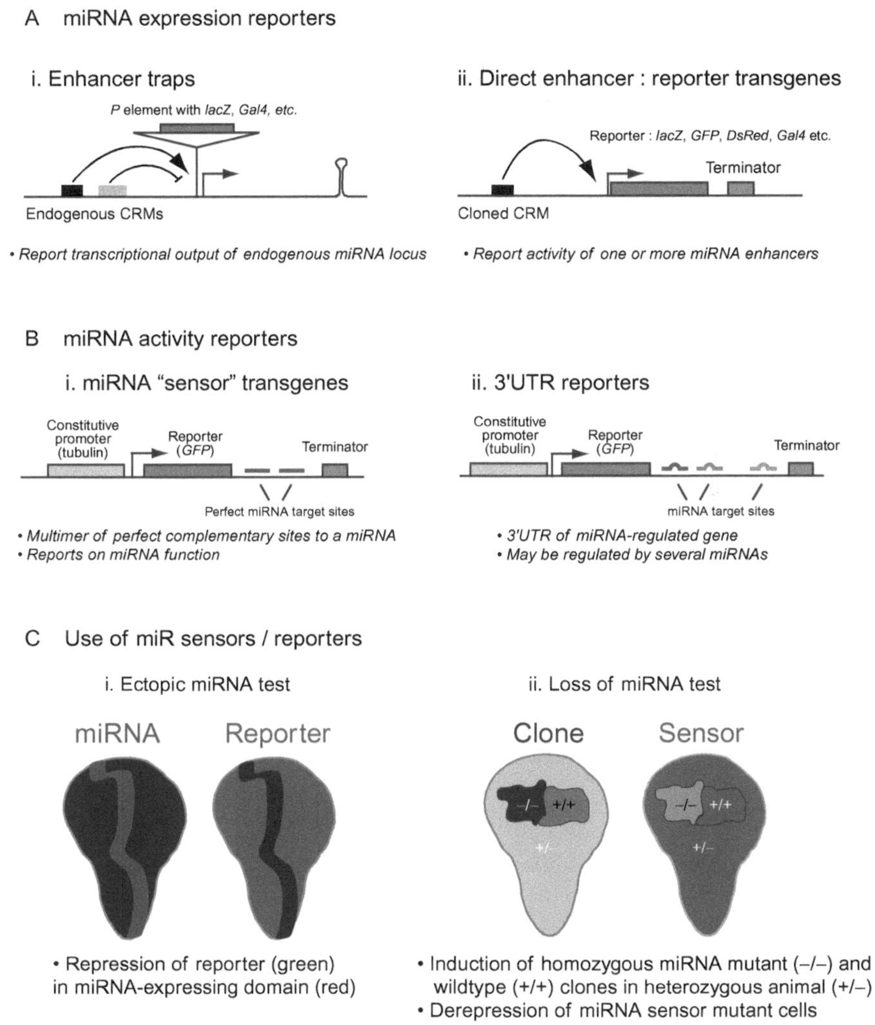

Figure 8.4 Reporters of miRNA expression and activity. miRNAs are transcribed as long primary transcripts, whose mature ~22 nt products guide Argonaute proteins complexes to target transcripts. Note that the transcription of a miRNA locus does not necessarily report directly on cells in which the miRNA is active. (A) miRNA expression reporters are the same as those used to report expression of protein-coding genes. (i) Putative regulatory sequences are identified by enhancer traps (insertion of transposon with a marker gene, often *lacZ* or *Gal4*) where the endogenous *cis*-regulatory modules (CRMs) drive expression of the marker. (ii) Cloned CRMs can also be used to analyze miRNA gene expression; however, these may only report on a subset of the full transcriptional control of a miRNA. (B) Sensors of miRNA activity typically use a marker gene (often *GFP*) bearing sequences complimentary to the miRNA of interest (i) or with the 3′UTR of a gene of interest (ii) cloned downstream. Sensor activity is inverse to that of the miRNA: low sensor expression equates to high

postembryonic expression of these miRNAs in the nervous system. In many of these examples, relevant binding sites for transcription factors have been elucidated. For example, analysis of *mir-1* CRMs indicated direct regulation by the transcription factors Dorsal, Twist, d-Mef2, and potentially SRF (Biemar *et al.*, 2005; Kwon *et al.*, 2005; Sokol and Ambros, 2005).

Even without prior knowledge of an endogenous transcript pattern, reporters can provide an entry point to elucidate spatial patterns. In this case, a critical concern is to include sufficient regulatory sequence to encompass relevant inputs. This might be addressed by inserting reporter sequences into large constructs (such as fosmids or BACs), or better yet, by inserting within the endogenous locus. One way this can be accomplished is to include a reporter, such as Gal4, within an HR targeting cassette. Combining a UAS reporter with a Gal4 knockin allele was first used to demonstrate that *mir-278* was highly active in the fat body, where it regulates energy homeostasis by limiting insulin pathway activity and levels of circulating sugar (Teleman *et al.*, 2006). Subsequent integrations of Gal4 into the *let-7/mir-100/mir-125* locus (Sokol *et al.*, 2008) and the *mir-263a* and *mir-263b* loci (Hilgers *et al.*, 2010) have been useful to report on their expression, as well as to manipulate gene expression within miRNA-expressing cells. Finally, it is worth recalling that as many miRNA loci are *P* element hotspots, many of these were hit as enhancer traps in the pre-miRNA era. As mentioned in Section 2.2, a *lacZ* insertion in *mir-263a/bereft* was critical to reveal its sensory organ-specific expression (Hardiman *et al.*, 2002).

5.3. Genetically encoded sensors of miRNA activity

The above methods have collectively been powerful in analyzing miRNA expression in the *Drosophila* system. However, one potential limitation is that these strategies do not directly report on miRNA activity. Given the growing appreciation of posttranscriptional regulation of miRNA processing and/or function (Kim *et al.*, 2010; Siomi and Siomi, 2010), it is useful to be able to monitor miRNA activity *in vivo*. Such an approach was

miRNA activity and vice versa. (C) Schematic representation of *Drosophila* wing imaginal discs, illustrating how miRNA reporters and sensors reflect miRNA activity in overexpression and loss-of-function scenarios. (i) Overexpression of the miRNA in a defined stripe of cells (red) leads to cell-autonomous repression of the reporter (green); neighboring non-miRNA-expressing cells serve as an internal control for the experiment. (ii) Mutant clones of a miRNA (negatively marked by absence of red staining) derepress the miRNA sensor (green), indicating loss of an endogenous miRNA:target repression event. Homozygous mutant (−/−) and wild-type cells (+/+) are generated within an otherwise heterozygous animal. Again, cell-autonomy of the sensor derepression provides stringent controls for the experiment. (See Color Insert.)

pioneered 9 years ago with the bantam miRNA (Brennecke *et al.*, 2003). The bantam "sensor" transgene consisted of a ubiquitously expressed GFP transcript with several perfectly complementary bantam sites in its 3′UTR (Fig. 8.4B); thus, the expression of GFP is lowest where the activity of bantam is highest (Fig. 8.4C). The bantam sensor exhibits spatially modulated activity in imaginal discs, reflecting its function in growth regulation.

The bantam sensor has been widely exploited as a proxy readout of its response to various signaling pathways, presumably at the transcriptional level. For example, a complement to the aforementioned epistastic tests between upstream Hippo pathway members and bantam was to examine the bantam sensor in Hippo pathway mutant clones. These tests revealed lower bantam sensor levels in these clones in wing imaginal discs, indicating increased bantam activity (Nolo *et al.*, 2006; Thompson and Cohen, 2006). The bantam sensor was later shown to be upregulated by activation of Notch signaling in a particular region of the wing disc (Herranz *et al.*, 2008), suggesting a repressive input of Notch pathway activity onto bantam. Finally, the bantam sensor helped reveal a direct input of the Homothorax transcription factor, in complex with Yorkie, that drives proliferation in the eye imaginal disc (Peng *et al.*, 2009).

Most recently, exploitation of the bantam sensor uncovered a novel and direct intersection of the Hippo and TGF-β signaling pathways that drives bantam expression in both wing and eye discs (Oh and Irvine, 2011). Indeed, protein–protein interactions between the mammalian Yki ortholog Yap and TGF-β pathway Smad transcription factors have been documented (Alarcon *et al.*, 2009; Ferrigno *et al.*, 2002), although the *in vivo* consequences of this for Hippo signaling or TGF-β signaling were not known. Taking a cue from the essential role for cell signaling via the *Drosophila* TGF-β ligand Dpp for tissue growth and patterning (Affolter and Basler, 2007), it was found that coexpression of Yorkie and with an activated Dpp receptor (Tkv^{QD}) synergistically promoted tissue growth and repression of the bantam sensor, indicating increased bantam activity. Reciprocally, blocking Dpp pathway activity increased bantam sensor expression, indicating repression of bantam function (Martin *et al.*, 2004; Oh and Irvine, 2011). (Note that these experiments involved keeping mutant cells alive using the antiapoptotic factor p35, since cells are otherwise succumb without Dpp signaling.)

These observations laid a foundation for multifaceted studies involving many other principles we have discussed in this review. First, epistatic tests (Section 4.2) showed that forced expression of bantam could partially rescue the failure of cells lacking Dpp signaling to survive and proliferate. Second, enhancer bashing (Section 5.2) identified a bantam CRM that was synergistically responsive to Yorkie and the Dpp pathway function in imaginal discs. Finally, it was shown that a novel transcriptional complex containing Yorkie and the Dpp pathway transcription factor Mad directly binds and activates this bantam CRM.

Altogether, this collection of studies of the bantam sensor helped reveal an extraordinary diversity of transcriptional inputs into this miRNA. Such complexity befits an essential growth locus such as bantam, presumably to coordinate precise tissue growth underlying appropriate development. Interestingly, bantam has additional roles independent of growth, such as in scaling growth of neural dendrites (Parrish *et al.*, 2009), maintenance of germline stem cells (Yang *et al.*, 2009), and control of circadian rhythm (Kadener *et al.*, 2009a). Presumably each of these other settings is associated with other transcriptional inputs to bantam, which remain to be elucidated. Finally, while the bantam sensor is the most well studied in *Drosophila*, other sensors for individual miRNAs [e.g., miR-9a (Bejarano *et al.*, 2010) or miR-14 (Varghese *et al.*, 2010)] or for miRNA-regulated 3′UTRs (Fig. 8.4B; Bejarano *et al.*, 2010; Brennecke *et al.*, 2003; Friggi-Grelin *et al.*, 2008) have been informative probes of miRNA activity *in vivo*. In many postembryonic tissues, one can induce clones that are homozygous for a mutant allele of interest. In combination with a miRNA sensor, induction of clones of a miRNA mutant can report on the activity of that miRNA in a particular tissue; derepression of a miRNA or 3′UTR sensor reports directly on miRNA activity (Fig. 8.4C).

6. Conclusions and Future Prospects

We have emphasized a broad selection of *Drosophila* genetic techniques used to manipulate miRNA activity, elucidate associated pathways and key target genes, and interrogate miRNA expression, all in the context of the intact animal. Overall, we wish to highlight that miRNAs are just like any other genes, in that they can be studied using most of the same techniques used to study "interesting" protein-coding genes. Although the field of miRNA biology is dominated by reverse genetics, many miRNAs are associated with palpable phenotypes, both loss- and gain-of-function. However, it can take some searching to find the right place, the right time, and the right markers with which to characterize miRNA activities. However, we also highlight that manifestations of miRNA dysfunction have existed in *Drosophila*, as with *C. elegans*, long before the formal recognition of miRNAs as a unified class of regulatory molecule. We conclude with a few final points with a view to the future of miRNA genetics in *Drosophila*.

First, we are struck by the fact that overexpression of so many miRNAs induces compelling and interpretable phenotypes. This is not necessarily expected from the viewpoint that animal miRNAs predominantly mediate subtle repression of large groups of transcripts, which might imply that manipulation of miRNAs might either yield few phenotypes or alternatively elicit nonspecific toxic effects caused by simultaneous misregulation of hundreds of transcripts. As further examples, misexpression of many miRNAs elicits specific developmental phenotypes reflecting gain- or loss-of-function of the core cell signaling pathways (Hagen and Lai, 2008), such as Notch (Lai and Posakony, 1997; Lai *et al.*, 1998, 2005; Stark *et al.*, 2003), Hippo (Brennecke *et al.*, 2003), Hedgehog (Friggi-Grelin *et al.*, 2008), and Wnt (Silver *et al.*, 2007). Efforts are underway to extend plasmid-based UAS-miRNA libraries used to screen tissue culture cells (Silver *et al.*, 2007) into transgenes, for systematic *in vivo* phenotypic screening in the animal (Y. Chou, F. Bejarano, D. Bortolamiol-Becet, and E. C. Lai, Unpublished). We anticipate that a similar set of resources in the mouse would have a high probability of revealing a great diversity of disease-relevant miRNA activities, most of which would probably not be easily anticipated from extensive lists of miRNA target predictions.

Conversely, the generation and analysis of more miRNA deletion alleles are heavily anticipated. In contrast to the *C. elegans* system, where very few miRNA mutants have overt phenotypes (Miska *et al.*, 2007), even when examining multiple mutants of entire families (Alvarez-Saavedra and Horvitz, 2010), the extant literature indicates an impressive set of developmental and physiological defects associated with many different *Drosophila* miRNA mutants, under nonperturbed conditions. Perhaps this reflects the rich phenotypic assays available in *Drosophila*, and we suspect that mammalian systems (e.g., mice and humans) are similarly phenotypically rich. It is worth considering that many seemingly inert *C. elegans* miRNA deletions generate phenotypes in genetically sensitized backgrounds (Brenner *et al.*, 2010). Similarly, environmental stress can strongly enhance the phenotypes of the *Drosophila mir-7* mutant (Li *et al.*, 2009). Therefore, future miRNA deletion studies in flies should take advantage of both normal and sensitized backgrounds.

The flexibility of manipulating endogenous miRNA loci will be greatly increased by recent upgrades to HR technologies. These include the development of "ends-out" targeting for direct allele replacement (Gong and Golic, 2003) and various modifications that improve the ease and efficiency of detecting candidate targeting events (Huang *et al.*, 2009). Even more valuable are the development of "genomic engineering" strategies to permit rapid construction of allelic variants using a founder line modified to contain phage phiC31 integrase recognition sites (Fig. 8.1F; Choi *et al.*, 2009; Gao *et al.*, 2008; Huang *et al.*, 2009). This is probably the way forward for generation of reverse-engineered alleles in *Drosophila*, including miRNAs (Weng *et al.*, 2009). For example, one can imagine generating one founder mutant allele that deletes a miRNA locus, then quickly using that as a platform to reintroduce the miRNA to demonstrate rescue of the targeted allele, to insert a nuclear enhancer trap marker, a membrane-localized marker (e.g., useful with neural genes to delineate their projections), a

Gal4 driver line to express other genes in the miRNA pattern, to insert variant versions of the miRNA sequence for structure–function tests, and so forth. The near future should see broad utilization of these powerful technologies.

Finally, we envision that the ultimate proof of critical miRNA:target interactions will involve assessment of target transgenes specifically mutated for miRNA-binding sites. In fact, such evidence was generated in *Drosophila* in the "pre-miRNA era," with the observation that genomic transgenes for Notch pathway members *Bearded* and *E(spl)m8* specifically mutated for Brd box and K box miRNA-binding sites generated dominant gain-of-function phenotypes in the nervous system (Lai and Posakony, 1997; Lai *et al.*, 1998), echoing the dominant alleles involving 3′UTR aberrations that permitted the cloning of these genes (Klämbt *et al.*, 1989; Leviten *et al.*, 1997). Similarly, the very recognition of the miRNA pathway depended in part on the existence of *lin-14* gain-of-function mutants caused by 3′UTR deletions that removed lin-4 binding sites (Ruvkun and Giusto, 1989; Wightman *et al.*, 1991). Although there is good evidence that a large fraction of the *Drosophila* transcriptome is regulated by highly conserved miRNA-binding sites (Ruby *et al.*, 2007b), it is difficult to imagine that the specific loss of these regulatory interactions will result in noticeable phenotypes. Greater knowledge of some of these "critical" targets, for which loss of miRNA-mediated regulation is associated with developmental or physiological abnormalities, will not only be interesting to insect fanciers, but also should be valuable for understanding how miRNA dysfunction contributes to human disease.

ACKNOWLEDGMENTS

Q. D. was supported by a fellowship from the Swedish Research Council. Work in E. C. L.'s group was supported by the Burroughs Wellcome Fund, the Alfred Bressler Scholars Fund, the Starr Cancer Consortium (I3-A139), and the NIH (R01-GM083300).

REFERENCES

Abbott, A. L., Alvarez-Saavedra, E., Miska, E. A., Lau, N. C., Bartel, D. P., Horvitz, H. R., and Ambros, V. (2005). *Dev. Cell* **9,** 403–414.

Abdelilah-Seyfried, S., Chan, Y. M., Zeng, C., Justice, N. J., Younger-Shepherd, S., Sharp, L. E., *et al.* (2000). *Genetics* **155,** 733–752.

Aboobaker, A. A., Tomancak, P., Patel, N., Rubin, G. M., and Lai, E. C. (2005). *Proc. Natl. Acad. Sci. USA* **102,** 18017–18022.

Affolter, M., and Basler, K. (2007). *Nat. Rev. Genet.* **8,** 663–674.

Alarcon, C., Zaromytidou, A. I., Xi, Q., Gao, S., Yu, J., Fujisawa, S., *et al.* (2009). *Cell* **139,** 757–769.

Alvarez-Saavedra, E., and Horvitz, H. R. (2010). *Curr. Biol.* **20,** 367–373.

Ambros, V., Lee, R. C., Lavanway, A., Williams, P. T., and Jewell, D. (2003). *Curr. Biol.* **13,** 807–818.

Aravin, A., Lagos-Quintana, M., Yalcin, A., Zavolan, M., Marks, D., Snyder, B., *et al.* (2003). *Dev. Cell* **5,** 337–350.

Baek, D., Villen, J., Shin, C., Camargo, F. D., Gygi, S. P., and Bartel, D. P. (2008). *Nature* **455,** 64–71.

Bartel, D. P. (2009). *Cell* **136,** 215–233.

Bejarano, F., Smibert, P., and Lai, E. C. (2010). *Dev. Biol.* **338,** 63–73.

Bellen, H. J., Levis, R. W., Liao, G., He, Y., Carlson, J. W., Tsang, G., *et al.* (2004). *Genetics* **167,** 761–781.

Bellen, H. J., Levis, R. W., He, Y., Carlson, J. W., Evans-Holm, M., Bae, E., *et al.* (2011). *Genetics* **188,** 731–743.

Bender, W. (2008). *Genes Dev.* **22,** 14–19.

Berezikov, E., Robine, N., Samsonova, A., Westholm, J. O., Naqvi, A., Hung, J. H., *et al.* (2011). *Genome Res.* **21,** 203–215.

Biemar, F., Zinzen, R., Ronshaugen, M., Sementchenko, V., Manak, J. R., and Levine, M. S. (2005). *Proc. Natl. Acad. Sci. USA* **102,** 15907–15911.

Bier, E., Väessin, H., Shepherd, S., Lee, K., McCall, K., Barbel, S., *et al.* (1989). *Genes Dev.* **3,** 1273–1287.

Birney, E., Stamatoyannopoulos, J. A., Dutta, A., Guigo, R., Gingeras, T. R., Margulies, E. H., *et al.* (2007). *Nature* **447,** 799–816.

Biryukova, I., Asmar, J., Abdesselem, H., and Heitzler, P. (2009). *Dev. Biol.* **327,** 487–496.

Brand, A. H., and Perrimon, N. (1993). *Development* **118,** 401–415.

Brennecke, J., Hipfner, D. R., Stark, A., Russell, R. B., and Cohen, S. M. (2003). *Cell* **113,** 25–36.

Brennecke, J., Stark, A., Russell, R. B., and Cohen, S. M. (2005). *PLoS Biol.* **3,** e85.

Brenner, J. L., Jasiewicz, K. L., Fahley, A. F., Kemp, B. J., and Abbott, A. L. (2010). *Curr. Biol.* **20,** 1321–1325.

Brodersen, P., and Voinnet, O. (2009). *Nat. Rev. Mol. Cell Biol.* **10,** 141–148.

Bushati, N., Stark, A., Brennecke, J., and Cohen, S. M. (2008). *Curr. Biol.* **18,** 501–506.

Caygill, E. E., and Johnston, L. A. (2008). *Curr. Biol.* **18,** 943–950.

Cayirlioglu, P., Kadow, I. G., Zhan, X., Okamura, K., Suh, G. S., Gunning, D., *et al.* (2008). *Science* **319,** 1256–1260.

Chalfie, M., Horvitz, H. R., and Sulston, J. E. (1981). *Cell* **24,** 59–69.

Choi, C. M., Vilain, S., Langen, M., Van Kelst, S., De Geest, N., Yan, J., *et al.* (2009). *Science* **324,** 54.

Chung, W. J., Agius, P., Westholm, J. O., Chen, M., Okamura, K., Robine, N., *et al.* (2011). *Genome Res.* **21,** 286–300.

Czech, B., and Hannon, G. J. (2010). *Nat. Rev. Genet.* **12,** 19–31.

Deak, P., Omar, M. M., Saunders, R. D., Pal, M., Komonyi, O., Szidonya, J., *et al.* (1997). *Genetics* **147,** 1697–1722.

Dietzl, G., Chen, D., Schnorrer, F., Su, K. C., Barinova, Y., Fellner, M., *et al.* (2007). *Nature* **448,** 151–156.

Doench, J. G., and Sharp, P. A. (2004). *Genes Dev.* **18,** 504–511.

Ebert, M. S., Neilson, J. R., and Sharp, P. A. (2007). *Nat. Methods* **4,** 721–726.

Ejsmont, R. K., Sarov, M., Winkler, S., Lipinski, K. A., and Tomancak, P. (2009). *Nat. Methods* **6,** 435–437.

Enderle, D., Beisel, C., Stadler, M. B., Gerstung, M., Athri, P., and Paro, R. (2011). *Genome Res.* **21,** 216–226.

Ferrigno, O., Lallemand, F., Verrecchia, F., L'Hoste, S., Camonis, J., Atfi, A., and Mauviel, A. (2002). *Oncogene* **21,** 4879–4884.

Flynt, A. S., and Lai, E. C. (2008). *Nat. Rev. Genet.* **9,** 831–842.

Flynt, A. S., Chung, W. J., Greimann, J. C., Lima, C. D., and Lai, E. C. (2010). *Mol. Cell* **38,** 900–907.

Friedman, R. C., Farh, K. K., Burge, C. B., and Bartel, D. P. (2009). *Genome Res.* **19,** 92–105.

Friggi-Grelin, F., Lavenant-Staccini, L., and Therond, P. (2008). *Genetics* **179,** 429–439.

Gao, G., McMahon, C., Chen, J., and Rong, Y. S. (2008). *Proc. Natl. Acad. Sci. USA* **105,** 13999–14004.

Gentner, B., Schira, G., Giustacchini, A., Amendola, M., Brown, B. D., Ponzoni, M., and Naldini, L. (2009). *Nat. Methods* **6,** 63–66.

Gerstein, M. B., Lu, Z. J., Van Nostrand, E. L., Cheng, C., Arshinoff, B. I., Liu, T., *et al.* (2010). *Science* **330,** 1775–1787.

Gong, W. J., and Golic, K. G. (2003). *Proc. Natl. Acad. Sci. USA* **100,** 2556–2561.

Graveley, B. R., Brooks, A. N., Carlson, J. W., Duff, M. O., Landolin, J. M., Yang, L., *et al.* (2011). *Nature* **471,** 473–479.

Guo, H., Ingolia, N. T., Weissman, J. S., and Bartel, D. P. (2010). *Nature* **466,** 835–840.

Hagen, J. W., and Lai, E. C. (2008). *Cell Cycle* **7,** 2327–2332.

Hardiman, K. E., Brewster, R., Khan, S. M., Deo, M., and Bodmer, R. (2002). *Genetics* **161,** 231–247.

He, Q., Bardet, A. F., Patton, B., Purvis, J., Johnston, J., Paulson, A., *et al.* (2011). *Nat. Genet.* **43,** 414–420.

Herranz, H., Perez, L., Martin, F. A., and Milan, M. (2008). *EMBO J.* **27,** 1633–1645.

Hilgers, V., Bushati, N., and Cohen, S. M. (2010). *PLoS Biol.* **8,** e1000396.

Hipfner, D. R., Weigmann, K., and Cohen, S. M. (2002). *Genetics* **161,** 1527–1537.

Huang, J., Zhou, W., Dong, W., Watson, A. M., and Hong, Y. (2009). *Proc. Natl. Acad. Sci. USA* **106,** 8284–8289.

Hyun, S., Lee, J. H., Jin, H., Nam, J., Namkoong, B., Lee, G., *et al.* (2009). *Cell* **139,** 1096–1108.

Iovino, N., Pane, A., and Gaul, U. (2009). *Dev. Cell* **17,** 123–133.

Jan, C. H., Friedman, R. C., Ruby, J. G., and Bartel, D. P. (2011). *Nature* **469,** 97–101.

Johnston, R. J., and Hobert, O. (2003). *Nature* **426,** 845–849.

Kadener, S., Menet, J. S., Sugino, K., Horwich, M. D., Weissbein, U., Nawathean, P., *et al.* (2009a). *Genes Dev.* **23,** 2179–2191.

Kadener, S., Rodriguez, J., Abruzzi, K. C., Khodor, Y. L., Sugino, K., Marr, M. T., 2nd, *et al.* (2009b). *RNA* **15,** 537–545.

Karim, F. D., Chang, H. C., Therrien, M., Wassarman, D. A., Laverty, T., and Rubin, G. M. (1996). *Genetics* **143,** 315–329.

Karr, J., Vagin, V., Chen, K., Ganesan, S., Olenkina, O., Gvozdev, V., and Featherstone, D. E. (2009). *J. Cell Biol.* **185,** 685–697.

Karres, J. S., Hilgers, V., Carrera, I., Treisman, J., and Cohen, S. M. (2007). *Cell* **131,** 136–145.

Kennerdell, J. R., and Carthew, R. W. (2000). *Nat. Biotechnol.* **18,** 896–898.

Kim, V. N., Han, J., and Siomi, M. C. (2009). *Nat. Rev. Mol. Cell Biol.* **10,** 126–139.

Kim, Y. K., Heo, I., and Kim, V. N. (2010). *Cell* **143,** 703–709.

King, I. N., Qian, L., Liang, J., Huang, Y., Shieh, J. T., Kwon, C., and Srivastava, D. (2011). *Dev. Cell* **20,** 497–510.

Klämbt, C., Knust, E., Tietze, K., and Campos-Ortega, J. (1989). *EMBO J.* **8,** 203–210.

Kosman, D., Mizutani, C. M., Lemons, D., Cox, W. G., McGinnis, W., and Bier, E. (2004). *Science* **305,** 846.

Kraut, R., Menon, K., and Zinn, K. (2001). *Curr. Biol.* **11,** 417–430.

Krutzfeldt, J., Rajewsky, N., Braich, R., Rajeev, K. G., Tuschl, T., Manoharan, M., and Stoffel, M. (2005). *Nature* **438,** 685–689.

Kwon, C., Han, Z., Olson, E. N., and Srivastava, D. (2005). *Proc. Natl. Acad. Sci. USA* **102,** 18986–18991.
Lagos-Quintana, M., Rauhut, R., Lendeckel, W., and Tuschl, T. (2001). *Science* **294,** 853–858.
Lai, E. C. (2002). *Nat. Genet.* **30,** 363–364.
Lai, E. C., and Posakony, J. W. (1997). *Development* **124,** 4847–4856.
Lai, E. C., and Posakony, J. W. (1998). *Cell* **93,** 1103–1104.
Lai, E. C., Burks, C., and Posakony, J. W. (1998). *Development* **125,** 4077–4088.
Lai, E. C., Tomancak, P., Williams, R. W., and Rubin, G. M. (2003). *Genome Biol.* **4,** R42.1–R42.20.
Lai, E. C., Tam, B., and Rubin, G. M. (2005). *Genes Dev.* **19,** 1067–1080.
Leaman, D., Chen, P. Y., Fak, J., Yalcin, A., Pearce, M., Unnerstall, U., *et al.* (2005). *Cell* **121,** 1097–1108.
Lee, R. C., Feinbaum, R. L., and Ambros, V. (1993). *Cell* **75,** 843–854.
Lee, Y., Kim, M., Han, J., Yeom, K. H., Lee, S., Baek, S. H., and Kim, V. N. (2004). *EMBO J.* **23,** 4051–4060.
Lee, Y., Lee, J., Bang, S., Hyun, S., Kang, J., Hong, S. T., *et al.* (2005). *Nat. Genet.* **37,** 305–310.
Leviten, M. W., Lai, E. C., and Posakony, J. W. (1997). *Development* **124,** 4039–4051.
Lewis, E. B. (1978). *Nature* **276,** 565–570.
Li, X., and Carthew, R. W. (2005). *Cell* **123,** 1267–1277.
Li, Y., Wang, F., Lee, J. A., and Gao, F. B. (2006). *Genes Dev.* **20,** 2793–2805.
Li, X., Cassidy, J. J., Reinke, C. A., Fischboeck, S., and Carthew, R. W. (2009). *Cell* **137,** 273–282.
Lifschytz, E., and Green, M. M. (1979). *Mol. Gen. Genet.* **171,** 153–159.
Lim, L. P., Lau, N. C., Garrett-Engele, P., Grimson, A., Schelter, J. M., Castle, J., *et al.* (2005). *Nature* **433,** 769–773.
Lindsley, D. L., Sandler, L., Baker, B. S., Carpenter, A. T. C., Denell, R. F., Hall, J. C., *et al.* (1972). *Genetics* **71,** 157–184.
Loya, C. M., Lu, C. S., Van Vactor, D., and Fulga, T. A. (2009). *Nat. Methods* **6,** 897–903.
Macquarrie, K. L., Fong, A. P., Morse, R. H., and Tapscott, S. J. (2011). *Trends Genet.* **27,** 141–148.
Manak, J. R., Dike, S., Sementchenko, V., Kapranov, P., Biemar, F., Long, J., *et al.* (2006). *Nat. Genet.* **38,** 1151–1158.
Mangone, M., Manoharan, A. P., Thierry-Mieg, D., Thierry-Mieg, J., Han, T., Mackowiak, S. D., *et al.* (2010). *Science* **329,** 432–435.
Martin, F. A., Perez-Garijo, A., Moreno, E., and Morata, G. (2004). *Development* **131,** 4921–4930.
Miska, E. A., Alvarez-Saavedra, E., Abbott, A. L., Lau, N. C., Hellman, A. B., McGonagle, S. M., *et al.* (2007). *PLoS Genet.* **3,** e215.
Moss, E. G., Lee, R. C., and Ambros, V. (1997). *Cell* **88,** 637–646.
Nairz, K., Rottig, C., Rintelen, F., Zdobnov, E., Moser, M., and Hafen, E. (2006). *Dev. Biol.* **291,** 314–324.
Nolo, R., Abbott, L., and Bellen, H. J. (2001). *Genetics* **157,** 307–315.
Nolo, R., Morrison, C. M., Tao, C., Zhang, X., and Halder, G. (2006). *Curr. Biol.* **16,** 1895–1904.
Oh, H., and Irvine, K. D. (2011). *Dev. Cell* **20,** 109–122.
Okamura, K., Hagen, J. W., Duan, H., Tyler, D. M., and Lai, E. C. (2007). *Cell* **130,** 89–100.
O'Keefe, L. V., Smibert, P., Colella, A., Chataway, T. K., Saint, R., and Richards, R. I. (2007). *Trends Genet.* **23,** 238–242.
Pan, D. (2010). *Dev. Cell* **19,** 491–505.

Parks, A. L., Cook, K. R., Belvin, M., Dompe, N. A., Fawcett, R., Huppert, K., *et al.* (2004). *Nat. Genet.* **36,** 288–292.

Parrish, J. Z., Xu, P., Kim, C. C., Jan, L. Y., and Jan, Y. N. (2009). *Neuron* **63,** 788–802.

Pena-Rangel, M. T., Rodriguez, I., and Riesgo-Escovar, J. R. (2002). *Genetics* **160,** 1035–1050.

Peng, H. W., Slattery, M., and Mann, R. S. (2009). *Genes Dev.* **23,** 2307–2319.

Raisin, S., Pantalacci, S., Breittmayer, J. P., and Leopold, P. (2003). *Genetics* **164,** 1015–1025.

Reinhart, B. J., Slack, F., Basson, M., Pasquinelli, A., Bettinger, J., Rougvie, A., *et al.* (2000). *Nature* **403,** 901–906.

Roegiers, F., Kavaler, J., Tolwinski, N., Chou, Y. T., Duan, H., Bejarano, F., *et al.* (2009). *Genetics* **182,** 407–410.

Rong, Y. S., and Golic, K. G. (2000). *Science* **288,** 2013–2018.

Rong, Y. S., and Golic, K. G. (2001). *Genetics* **157,** 1307–1312.

Rong, Y. S., Titen, S. W., Xie, H. B., Golic, M. M., Bastiani, M., Bandyopadhyay, P., *et al.* (2002). *Genes Dev.* **16,** 1568–1581.

Ronshaugen, M., Biemar, F., Piel, J., Levine, M., and Lai, E. C. (2005). *Genes Dev.* **19,** 2947–2952.

Rørth, P., Szabo, K., Bailey, A., Laverty, T., Rehm, J., Rubin, G. M., *et al.* (1998). *Development* **125,** 1049–1057.

Roy, S., Ernst, J., Kharchenko, P. V., Kheradpour, P., Negre, N., Eaton, M. L., *et al.* (2010). *Science* **330,** 1787–1797.

Ruby, J. G., Jan, C. H., and Bartel, D. P. (2007a). *Nature* **448,** 83–86.

Ruby, J. G., Stark, A., Johnston, W. K., Kellis, M., Bartel, D. P., and Lai, E. C. (2007b). *Genome Res.* **17,** 1850–1864.

Ruvkun, G., and Giusto, J. (1989). *Nature* **338,** 313–319.

Ryder, E., Ashburner, M., Bautista-Llacer, R., Drummond, J., Webster, J., Johnson, G., *et al.* (2007). *Genetics* **177,** 615–629.

Silver, S. J., Hagen, J. W., Okamura, K., Perrimon, N., and Lai, E. C. (2007). *Proc. Natl. Acad. Sci. USA* **104,** 18151–18156.

Simon, M. A., Bowtell, D. D., Dodson, G. S., Laverty, T. R., and Rubin, G. M. (1991). *Cell* **67,** 701–716.

Siomi, H., and Siomi, M. C. (2010). *Mol. Cell* **38,** 323–332.

Smibert, P., and Lai, E. C. (2008). *Cell Cycle* **7,** 2500–2508.

Smibert, P., and Lai, E. C. (2010). *Semin. Cell Dev. Biol.* **21,** 745–753.

Sokol, N. S., and Ambros, V. (2005). *Genes Dev.* **19,** 2343–2354.

Sokol, N. S., Xu, P., Jan, Y. N., and Ambros, V. (2008). *Genes Dev.* **22,** 1591–1596.

Spradling, A. C., Stern, D., Beaton, A., Rhem, E. J., Laverty, T., Mozden, N., *et al.* (1999). *Genetics* **153,** 135–177.

Stark, A., Brennecke, J., Russell, R. B., and Cohen, S. M. (2003). *PLoS Biol.* **1,** E60.

Stark, A., Brennecke, J., Bushati, N., Russell, R. B., and Cohen, S. M. (2005). *Cell* **123,** 1133–1146.

Stark, A., Lin, M. F., Kheradpour, P., Pedersen, J. S., Parts, L., Carlson, J. W., *et al.* (2007). *Nature* **450,** 219–232.

Stark, A., Bushati, N., Jan, C. H., Kheradpour, P., Hodges, E., Brennecke, J., *et al.* (2008). *Genes Dev.* **22,** 8–13.

Teleman, A. A., Maitra, S., and Cohen, S. M. (2006). *Genes Dev.* **20,** 417–422.

Thibault, S. T., Singer, M. A., Miyazaki, W. Y., Milash, B., Dompe, N. A., Singh, C. M., *et al.* (2004). *Nat. Genet.* **36,** 283–287.

Thompson, B. J., and Cohen, S. M. (2006). *Cell* **126,** 767–774.

Tsurudome, K., Tsang, K., Liao, E. H., Ball, R., Penney, J., Yang, J. S., *et al.* (2010). *Neuron* **68,** 879–893.

Tyler, D. M., Okamura, K., Chung, W. J., Hagen, J. W., Berezikov, E., Hannon, G. J., and Lai, E. C. (2008). *Genes Dev.* **22,** 26–36.
Varghese, J., and Cohen, S. M. (2007). *Genes Dev.* **21,** 2277–2282.
Varghese, J., Lim, S. F., and Cohen, S. M. (2010). *Genes Dev.* **24,** 2748–2753.
Venken, K. J., Carlson, J. W., Schulze, K. L., Pan, H., He, Y., Spokony, R., *et al.* (2009). *Nat. Methods* **6,** 431–434.
Weng, R., Chen, Y. W., Bushati, N., Cliffe, A., and Cohen, S. M. (2009). *Genetics* **183,** 399–402.
Wightman, B., Burglin, T. R., Gatto, J., Arasu, P., and Ruvkun, G. (1991). *Genes Dev.* **5,** 1813–1824.
Wightman, B., Ha, I., and Ruvkun, G. (1993). *Cell* **75,** 855–862.
Xu, P., Vernooy, S. Y., Guo, M., and Hay, B. A. (2003). *Curr. Biol.* **13,** 790–795.
Xu, X. L., Li, Y., Wang, F., and Gao, F. B. (2008). *J. Neurosci.* **28,** 11883–11889.
Yang, J. S., and Lai, E. C. (2011). *Mol. Cell* **43,** 892–903.
Yang, Y., Xu, S., Xia, L., Wang, J., Wen, S., Jin, P., and Chen, D. (2009). *PLoS Genet.* **5,** e1000444.

CHAPTER NINE

Biological Robustness and the Role of MicroRNAs: A Network Perspective

Nicolás Peláez[*,†,‡] *and* Richard W. Carthew[†,‡]

Contents

Abstract

Over the past decade, microRNA molecules have emerged as critical regulators in the expression and function of animal genomes. This review discusses the relationship between microRNA-mediated regulation and the robustness of biochemical networks that contain microRNAs. Most biochemical networks are robust; they are relatively insensitive to the precise values of reaction constants and concentrations of molecules acting within the network. MicroRNAs involved in network robustness may appear to be nonessential under favourable uniform conditions used in conventional laboratory experiments. However, the function of these molecules can be revealed under environmental and genetic perturbations. Recent advances have revealed unexpected features of microRNA organization in networks that help explain their promotion of robustness.

[*] Interdepartmental Program in Biological Sciences, Northwestern University, Evanston, Illinois, USA
[†] Department of Molecular Biosciences, Northwestern University, Evanston, Illinois, USA
[‡] Chicago Center for Systems Biology, Chicago, Illinois, USA

Current Topics in Developmental Biology, Volume 99
ISSN 0070-2153, DOI: 10.1016/B978-0-12-387038-4.00009-4

Biological activities such as development exhibit a property known as robustness. This term has been used to mean many things, and we define robustness as an event that happens reproducibly and uniformly even in the face of variability that can be induced by the environment, informatic (genetic and epigenetic) variation, local effects, and random chance. A few biological processes are quite variable and hence do not require robustness. However, most biological processes, particularly irreversible ones such as differentiation, are strongly robust to ensure a minimal impact of error. This review discusses the particular role that microRNAs (miRNAs) play in biological robustness.

1. MicroRNAs and Large-Scale Networks

Many studies have looked at miRNAs by first-order relationships: what molecules regulate a miRNA and what mRNA transcripts are regulated by a miRNA. Genome-wide studies have shown that, in plants and mammals, mRNAs targeted by miRNAs are overrepresented by GO terms associated with regulation of development (Rhoades *et al.*, 2002; Shalgi *et al.*, 2007). Clearly, miRNAs are important for development (Carthew, 2006). However, a paradox for many miRNAs is their lack of strong phenotypic consequences on development when individually mutated, and yet they are evolutionarily conserved. Some have argued that this paradox is due to the weak repression of target gene expression elicited by most miRNAs. Weak and tunable repression by miRNAs can generally elicit three distinct effects on their targets. MiRNAs can (i) dampen, (ii) denoise, and (iii) set thresholds to the levels of their targets (Bartel and Chen, 2004; Bushati and Cohen, 2007; Cohen *et al.*, 2006; Inui *et al.*, 2010). In the first type of effect, a miRNA reduces the level of target below an activity threshold acting like a switch; in the second type of effect, a miRNA buffers fluctuations in the target, limiting undesired signal propagation; in the third type of effect, a miRNA raises or changes the level of activation at which the target has to be induced to actively regulate a process. Each of these effects can potentially be harnessed to provide robustness to target gene regulation.

However, miRNAs also exist in higher-order relationships, and less emphasis has been placed on the biochemical networks that include miRNAs. Yet this is fundamental to understand what roles miRNAs play in robustness. Most biochemical networks are robust; they are relatively insensitive to the precise values of reaction constants and concentrations of molecules acting within the network (Barkai and Leibler, 1997; Eldar *et al.*, 2002). Our point of view is that miRNAs help to generate the robustness of biochemical networks. One means that they might generate network robustness is to dampen, denoise, and set thresholds for direct targets in a network. But as we shall see, miRNAs also generate network robustness because of biases in the kinds of targets that they regulate. Understanding how miRNAs do so will provide insight about how higher-level biological processes such as development are also made robust.

1.1. Network hubs and miRNAs

Molecular regulatory systems can be represented as networks composed of nodes and links (Fig. 9.1A). Nodes can be genes, sequence elements, or molecules such as proteins, metabolites, RNAs, etc. Links are the molecular interactions between the nodes. The degree of a node is the number of links that a node has with other nodes in the network. The collection of degrees for each node in the network is the degree distribution and is frequently represented as a graph of the frequency of each node degree type (Fig. 9.1B). While some links such as protein–protein interactions do not necessarily have an associated direction (undirected link), many links such as between a transcription factor (TF) and its gene target, or a miRNA and its mRNA target, are directed (i.e., a miRNA represses a target mRNA when the two are bound together, and not the other way around). In directed

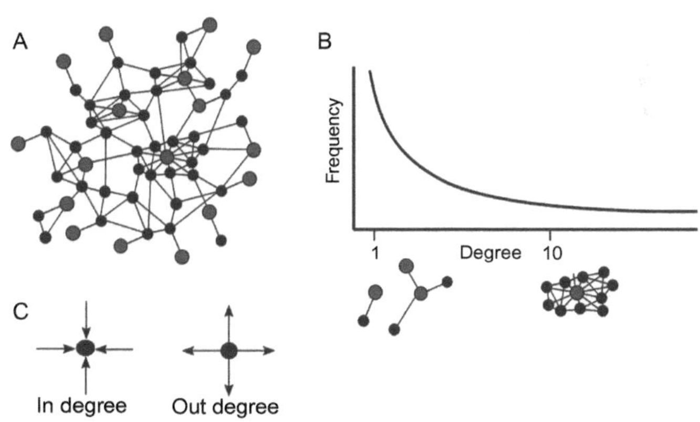

Figure 9.1 Biochemical network organization. (A) Schematized network of nodes (circles) and links (lines). Different molecular classes of nodes are highlighted in green and blue. (B) A typical degree distribution for a network, illustrating that most nodes have few links and a few nodes have many links. This organization gives rise to a power law distribution that has a long tail. (C) Each link between nodes can be directed in terms of cause–effect relationship. Links directed into a node effect that node, while links directed out from a node effect the other node connected by the link. (See Color Insert.)

networks where links have a directionality associated to them, the degree of a node can be further subdivided into links going in or out of the node (in and out degrees, respectively) (Fig. 9.1C).

The degree distribution is an important property of large-scale network organization, and it measures how the connectivity is distributed overall in the network. Most molecular networks studied to date such as protein–protein, signaling, or TF-gene networks have a degree distribution that deviates from random (i.e., a normal distribution) and follows instead an exponential distribution or a power law (Barabási and Oltvai, 2004; Martinez and Walhout, 2009). In such networks, the majority of nodes have a low to medium number of links, while a few nodes, called hubs, are highly connected to other nodes (Fig. 9.1B). Hubs mediate interactions among numerous and less connected nodes, allowing rapid coordination between different parts of the network. This type of system is robust to random loss of the less connected nodes but is sensitive to deletion of the hubs (Albert *et al.*, 2000). For example, in protein–protein interaction networks, there is a lethality–centrality relationship where highly connected components induce lethality when lost (Jeong *et al.*, 2001; Zotenko *et al.*, 2008).

Do miRNAs exhibit this lethality–centrality relationship? Systematic mutagenesis of many individual miRNA genes (Miska *et al.*, 2007) and paralogous families of miRNA genes (Alvarez-Saavedra and Horvitz, 2010) were performed in the nematode *Caenorhabditis elegans*. Most miRNAs either individually or in collective families are not essential for viability or development. Only miRNAs like *let-7*, one of the most highly connected miRNAs in the animal (Martinez *et al.*, 2008), elicit observable lethal phenotypes when knocked out (Reinhart *et al.*, 2000). The degree distribution of *C. elegans* miRNA-TF networks could explain why most miRNAs trigger subtle or nondetectable mutant phenotypes. If a miRNA node has only one or a few links to target within a given network, then loss of this node would generally have a small impact on network behavior. In *C. elegans*, there is a significant difference between nodes composed of TFs and nodes composed of miRNAs. TFs bind promoters in a scale-free manner, that is, the TF-out degree distribution follows a power law, and there are clear TF hubs binding many promoters. In contrast, the miRNA-out degree distribution follows an exponential distribution, that is, there are no clear miRNA hubs even though some miRNAs are more connected than others (Martinez and Walhout, 2009; Martinez *et al.*, 2008).

Although miRNAs do not exhibit hub-like properties, frequently the direct targets of miRNAs behave as network hubs. These target hubs often contain many in-links from different miRNAs (Fig. 9.2). In the *C. elegans* and human miRNA–mRNA target networks, the target in-links follow a power law distribution (Martinez *et al.*, 2008; Mookherjee *et al.*, 2009a).

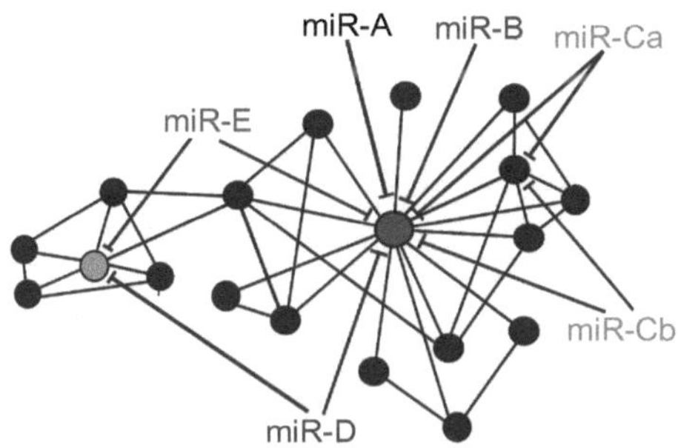

Figure 9.2 MicroRNAs regulate network hubs. Two highly connected nodes are highlighted in green and yellow, and these have hub-like features. Various miRNAs preferentially regulate the network through these hubs. (See Color Insert.)

In such networks, target hubs exist that are linked by 15 or more miRNAs. Strikingly, these target hubs are enriched for TFs (Martinez *et al.*, 2008) and factors involved in regulation of development (Shalgi *et al.*, 2007). Target hubs also tend to be more connected by protein–protein interaction links than lowly connected targets. A positive correlation has been seen between the number of miRNA-binding sites in the 3′UTR of a gene and the connectivity of its protein product to other proteins (Liang and Li, 2007). This propensity is not a side result of longer or evolutionarily conserved 3′UTRs (Liang and Li, 2007; Shalgi *et al.*, 2007). In the case of *C. elegans*, experimental evidence indicates such hubs are important for miRNA function. Combined knockout of individual miRNA genes and TF hubs led to synthetic phenotypes that were otherwise undetectable when either miRNA or hub was knocked out alone (Brenner *et al.*, 2010).

Target hubs can be more connected in other ways as well. miRNAs were found to preferentially target genes encoding enzymes that are metabolic hubs or cut point enzymes (Tibiche and Wang, 2008). These are capable of regulating metabolic mass flow at global and local scales, respectively (Tibiche and Wang, 2008). Highly connected scaffold proteins in signaling networks are also preferentially linked to miRNAs (Cui *et al.*, 2007). Scaffold proteins are important components of signaling pathways that lack enzyme activity but physically interact with upstream and downstream components of the pathways, often simultaneously. It has been found that miRNAs more frequently target highly connected scaffold proteins than less connected nodes of the same pathways (Cui *et al.*, 2007). Altogether these results suggest that miRNAs preferentially regulate highly connected nodes in various types of networks.

1.2. Signal flow and miRNAs

Signaling networks are crucial to establish expression patterns that lead to cell decisions. In such systems, signal transduction has a directional flow; it often begins with membrane-bound receptors that bind ligands to intracellular transduction proteins interacting with other proteins, to translocation of effectors to the nucleus resulting in altered gene expression. Cui *et al.* (2007) found that miRNAs target signaling proteins of a human network more frequently than what would be expected by chance. Importantly, the distribution of miRNA targets in the signaling network is correlated with signal flow and the position of the factors within the signaling network. The propensity for a factor to be regulated by miRNAs increases in the direction of signal flow, from ligands (9.1%) to cell surface receptors (18.8%), to intracellular transducers (31.2%), and to nuclear proteins (50%) (Cui *et al.*, 2007).

1.3. Network modules and miRNAs

Biochemical networks are frequently organized into a set of distinct subnetworks called modules. A module exists as a group of nodes that are more highly connected to each other than to the rest of the network. Modules are interconnected, typically through nodes called bottlenecks (Fig. 9.3). Bottlenecks can in some cases be hubs. Since bottlenecks connect modules, they have a high betweenness centrality, that is, are nodes with many "shortest paths" going through them. They are analogous to major bridges and tunnels with multiple parallel links between

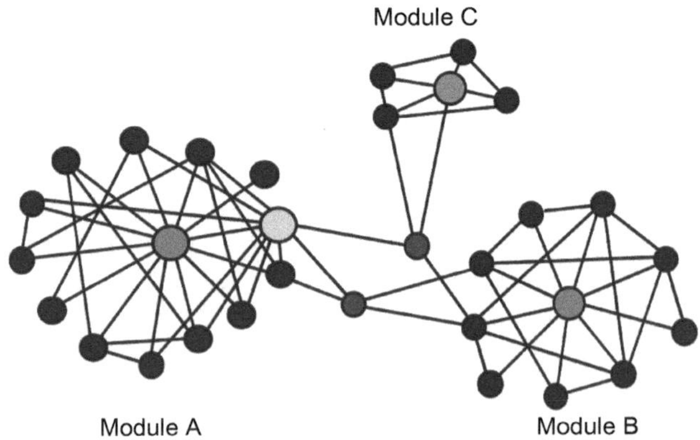

Figure 9.3 Modular design of biochemical networks. A network with three modules. Modules A and B have hubs (green) and bottlenecks (yellow and red). Module C has a hub–bottleneck (blue). (See Color Insert.)

two nodes that themselves are part of two distinct modules. Given their central position in networks, bottlenecks are often essential proteins (Yu *et al.*, 2007). Modularity can impart robustness to networks, as it provides the ability for certain functions to be carried out in a semiautonomous manner through the coordinated interactions of relatively small subsets of molecules more densely connected to themselves than to other network components (Hartwell *et al.*, 1999). Modularity can make systems more evolvable and in some cases more tolerant to the random loss or modification of a module's component. Modules have been mapped for developmental processes such as early embryogenesis of *C. elegans* (Gunsalus *et al.*, 2005) and have been in some cases shown to operate as robust entities despite variations in input signals and kinetic constants that govern their behavior (von Dassow *et al.*, 2000).

MicroRNAs frequently regulate module bottlenecks in networks analyzed on the genome scale (Hsu *et al.*, 2008) (Fig. 9.4A). More detailed

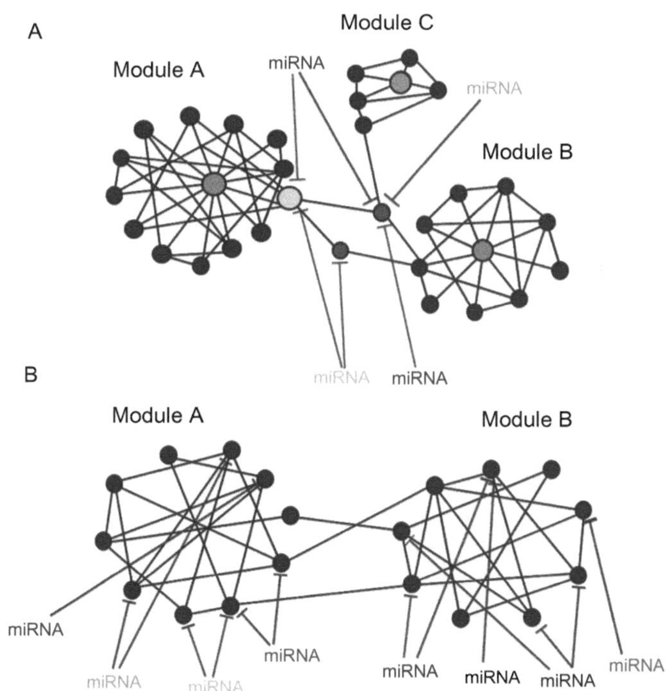

Figure 9.4 MicroRNAs selectively regulate modules. (A) Various miRNAs preferentially regulate networks through module bottlenecks. (B) Families or classes of miRNAs selectively regulate certain modules in a network. Each miRNA frequently regulates more than one node in a particular module. (See Color Insert.)

studies have found corroboration for this bias. One study focused on a protein–protein network regulated by miR-204, a miRNA that was shown to function as a tumor suppressor (Lee *et al.*, 2010). Within the network, two distinct modules (cell adhesion and cell cycle) are found, and each module is connected to the other via bottlenecks. Within each module, miR-204 preferentially targets mRNAs encoding hub and bottleneck proteins (Lee *et al.*, 2010). Interestingly, miR-204 suppression significantly augments cell cycle and extracellular matrix remodeling *in vitro* and *in vivo* (Lee *et al.*, 2010). Another study showed that several miRNAs are predicted as regulators for various modules of tightly coexpressed genes (Bonnet *et al.*, 2010). miR-200a is the top regulator of a small module of nine genes that is part of a larger network, such that this miRNA regulates the module via the TF ZEB1. Interestingly, this module is most likely involved in epithelial homeostasis, and its dysregulation could contribute to the malignant process in cancer cells (Bonnet *et al.*, 2010).

A second modularity-related property of miRNAs is that they cotarget molecules belonging to the same module (Fig. 9.4B). Cotargeting occurs by subsets of either unrelated miRNAs or a particular miRNA family. For example, there is pervasive regulation of several related transcription repressors that function in the Notch signaling network by three different classes of miRNAs that each recognizes a similar seed sequence in their targets (Lai *et al.*, 2005). Further support for this property has come from computational studies. Basu *et al.* (2011) found that coregulated targets tend to be organized within network modules. More than half of modules with prevalent coregulated targets are not simply explained by seed similarity (Mookherjee *et al.*, 2009b). Two other computational analyses showed that cotargeted genes and their interacting neighbors jointly show significantly higher modularity, and clustered miRNAs jointly regulate proteins in close proximity within a protein–protein interaction network (Hsu *et al.*, 2008; Yuan *et al.*, 2009). These analyses support the notion that coregulation of targets within modules is a prevalent phenomenon.

As factors belonging to a module are corepressed, modular regulation increases the redundancy of the cotarget network, making it robust to the individual loss or rewiring of some of the miRNAs. Consistent with this feature, miRNAs are individually not essential (Miska *et al.*, 2007). The concept of miRNA regulative modularity is also in accordance with observations that relate specific miRNAs to cancer development. Up- or downregulation of miRNAs repressing particular modules is associated with loss of robustness that is linked with cancer progression (Bandyopadhyay *et al.*, 2010). For example, five miRNAs (miR-19b, miR-20a, miR-26a, miR-92, and miR-223), which are capable of promoting T-cell acute lymphoblastic leukemia (T-ALL) in a mouse model, account for the majority of miRNA expression in human T-ALL. This small set of miRNAs is responsible for the cooperative suppression of several tumor suppressor factors, achieved through an overlapping and cooperative regulation of these miRNAs (Mavrakis *et al.*, 2011).

1.4. How do these properties affect network robustness?

Nonuniform perturbation in network activity is most keenly "felt" by critical nodes such as hubs and bottlenecks since their extensive in-links act to amplify any perturbation. This amplification must be dampened in order to maintain these nodes working in synchrony with less connected nodes that do not experience such a degree of amplification. miRNAs weakly repress protein expression and thus are well suited to dampen hubs or bottlenecks in times of perturbation. Another reason critical nodes are targeted is one of impact. If a hub or bottleneck is perturbed, it has the most impact on the network's stability. Therefore, miRNAs that dampen such perturbation will contribute more greatly to network stability than miRNAs that target other types of nodes. A flip side of the impact theme is that perturbation of one section of a network is transduced most strongly to the rest of the network by hubs and bottlenecks. Dampening the transduced perturbation at these central nodes has both the broadest and swiftest effect on stabilizing the entire network and achieving synchronization. It also has the advantage of being able to respond to a wider variety of perturbations.

The modularity of miRNA targeting is also well suited to providing robustness. As a module is semiautonomous with distinctive biochemical properties, it specifically responds to perturbation relative to the rest of the network. A perturbation can thus be contained within a module by exact and coordinated regulation of the handful of nodes acting within a module or by regulation of bottlenecks. Clearly, miRNAs with tunable and parallel (redundant) regulatory capabilities are well suited to dampen perturbations within a module. This activity then would help prevent destabilization of the remainder of the network.

The pattern of miRNA regulation in signaling networks further provides robustness. Preferential regulation of the most downstream nodes in signal flow facilitates rapid responses with minimal lag when upstream nodes are perturbed. This pattern of regulation also makes signaling networks less prone to respond to noise resulting from signal propagation, since amplification of upstream noise would be dampened downstream.

2. MicroRNAs and Circuits

In this section, we review how small-scale circuits can provide robustness to biological processes, associating miRNA function with particular features of networks.

2.1. Circuits are recurrent patterns in large-scale networks

Large-scale networks can be deconstructed into circuits composed of smaller groups of nodes. Analysis of prokaryotic TF-target gene networks led to the discovery that certain types of circuits occur inside networks at frequencies much higher than in randomized control networks (Shen-Orr *et al.*, 2002). Such overrepresented circuits are called network motifs, and they constitute "building blocks" of larger networks that preserve their functions independent of the network environment in which they are embedded (Alon, 2007; Milo *et al.*, 2002).

One class of network motif is a circuit called a feedforward loop (FFL). FFLs have two paths or arms of regulation, one direct (short arm) and one indirect (long arm). The upstream node in the loop regulates the downstream node directly and indirectly through an intermediate node (Fig. 9.5). Eight different FFL configurations exist, and four variables determine which combination of dynamic behaviors emerges in a particular FFL (Box 9.1). Computational modeling and experimental studies of prokaryotic FFLs show that such loops have specific information processing properties that differ from direct circuits (Goentoro *et al.*, 2009; Kaplan *et al.*, 2008; Mangan and Alon, 2003; Mangan *et al.*, 2003; Shen-Orr *et al.*, 2002). Such properties include acceleration or delay of a response, generation of signal pulses, and the ability to buffer the downstream node against fluctuations in the upstream node such that only persistent changes in the upstream node are transduced through the loop. FFLs provide robustness against stochastic fluctuations in the upstream node of the circuit.

Another type of circuit is the feedback loop (FBL). Positive and negative FBLs are known to be of central importance in biological processes (Fig. 9.5). Positive FBLs can amplify signals, create ultrasensitivity, and enable irreversible states of gene expression to occur (Brandman and Meyer, 2008; Chang *et al.*, 2010; Ferrell, 2002). Positive FBLs can give rise to bistable switches, that is, two alternative stable states without stable intermediates in between them (Ferrell, 2002). Double-negative FBLs can also stabilize gene expression in one state, though simulations have shown that double-negative FBLs are not sufficient to create bistable switches. Other features such as nonlinear positive feedback or balanced link strength are needed for a double-negative FBL to generate bistable behavior (Ferrell, 2002; Graham *et al.*, 2010). Single-negative FBLs are associated with homeostasis and desensitization (Ferrell, 2002). Thus, in different ways, positive and negative FBLs provide robustness of a circuit against fluctuation or perturbation.

We review the role of circuits that contain or are regulated by miRNAs, focusing on their possible roles in providing robustness. We refer the reader to four reviews on the topic that relate these circuits to developmental canalization (Hornstein and Shomron, 2006; Wu *et al.*, 2009), noise (Herranz and Cohen, 2010), and signal transduction (Inui *et al.*, 2010).

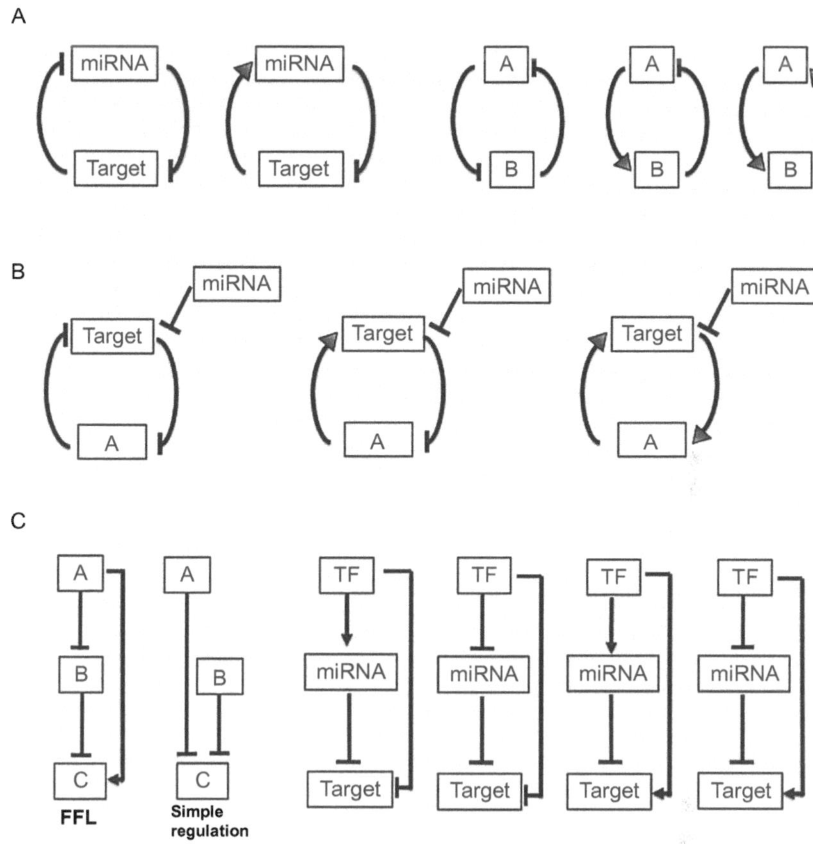

Figure 9.5 MicroRNAs are components of network circuits. (A) Feedback loops (FBLs) can be double-negative, single-negative, or positive in sign (right). Since miRNAs repress their targets, then miRNAs can form double-negative or single-negative FBLs with targets (left). (B) FBLs of any sign can be regulated by a miRNA that exists outside of the loop. (C) Feedforward loops (FFLs) are distinct from simpler regulatory circuits (left). MicroRNAs can be components of FFLs, as shown on the right, where a miRNA is an intermediate in four different FFLs. Two of the FFLs are coherent, in which each arm of the loop has the same sign. Two of the FFLs are incoherent since each arm of the loop has the opposite sign. (For color version of this figure, the reader is referred to the Web version of this chapter.)

2.2. Feedback loops and miRNAs

Genome-wide studies have shown that FBLs containing miRNAs are network motifs in *C. elegans* and mammals (Martinez *et al.*, 2008; Tsang *et al.*, 2007). FBLs containing miRNAs can be double-negative, where the miRNA represses a repressor of the miRNA (Fig. 9.5). The double-negative FBLs containing miRNAs can be associated with bistable dynamics that give rise to

Box 9.1 Feedforward loops are network motifs

In a FFL, the first factor of the loop regulates the last factor directly (short arm) and indirectly (long arm) through an intermediate factor. Four variables determine the dynamic behavior of a FFL. These variables are (i) the overall regulatory effect of the arms. The long arm can be positive or negative. (ii) Whether arms are coherent or incoherent (the same or opposite sign). (iii) The logic gate that integrates the signals from the two upstream nodes. This determines whether the downstream node can be regulated by both arms (AND gate) or by either arm (OR gate). (iv) The type of change in signal that passes through the loop. Does the change in the signal directed from an active to a repressed state (ON-to-OFF), or its inverse? Depending on the combination of these four variables, FFLs can act as pulsers, persistence detectors of change, or accelerators of response time. For example, consider the Type 3 coherent FFL where the upstream node directly and indirectly represses the downstream node through a second inhibitor activated by the first. When this specific architecture operates with an OR Gate, the resulting FFL produces a delayed response to a decrease (ON-to-OFF step) in the upstream node. As result, this type of FFL can filter ON-to-OFF pulses and only respond to persistent changes in the upstream node. This type of architecture exists in the Yan network linking *Pnt-P1*, *miR-7*, and *YAN* and is predicted to buffer YAN expression from stochastic decreases of *Pnt-P1*.

mutually exclusive expression of the miRNAs and their targets. miRNAs can also be found in single-negative FBLs, where the miRNA represses an activator of the miRNA. A limitation of these genome-wide studies has been their ability to experimentally verify the existence of computationally predicted links. For example, Martinez *et al.* (2008) derived their network from a combination of computational predictions of miRNA targets and Y1H experiments that established which TFs bound a library of DNA elements containing predicted promoters for miRNA genes.

Several experimentally verified examples of miRNA-containing FBLs have been described. Some of these are listed in Table 9.1 and have been reviewed extensively. Instead, we focus on two examples where a role for miRNA-mediated robustness has been shown. In *Drosophila melanogaster*, the transcription repressor YAN binds and represses the transcriptional enhancer of the *miR-7* gene. In turn, miR-7 binds and represses the protein expression of *YAN*. YAN and miR-7 are part of a network that regulates the transition from multipotent retinal progenitor cells to differentiated photoreceptors (Li and Carthew, 2005; Li *et al.*, 2009). The YAN network is a bistable system that transitions from a high YAN/low miR-7

Table 9.1 MicroRNAs and network motifs: Experimental evidence

System	Factors	Reference
FBLs		
Human granulocytes	miR-223; NFI-A; CEBP	Fazi *et al.* (2005)
C. elegans ASE neurons	lsy-6; miR-273; die-1; cog-1	Johnston *et al.* (2005)
C. elegans VPCs	miR-61; lin-12; vav-1	Yoo and Greenwald (2005)
Drosophila R cells	miR-7; Yan; Pnt-P1	Li and Carthew (2005)
Drosophila SOPs	miR-9a; Sens; Proneural	Li *et al.* (2006)
Cardiomyocytes	miR-1; miR-133l SRF; MEF-2	Chen *et al.* (2006)
FFLs		
Human B cells	miR-17-5p; c-myc; E2F	O'Donnell *et al.* (2005)
Drosophila R cells	miR-7; Yan; Pnt-P1	Li *et al.* (2009)

to a high miR-7/low YAN stable state (Graham *et al.*, 2010). These two states are stabilized through double-negative FBLs between YAN and its repressors,and dictate whether a cell remains multipotent or differentiated (Graham *et al.*, 2010). EGF receptor signaling induces the phosphorylation of YAN and a switching from the multipotent to the differentiated state. Nevertheless, the FBL between YAN and miR-7 is essential neither for the switch nor for the stable maintenance of the cells' states. Instead, the role of this FBL is probably to generate robustness to the network. When development is perturbed in a *miR-7* mutant by oscillating temperature, the switch occurs less robustly and errors in cell fate are observed (Li *et al.*, 2009). These errors are undetectable under uniform temperature conditions. Thus, this FBL can make differentiation robust to environmental perturbation.

A different mechanism is found in the differentiation of the *Drosophila* sensory organ precursors (SOPs; Li *et al.*, 2006). In this case, the miRNA is not itself part of a FBL but rather it regulates the responsiveness of the FBL. *miR-9a* represses expression of *Senseless*, which encodes a TF that induces SOP differentiation. The SOP cell fate choice is made by coupling Notch signaling to a positive FBL between various TFs, including Senseless. This choice does not fundamentally depend on *miR-9a* repression of *Senseless*, as *miR-9a* mutants are still capable of specifying their SOPs. Nevertheless, up to 40% of mutant animals make extra sensory organs (Li *et al.*, 2006). *miR-9a* thresholds *Senseless* expression such that, unless a threshold of *Senseless* expression is achieved, the TF FBL is not engaged and SOPs are not specified (Cohen *et al.*, 2006). Thus, *miR-9a* buffers SOP differentiation against fluctuations of Senseless.

2.3. Feedforward loops and miRNAs

Genome-wide studies have found FFLs that contain miRNAs are network motifs (Re *et al.*, 2009; Shalgi *et al.*, 2007; Tsang *et al.*, 2007). Thus, miRNAs could provide robustness by participating in FFLs (Fig. 9.5C). Nevertheless, the existence of a FFL topology is no guarantee that the processing property predicted for a FFL containing a miRNA follows the principles established for other kinds of FFLs. This is because the properties of such loops not only depend on the pattern of interactions but also on the molecular stoichiometry of the nodes and relative kinetics of the links. Finally, it is worth noting that not every functional FFL containing a miRNA might necessarily satisfy the computational assumptions made by Mangan and Alon (2003) for protein-based FFLs.

Several experimental examples of miRNA-containing FFLs have been described. One such example involves *Drosophila* miR-7 and its action in the eye differentiation network. Three FFLs are contained within the YAN network (Graham *et al.*, 2010; Li *et al.*, 2009) (Fig. 9.6). The first of these has a topology where miR-7 is in the middle of the long arm of the loop; the TF *Pnt-P1* directly represses *YAN* and indirectly represses *YAN* through miR-7 (Li and Carthew, 2005; Li *et al.*, 2009). This FFL was predicted to buffer variations of *Pnt-P1*, only accepting persistent *Pnt-P1* changes. Interlocked in the opposite direction is another FFL, where YAN directly represses *miR-7* and indirectly represses *miR-7* by preventing *Pnt-P1* transcription. This FFL is predicted to buffer *miR-7* expression from sudden nonpersistent changes in YAN abundance. In a third FFL, *Pnt-P1* directly activates *miR-7* and indirectly activates *miR-7* by repressing *YAN*. This last FFL is predicted to buffer *miR-7* expression from sudden nonstable changes in *Pnt-P1*. Collectively, these three interlocked FFLs could buffer sudden and nonpersistent changes in the abundance of two key regulatory factors, imparting robustness to the network.

The roles of these FFLs in providing robustness were tested in *miR-7* mutant flies (Li *et al.*, 2009). When mutant animals were subjected to temperature fluctuations, *YAN* showed abnormal overexpression and there were errors in differentiation. Under uniform temperature, *YAN* expression was normal. This result suggests that miR-7 acting in FBL and FFLs plays a specific role in buffering the network against environmental perturbation.

FFLs with a TF at the beginning of the loop, a miRNA in the middle of the long path, and a target gene at the end of the loop could provide robustness through several mechanisms. If both TF and miRNA repress the target, they would augment target repression asynchronously since the kinetics of synthesis and action between TFs and miRNAs are different. This could cause a delay or an acceleration in the expression of a target, thus reducing or increasing the time necessary to trigger a response. Second,

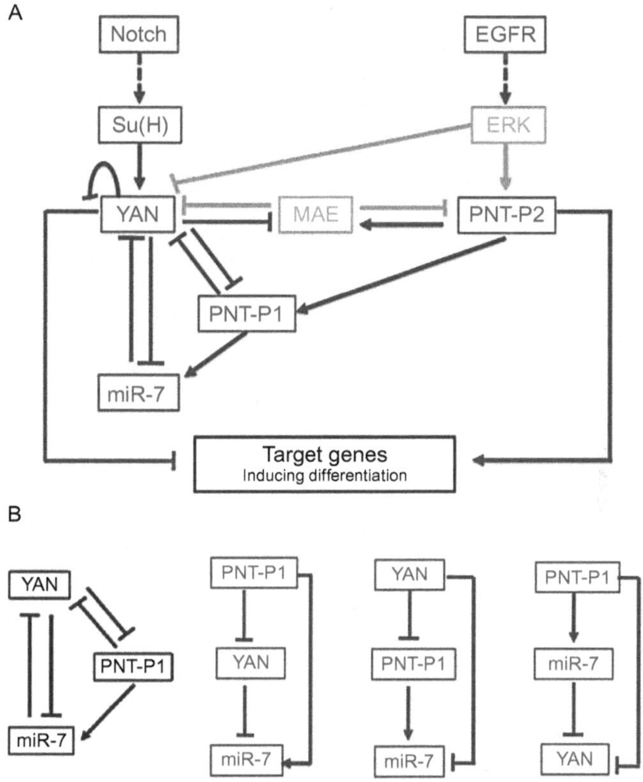

Figure 9.6 The YAN network. (A) The network is constituted by transcription factors (blue) *YAN, Pnt-P1, Pnt-2,* the miRNA *miR-7* (red), protein-interactors (green) *MAE* and *ERK,* and inputs from the *Notch* and *EFGR* signaling pathways (purple). The output of the network is the transcriptional regulation of target genes (box) required for differentiation. (B) Coupled FFLs and double-negative FFLs exist in the network. Three three-node FFLs linking *YAN, Pnt-P1,* and *miR-7* are interlocked (left, in black). These three FFLs are shown separately (in blue). (See Color Insert.)

compared to simple regulation circuits, such FFLs provide redundancy. Third, they would act as persistence detectors that only transduce stable changes in the activity of the upstream node. If the FFL is structured where the upstream TF activates and the miRNA represses the target, then the miRNA would set a threshold, assuring that unless a given level of TF activity is achieved, the downstream target is not affected.

Theoretical studies support the notion that FFLs containing miRNAs could provide robustness to the expression of miRNA targets (Hornstein and Shomron, 2006). A quantitative comparison of small RNA-based and protein-based regulation showed that small RNA-based mechanisms are better at filtering noise in input signals (Mehta *et al.*, 2008). Osella *et al.* (2011) computationally analyzed the ability of FFLs containing miRNAs to buffer noise, and they found that miRNAs can confer efficient noise control in the face of fluctuations of the upstream nodes. Thus, miRNA FFLs are predicted to buffer noise, as Ghosh *et al.* (2005) found for protein-only FFL models. Interestingly, Osella *et al.* (2011) found that optimal noise filtering does not necessarily require strong repression. Indeed, weak repression is a common feature of most miRNA–target interactions. It is possible that the networks that use miRNA-containing FFLs for buffering might have selected miRNAs for modest repression. This could explain a frequent paradox existing for many miRNAs, that is, an apparent dispensability and lack of strong phenotypic consequences when individually knocked out but a strong evolutionary conservation.

3. Conclusion

Robustness triggered by miRNAs is generally thought to be a consequence of the way in which miRNAs act upon their gene targets. While this principle is no doubt at work, we have attempted to discuss miRNAs in the context of simple and complex networks of regulation. miRNAs regulate circuits that can provide robustness to networks. Within networks, miRNAs favor regulation of central hubs and bottlenecks. Their regulation is frequently module-centric, and targeting propensity increases toward the downstream effectors of signaling networks. These biases in miRNA targeting are, in and of themselves, other means by which miRNAs more effectively generate robustness. These biases would then imply that the acquisition of targets by miRNAs is not necessarily to generate novel gene regulation but to stabilize gene networks. This idea would explain why experimentalists frequently observe few phenotypic changes when highly conserved miRNAs are mutated. It also begs the question as to whether the extraordinary high birth and death rates of animal miRNA genes (Lu *et al.*, 2008) might reflect dynamic buffering of gene expression prior to and subsequent to speciation.

ACKNOWLEDGMENTS

We thank Justin Cassidy, Adam Pah, and Patrick McMullen for helpful discussions. This work was funded by the NIH (GM077581), the Chicago Center for Systems Biology (CCSB), the Chicago Biomedical Consortium (CBC) with support from The Searle Funds at The Chicago Community Trust, and the Malkin Scholars Program from the Robert H. Lurie Comprehensive Cancer Center of Northwestern University.

REFERENCES

Albert, R., Jeong, H., and Barabasi, A. L. (2000). Error and attack tolerance of complex networks. *Nature* **406,** 378–382.

Alon, U. (2007). Network motifs: Theory and experimental approaches. *Nat. Rev. Genet.* **8,** 450–461.

Alvarez-Saavedra, E., and Horvitz, H. R. (2010). Many families of *C. elegans* microRNAs are not essential for development or viability. *Curr. Biol.* **20,** 367–373.

Bandyopadhyay, S., Mitra, R., Maulik, U., and Zhang, M. Q. (2010). Development of the human microRNA cancer network. *Silence* **1,** 6.

Barabási, A. L., and Oltvai, Z. N. (2004). Network biology: Understanding the cell's functional organization. *Nat. Rev. Genet.* **5,** 101–113.

Barkai, N., and Leibler, S. (1997). Robustness in simple biochemical networks. *Nature* **387,** 913–917.

Bartel, D. P., and Chen, C. Z. (2004). Micromanagers of gene expression: The potentially widespread influence of metazoan microRNAs. *Nat. Rev. Genet.* **5,** 396–400.

Basu, M., Bhattacharyya, N. P., and Mohanty, P. K. (2011). Modules of human micro-RNA co-target network. *J. Phys. Conf. Ser.* **297,** 012002.

Bonnet, E., Tatari, M., Joshi, A., Michoel, T., Marchal, K., Berx, G., and Van de Peer, Y. (2010). Module network inference from a cancer gene expression data set identifies microRNA regulated modules. *PLoS One* **5,** e10162.

Brandman, O., and Meyer, T. (2008). Feedback loops shape cellular signals in space and time. *Science* **322,** 390–395.

Brenner, J. L., Jasiewicz, K. L., Fahley, A. F., Kemp, B. J., and Abbott, A. L. (2010). Loss of individual microRNAs causes mutant phenotypes in sensitized genetic backgrounds in *C. elegans. Curr. Biol.* **20,** 1321–1325.

Bushati, N., and Cohen, S. M. (2007). MicroRNA functions. *Annu. Rev. Cell Dev. Biol.* **23,** 175–205.

Carthew, R. W. (2006). Gene regulation by microRNAs. *Curr. Opin. Genet. Dev.* **16,** 203–208.

Chang, D. E., Leung, S., Atkinson, M. R., Reifler, A., Forger, D., and Ninfa, A. J. (2010). Building biological memory by linking positive feedback loops. *Proc. Natl. Acad. Sci. USA* **107,** 175–180.

Chen, J. F., Mandel, E. M., Thomson, J. M., Wu, Q., Callis, T. E., Hammond, S. M., Conlon, F. L., and Wang, D. Z. (2006). The role of microRNA-1 and microRNA-133 in skeletal muscle proliferation and differentiation. *Nat. Genet.* **38,** 228–233.

Cohen, S. M., Brennecke, J., and Stark, A. (2006). Denoising feedback loops by thresholding—A new role for microRNAs. *Genes Dev.* **20,** 2769–2772.

Cui, Q., Yu, Z., Pan, Y., Purisima, E. O., and Wang, E. (2007). MicroRNAs preferentially target the genes with high transcriptional regulation complexity. *Biochem. Biophys. Res. Commun.* **352,** 733–738.

Eldar, A., Dorfman, R., Weiss, D., Ashe, H., Shilo, B. Z., and Barkai, N. (2002). Robustness of the BMP morphogen gradient in Drosophila embryonic patterning. *Nature* **419,** 304–308.

Fazi, F., Rosa, A., Fatica, A., Gelmetti, V., De Marchis, M. L., Nervi, C., and Bozzoni, I. (2005). A minicircuitry comprised of microRNA-223 and transcription factors NFI-A and C/EBPalpha regulates human granulopoiesis. *Cell* **123,** 819–831.

Ferrell, J. E., Jr. (2002). Self-perpetuating states in signal transduction: Positive feedback, double-negative feedback and bistability. *Curr. Opin. Cell Biol.* **14,** 140–148.

Ghosh, B., Karmakar, R., and Bose, I. (2005). Noise characteristics of feed forward loops. *Phys. Biol.* **2,** 36–45.

Goentoro, L., Shoval, O., Kirschner, M. W., and Alon, U. (2009). The incoherent feedforward loop can provide fold-change detection in gene regulation. *Mol. Cell* **11,** 894–899.

Graham, T. G., Tabei, S. M., Dinner, A. R., and Rebay, I. (2010). Modeling bistable cell-fate choices in the Drosophila eye: Qualitative and quantitative perspectives. *Development* **137,** 2265–2278.

Gunsalus, K. C., Ge, H., Schetter, A. J., *et al.* (2005). Predictive models of molecular machines involved in *Caenorhabditis elegans* early embryogenesis. *Nature* **436,** 861–865.

Hartwell, L. H., Hopfield, J. J., Leibler, S., and Murray, A. W. (1999). From molecular to modular cell biology. *Nature* **402,** C47–C52.

Herranz, H., and Cohen, S. M. (2010). MicroRNAs and gene regulatory networks: Managing the impact of noise in biological systems. *Genes Dev.* **24,** 1339–1344.

Hornstein, E., and Shomron, N. (2006). Canalization of development by microRNAs. *Nat. Genet.* **38**(Suppl.), S20–S24.

Hsu, C. W., Juan, H. F., and Huang, H. C. (2008). Characterization of microRNA-regulated protein–protein interaction network. *Proteomics* **8,** 1975–1979.

Inui, M., Martello, G., and Piccolo, S. (2010). MicroRNA control of signal transduction. *Nat. Rev. Mol. Cell Biol.* **11,** 252–263.

Jeong, H., Mason, S. P., Barabási, A.-L., and Oltvai, Z. N. (2001). Lethality and centrality in protein networks. *Nature* **411,** 41–42.

Johnston, R. J., Jr., Chang, S., Etchberger, J. F., Ortiz, C. O., and Hobert, O. (2005). MicroRNAs acting in a double-negative feedback loop to control a neuronal cell fate decision. *Proc. Natl. Acad. Sci. USA* **102,** 12449–12454.

Kaplan, S., Bren, A., Dekel, E., and Alon, U. (2008). The incoherent feed-forward loop can generate non-monotonic input functions for genes. *Mol. Syst. Biol.* **4,** 203.

Lai, E. C., Tam, B., and Rubin, G. M. (2005). Pervasive regulation of *Drosophila* Notch target genes by GY-box-, Brd-box-, and K-box-class microRNAs. *Genes Dev.* **19,** 1067–1080.

Lee, Y., Yang, X., Huang, Y., *et al.* (2010). Network modeling identifies molecular functions targeted by miR-204 to suppress head and neck tumor metastasis. *PLoS Comput. Biol.* **6,** e1000730.

Li, X., and Carthew, R. W. (2005). A microRNA mediates EGF receptor signaling and promotes photoreceptor differentiation in the *Drosophila* eye. *Cell* **123,** 1267–1277.

Li, Y., Wang, F., Lee, J. A., and Gao, F. B. (2006). MicroRNA-9a ensures the precise specification of sensory organ precursors in *Drosophila*. *Genes Dev.* **20,** 2793–2805.

Li, X., Cassidy, J. J., Reinke, C. A., Fischboeck, S., and Carthew, R. W. (2009). A microRNA imparts robustness against environmental fluctuation during development. *Cell* **137,** 273–282.

Liang, H., and Li, W. H. (2007). MicroRNA regulation of human protein-protein interaction network. *RNA* **13,** 1402–1408.

Lu, J., Shen, Y., Wu, Q., Kumar, S., He, B., Carthew, R. W., Wang, S., and Wu, C. I. (2008). The birth and death of microRNA genes in *Drosophila*. *Nat. Genet.* **40,** 351–355.

Mangan, S., and Alon, U. (2003). Structure and function of the feed-forward loop network motif. *Proc. Natl. Acad. Sci. USA* **100,** 11980–11985.

Mangan, S., Zaslaver, A., and Alon, U. (2003). The coherent feed forward loop serves as a sign-sensitive delay element in transcription networks. *J. Mol. Biol.* **334,** 197–204.

Martinez, N. J., and Walhout, A. J. (2009). The interplay between transcription factors and microRNAs in genome-scale regulatory networks. *Bioessays* **31,** 435–445. Review.

Martinez, N. J., Ow, M. C., Barrasa, M. I., Hammell, M., Sequerra, R., Doucette-Stamm, L., Roth, F. P., Ambros, V., and Walhout, A. J. M. (2008). A *C. elegans* genome-scale microRNA network contains composite feedback loops with high flux capacity. *Genes Dev.* **22,** 2535–2549.

Mavrakis, K. J., Van Der Meulen, J., Wolfe, A. L., *et al.* (2011). A cooperative microRNA-tumor suppressor gene network in acute T-cell lymphoblastic leukemia (T-ALL). *Nat. Genet.* **43,** 673–678.

Mehta, P., Goyal, S., and Wingreen, N. S. (2008). A quantitative comparison of sRNA-based and protein-based gene regulation. *Mol. Syst. Biol.* **4,** 221, Published online 2008 October 14.

Milo, R., Shen-Orr, S., Itzkovitz, S., Kashtan, N., Chklovskii, D., and Alon, U. (2002). Network motifs: Simple building blocks of complex networks. *Science* **298,** 824–827.

Miska, E. A., Alvarez-Saavedra, E., Abbott, A. L., Lau, N. C., Hellman, A. B., McGonagle, S. M., Bartel, D. P., Ambros, V. R., and Horvitz, H. R. (2007). Most *Caenorhabditis elegans* microRNAs are individually not essential for development or viability. *PLoS Genet.* **3,** e215.

Mookherjee, S., Sinha, M., Mukhopadhyay, S., Bhattacharyya, N. P., and Mohanty, P. K. (2009a). MicroRNA interaction network in human: Implications of clustered microRNAs in biological pathways and genetic disease. *Online J. Bioinform.* **10,** 280.

Mookherjee, S., Sinha, M., Mukhopadhyay, S., Bhattacharyya, N. P., and Mohanty, P. K. (2009b). Analysis of clustered microRNAs in biological pathways. *Online J. Bioinform.* **10,** 296.

O'Donnell, K. A., Wentzel, E. A., Zeller, K. I., Dang, C. V., and Mendell, J. T. (2005). c-Myc-regulated microRNAs modulate E2F1 expression. *Nature* **435,** 839–843.

Osella, M., Bosia, C., Corá, D., and Caselle, M. (2011). The role of incoherent microRNA-mediated feedforward loops in noise buffering. *PLoS Comput. Biol.* **7,** e1001101.

Re, A., Corá, D., Taverna, D., and Caselle, M. (2009). Genome-wide survey of microRNA-transcription factor feed-forward regulatory circuits in human. *Mol. Biosyst.* **5,** 854–867.

Reinhart, B. J., Slack, F. J., Basson, M., Pasquinelli, A. E., Bettinger, J. C., Rougvie, A. E., Horvitz, H. R., and Ruvkun, G. (2000). The 21-nt let-7 RNA regulates developmental timing in *C. elegans*. *Nature* **403,** 901–906.

Rhoades, M. W., Reinhart, B. J., Lim, L. P., Burge, C. B., Bartel, B., and Bartel, D. P. (2002). Prediction of plant microRNA targets. *Cell* **110,** 513–520.

Shalgi, R., Lieber, D., Oren, M., and Pilpel, Y. (2007). Global and local architecture of the mammalian microRNA-transcription factor regulatory network. *PLoS Comput. Biol.* **3,** e131.

Shen-Orr, S. S., Milo, R., Mangan, S., and Alon, U. (2002). Network motifs in the transcriptional regulation network of *Escherichia coli*. *Nat. Genet.* **31,** 64–68.

Tibiche, C., and Wang, E. (2008). MicroRNA regulatory patterns on the human metabolic network. *Open Syst. Biol. J.* **1,** 1–8.

Tsang, J., Zhu, J., and van Oudenaarden, A. (2007). MicroRNA-mediated feedback and feedforward loops are recurrent network motifs in mammals. *Mol. Cell* **26,** 753–767.

von Dassow, G., Meir, E., Munro, E. M., and Odell, G. M. (2000). The segment polarity network is a robust developmental module. *Nature* **406,** 188–192.

Wu, C. I., Shen, Y., and Tang, T. (2009). Evolution under canalization and the dual roles of microRNAs: A hypothesis. *Genome Res.* **19,** 734–743.

Yoo, A. S., and Greenwald, I. (2005). LIN-12/notch activation leads to MicroRNA-mediated down-regulation of Vav in *C. elegans*. *Science* **310,** 1330–1333.

Yu, H., Kim, P. M., Sprecher, E., Trifonov, V., and Gerstein, M. (2007). The importance of bottlenecks in protein networks: Correlation with gene essentiality and expression dynamics. *PLoS Comput. Biol.* **3,** e59.

Yuan, X., Liu, C., Yang, P., He, S., Liao, Q., Kang, S., and Zhao, Y. (2009). Clustered microRNAs' coordination in regulating protein-protein interaction network. *BMC Syst. Biol.* **3,** 65.

Zotenko, E., Mestre, J., O'Leary, D. P., and Przytycka, T. M. (2008). Why do hubs in the yeast protein interaction network tend to be essential: Reexamining the connection between the network topology and essentiality. *PLoS Comput. Biol.* **4,** e1000140.

Index

Note: Page numbers followed by "*f*" indicate figures, and "*t*" indicate tables.

Contents of Previous Volumes

Volume 47

Volume 48

Volume 50

Volume 51

Volume 52

Volume 53

Volume 54

Volume 55

Volume 56

Volume 57

Volume 62

Volume 63

Volume 64

Volume 65

Volume 68

Volume 69

Volume 70

Volume 71

Volume 72

Volume 73

Volume 74

Volume 75

Volume 76

Volume 77

Volume 80

Volume 81

Volume 82

Volume 83

Volume 84

Volume 85

Volume 86

Volume 87

Volume 88

Volume 89

Volume 90

Volume 91

Volume 92

Volume 93

Volume 94

Volume 95

Volume 96

FSC
www.fsc.org
MIX
Papier | Fördert
gute Waldnutzung
FSC® C083411

Kolibri 360 GmbH
Ferdinand-Jühlke-Straße 7
99095 Erfurt, Deutschland
produktsicherheit@kolibri360.de